AF352920

Bistatic HF Radar

Bistatic HF Radar

Editor

Stuart Anderson

MDPI • Basel • Beijing • Wuhan • Barcelona • Belgrade • Manchester • Tokyo • Cluj • Tianjin

Editor
Stuart Anderson
University of Adelaide
Australia

Editorial Office
MDPI
St. Alban-Anlage 66
4052 Basel, Switzerland

This is a reprint of articles from the Special Issue published online in the open access journal *Remote Sensing* (ISSN 2072-4292) (available at: http://www.mdpi.com).

For citation purposes, cite each article independently as indicated on the article page online and as indicated below:

LastName, A.A.; LastName, B.B.; LastName, C.C. Article Title. *Journal Name* **Year**, *Article Number*, Page Range.

ISBN 978-3-03943-330-8 (Hbk)
ISBN 978-3-03943-331-5 (PDF)

Contents

About the Editor

Stuart Anderson received the B.Sc. and Ph.D. degrees in physics from the University of Western Australia, Perth, Australia, in 1968 and 1972, respectively. In 1974, he was invited to join the team being assembled in the Australian Defence Science and Technology Organization to develop the Jindalee over-the-horizon radar system, where he assumed responsibility for ocean surveillance and remote sensing, leading to the world's first fully operational OTHR wide-area ocean surveillance system. He has worked as a Visiting Scientist in a number of countries, particularly the U.S., the U.K., and France, as a consultant to their national HF radar programs. Dr. Anderson holds or has held Adjunct and Visiting Professor appointments at numerous universities in Australia and overseas, including University College London, Université Paris VI, and Université Rennes I, which, in 2005, awarded him an honorary doctorate for his contributions to radar science. In 2014 he retired from DSTO and took up the position of Adjunct Professor of Physics at the University of Adelaide. His research interests span ionospheric physics, radiowave propagation, radio oceanography, electromagnetic scattering, inverse problems, signal processing, passive coherent location, and microwave polarimetry. He has published over 350 journal papers, conference papers, book chapters, and reports in these fields. Dr. Anderson was the recipient of the 1992 Australian Minister of Defence Science Award for Research Achievement for his pioneering contributions to over-the-horizon radar in both skywave and surface wave forms. He is the principal author of the chapter on OTH radar in the authoritative *Radar Handbook*.

Preface to "Bistatic HF Radar"

The proliferation of HF radar systems for ocean remote sensing and maritime surveillance continues apace, with hundreds of such radars now deployed around the world. The overwhelming majority of these radars operate in the conventional monostatic configuration, with the transmitting and receiving systems collocated or closely spaced (the term "quasi-monostatic" is often used in this case). This simple geometry has obvious advantages in terms of cost, siting requirements, communications, maintenance, signal processing, and echo interpretation, and it has been adopted by HF radars exploiting line-of-sight, surface wave, and skywave propagation modalities.

All these considerations notwithstanding, in some circumstances there can be compelling reasons to implement bistatic configurations, defined as geometries in which the separation between transmitter and receiver is comparable with the range to the zones being interrogated. Factors that can drive this decision include energy budget, desire to exploit hybrid propagation modes, scattering characteristics of the targets of interest, properties of the clutter, survivability, and covertness.

While the literature on the design and application of monostatic HF radars continues to thrive, the same does not hold for the literature on bistatic configurations. Motivated by our desire to expand the palette of missions that can be addressed by HF radar, especially some that cannot be addressed by monostatic radars, we have compiled this Special Issue of Remote Sensing.

The issue contains nine papers, embracing contributions from authors in a dozen centers of HF radar research in Australia, Canada, China, the UK, and the USA. The opening paper, by Anderson, catalogs the many possible bistatic configurations according to the propagation modes involved and describes a number of radar missions where the bistatic geometry yields enhanced radar capability. Next, there are three papers dealing with generalizations of well-known perturbation-theoretic methods of HF scatter from the sea surface. Chen et al. treat the case of signals incident at grazing incidence from a shore-based transmitter and scattered upwards to be received by an airborne receiver; they compute the spectra to second order. Yao et al. consider the situation where the radar transmitter is mounted on a floating platform subject to motion with 6 degrees of freedom and explore different options for receiver placement and the resulting impact on the echo spectral structure. Silva et al. address the problem of high sea states, where the standard perturbation-theoretic models break down, and derive expressions for the modified first-order spectrum under various conditions.

The following paper, by Hardman et al., deals with the inverse problem of estimating the directional wave spectrum from the HF radar Doppler spectrum. They generalize the Seaview monostatic inversion method to handle bistatic geometries and assess its performance on simulated data.

While remote sensing of ocean currents and sea state is often the primary mission of HF radar, ship detection and tracking are of increasing interest, and the next three papers focus on this surveillance mission. Ji et al. examine the effects of ship motion on the bistatic first-order clutter returns. They develop the relevant theory and present simulated results for various configurations, then support the modeling with measurements carried out with two radars, one mounted on a cooperating vessel. Next, Sun et al. describe a newly-developed multistatic HFSWR, one with a single transmitter but two receiving stations, and demonstrate the improved tracking performance that can be achieved with such a configuration. This immediately raises the question of a reciprocal design, one with multiple transmitters and a single receiver. Liu et al. explore this concept in their paper, reporting a passive radar system that uses multiple GPS satellites as illuminators.

Although this system operates in a much higher frequency band than HFSWR, it serves to illustrate some of the problems that arise when multiple transmitted signals need to be separated and processed at the receiving station; we anticipate that equivalent problems would arise with an analogous HFSWR configuration. Finally, Zhang et al. point out that, in practice, ship tracking is far from straightforward, with track fracture arising from a combination of many factors, including highly maneuverable vessels, dense channels, target occlusion, strong clutter/interference, long sampling intervals, and low detection probabilities. They describe a sophisticated tracking technique—an interacting multiple model extended Kalman filter combined with a machine learning architecture—and demonstrate its efficacy using real data from a stereoscopic HFSWR system.

The diversity of HF radar configurations represented in this Special Issue does not exhaust all the possibilities, as cataloged in the taxonomy shown in the first paper, but the impressive variety of bistatic HF radar systems now in operation, and the special capabilities that they offer, will ensure their continuing proliferation and the development of new concepts and missions.

Stuart Anderson
Editor

 *remote sensing*

Article

Bistatic and Stereoscopic Configurations for HF Radar

Stuart Anderson

Physics Department, University of Adelaide, Adelaide 5005, Australia; stuart.anderson@adelaide.edu.au

Received: 20 January 2020; Accepted: 18 February 2020; Published: 20 February 2020

Abstract: Most HF radars operate in a monostatic or quasi-monostatic configuration. The collocation of transmit and receive facilities simplifies testing and maintenance, reduces demands on communications networks, and enables the use of established and relatively straightforward signal processing and data interpretation techniques. Radars of this type are well-suited to missions such as current mapping, waveheight measurement, and the detection of ships and aircraft. The high scientific, defense, and economic value of the radar products is evident from the fact that hundreds of HF radars are presently in operation, the great majority of them relying on the surface wave mode of propagation, though some systems employ line-of-sight or skywave modalities. Yet, notwithstanding the versatility and proven capabilities of monostatic HF radars, there are some types of observations for which the monostatic geometry renders them less effective. In these cases, one must turn to more general radar configurations, including those that employ a multiplicity of propagation modalities to achieve the desired illumination, scattering selectivity, and echo reception. In this paper, we survey some of the considerations that arise with bistatic HF radar configurations, explore some of the missions for which they are optimal, and describe some practical techniques that can guide their design and deployment.

Keywords: HF radar; bistatic radar; HFSWR; OTH radar

---

1. Introduction

Remote sensing of our geophysical environment by means of radio waves in the HF band is now a truly global activity, with decametric radars operating in scores of countries, and on every continent [1]. In a number of instances, international collaborations facilitate the integration of the outputs from individual radars to yield regional or even basin-scale products, thereby increasing the quality, diversity, and utility of the derived information [2].

The overwhelming majority of these radars operate in the conventional monostatic configuration, with the transmitting and receiving systems collocated or closely spaced (the term quasi-monostatic is often used in this case). This simple geometry has obvious advantages in terms of cost, siting requirements, communications, maintenance, signal processing, and echo interpretation, and has been adopted by HF radars exploiting line-of-sight, surface wave, and skywave propagation modalities.

All these considerations notwithstanding, in some circumstances, there can be compelling reasons to implement bistatic configurations, often defined as geometries in which the separation between transmitter and receiver is comparable with the range to the zones being interrogated. Factors that can drive this decision include energy budget, desire to exploit hybrid propagation modes, scattering characteristics of the targets of interest, properties of the clutter, survivability, and covertness. Bistatic HF radars with very specific missions have been deployed since the 1960s, predominantly in defense applications, but the convenience of monostatic designs and the adequacy of their standard remote sensing products have tended to discourage wider adoption of bistatic configurations.

Once we allow for the separation of transmit and receive facilities, many possible configurations emerge. Each of these subsystems can be located on land, at sea, in the air, or even in space, with a range of propagation mode combinations possible for the signal paths from transmitter to target and target to receiver. Of these, line-of-sight, ground wave (we shall use the term *surface wave* throughout this paper, though strictly it refers to only one component of the total field–the dominant one at over-the-horizon ranges), and skywave modes are by far the most common, though more exotic propagation mechanisms have been explored. Figure 1 presents a taxonomy of the main configurations; those that are understood to have been implemented, or at least reached the advanced design and experimentation phase [3], are indicated by the colored dots (E. Lyon, personal communication, May 19, 2015).

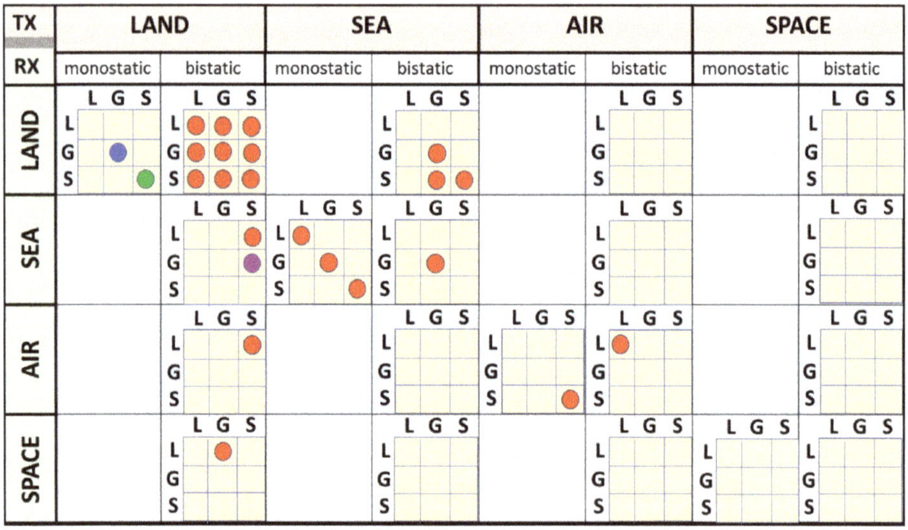

Figure 1. A taxonomy of HF radar configurations. The conventional monostatic surface wave and skywave radars are indicated with blue and green markers, respectively; the topical hybrid sky–surface wave configuration is shown by the magenta marker.

An obvious generalization of these single radar configurations is the deployment of multiple radars to interrogate a common area of interest. This is the standard modus operandi of current mapping HF surface wave radars (HFSWR) such as the CODAR SeaSonde [4] and the Helzel Messtechnik WERA [5], where two or more measurements of radial velocity are combined to yield a resultant vector. We note that measurements from these two distinct radar designs—based on direction-finding and beam-forming, respectively—can be combined to expand network coverage and reduce down-time [6]. Skywave radar networks with overlapping coverage have been operational in Australia (JORN) [7] and the United States (ROTHR) [8] for decades; not surprisingly, there are many issues to be taken into account when designing such configurations [9,10]. The term stereoscopic has been used to describe these multi-monostatic configurations; other applications include ship target dynamic signature analysis and excitation of nonlinear scattering mechanisms.

Another generalisation is the use of relay stations; that is, combined receive–transmit facilities that acquire the signal radiated by the primary radar transmitter, amplify it, possibly with additional modulation, and then reradiate it, thereby extending the range of the system or facilitating other radar functions.

This diverse array of system geometries offers many opportunities for remote sensing. In particular, the ability to extend the range of Bragg resonant scattering to lower wavenumbers opens the way to observing some environmental phenomena to which monostatic radars are insensitive. One example of this is the determination of sea ice parameters. Short sea waves are rapidly attenuated as they enter the marginal ice zone; only long waves penetrate to useful distances into the ice field. The sea ice properties are encoded in the radar Doppler spectrum, most visibly in the first-order peaks [11]. For a monostatic radar to observe these peaks, it would need to operate at a very low frequency, below those employed by present-day HF radars, but a bistatic geometry enables the returns from longer waves to be measured. Another example is the investigation of the physics of the ionosphere via analysis of impressed phase modulation [12,13], wavefront distortion [14], and polarization transformation [15] of oblique (bistatic) radar reflections; these are largely inaccessible to monostatic radars.

In this paper, we explore many of the issues that arise with bistatic HF radar configurations, basing our analysis on the formal radar process model presented in the following section. After examining the implications for the component elements of the radar observation process, we proceed to describe some specific radar missions that benefit from the physics of bistatic scattering and/or hybrid propagation modes. The term hybrid is often applied to configurations where the outbound and inbound propagation modalities are different; that is, they lie off the diagonals in the boxes of Figure 1. Along the way, we describe and illustrate some practical techniques that can serve as a guide to bistatic HF radar design and deployment. In particular, we look at the problem of site selection, a challenge that is compounded by the need to address multiple radar missions.

2. The General Radar Process Model

The radar process model formulation first introduced in [16] is ideally suited for our purpose as it makes explicit the temporal sequence of the signal trajectory and of this in mind, and noting that multizone scattering in the course of signal propagation (see below) has been observed to be significant for both skywave and surface wave HF radars [17,18], the formulation of the radar process is expressed as a concatenation of operators,

$$s = \sum_{n_B=1}^{N} \widetilde{R} \left[\prod_{j=1}^{n_B} \widetilde{M}_{S(j)}^{S\lfloor j+1 \rfloor} \widetilde{S}(j) \right] \widetilde{M}_T^{S(1)} \widetilde{T}w + \sum_{l=1}^{N_J} \sum_{m_B=1}^{M} \widetilde{R} \left[\prod_{k=1}^{n_B} \widetilde{M}_{S(k)}^{S\lfloor k+1 \rfloor} \widetilde{S}(k) \right] \widetilde{M}_N^{S(1)} n_l + m \tag{1}$$

where

w represents the selected waveform,

$\widetilde{T}$ represents the transmitting complex, including amplifiers and antennas,

$\widetilde{M}_T^{S(1)}$ represents propagation from transmitter to the first scattering zone,

$\widetilde{S}(j)$ represents all scattering processes in the j-th scattering zone,

$\widetilde{M}_{S(j)}^{S\lfloor j+1 \rfloor}$ represents propagation from the j-th scattering zone to the (j+1)-th zone,

n_B denotes the number of scattering zones that the signal visits on a specific route from the transmitter to the receiver,

N_J denotes the number of external noise sources or jammers,

$\widetilde{M}_N^{S(1)}$ represents propagation from the i-th noise source to its first scattering zone,

m_B denotes the number of scattering zones that the i-th noise emission visits on a specific route from its source to the receiver,

N, M denote the maximum number of zones visited by signal and external noise, respectively,

$\widetilde{R}$ represents the receiving complex, including antennas and receivers,

m represents internal noise,

s represents the signal delivered to the processing stage.

If the transmitter and/or receiver are in motion, as with shipborne radars, for example, a slight generalization is in order. Adopting the frame-hopping paradigm, we insert Lorenz transformation operators:

$$\widetilde{T} \rightarrow \widetilde{L}_T \widetilde{T} \tag{2}$$

and

$$\widetilde{R} \rightarrow \widetilde{R} \widetilde{L}_R \tag{3}$$

to take kinematic effects into account.

The effective design of bistatic HF radar systems requires decisions that involve all the terms in the process model, singly, pairwise, or collectively. Ultimately, the design problem is one of optimization; that is, finding the best combination of siting and radar parameters as measured by performance over the set of missions to be addressed. In general, this is a multi-objective problem as radars may be designed to perform air and surface surveillance as well as remote sensing of one or more geophysical variables. Later in this paper, we will describe tools for achieving this optimization, but first we examine some of the most important considerations associated with the individual operators.

3. Consequences of Bistatic Geometry on the Radar Process Model Operators

3.1. Waveform

Most HF radars nowadays employ a variant of the linear FMCW waveform, ranging from a continuous signal, through interrupted FMCW, to FM pulses with a low duty cycle. Interrupted versions include notched sweeps as well as frequency-hopping and spaced sweep formats. In addition to the FMCW class, phase-coded pulse waveforms can still be heard. For most of these options, MIMO (multiple input, multiple output) implementations are possible.

When one moves from monostatic or quasi-monostatic to the bistatic case, several considerations need to be kept in mind. First, the separation of transmit and receive facilities greatly reduces the problem of self-interference, thereby expanding the waveform parameter space. For the moment, we set aside the case where radars in a stereoscopic configuration are sharing a common transmission frequency band. Second, it is well known that range-folded echoes pose a serious hazard for monostatic radars, arising from the combination of long-range propagation of HF radiowaves and the abundance of ionospheric and terrestrial scatterers. As illustrated in Figure 2, bistatic configurations offer a greater freedom with choice of waveform repetition frequency because the range-ambiguous zones of illumination are displaced from those of the receiving system. We note here that the use of non-repetitive waveforms is another tool for reducing this threat, though few HF radars presently employ such signals.

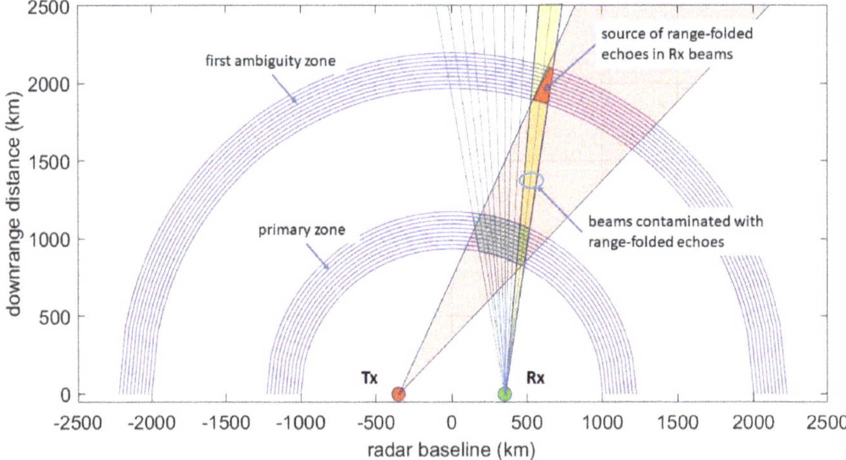

Figure 2. The problem of range-ambiguous echoes associated with periodic waveforms is greatly reduced with bistatic configurations. These enable one to steer the receiver beams over the desired ambiguity zone whilst rejecting unwanted returns.

Yet, even bistatic systems are advised to take account of the far-range illumination pattern as the magnitude of unwanted environmental echoes may be sufficient to disrupt through receiving array sidelobes. Third, an associated problem is the prospect of round-the-world (RTW) propagation—in powerful HF skywave radars, signals have been observed after three transits around the Earth. Of course, for low-power radars, background noise will almost invariably swamp RTW returns. Fourth, when pulsed waveforms are used, the fact that bistatic geometry couples time delay to angle-of-arrival may require that one implements a pulse-chasing capability [19], with its attendant penalties. Fifth, the spatial properties of bistatic resolution cells are well-known [20], but less attention has been paid to what we might call the Doppler sensitivity, $\frac{\partial \omega}{\partial v}$, where ω is the Doppler shift and v is the target speed. To quantify this, recall that the bistatic Doppler shift of a target with velocity $\vec{v}$ at location x given by

$$\omega = -\frac{2\pi}{\lambda} \frac{d}{dt}\left(r_T^x + r_x^R\right) = -k\left(\hat{r}_T^x \cdot \vec{v} + \vec{r}_x^R \cdot \vec{v}\right) = -k\left(\hat{r}_T^x + \hat{r}_x^R\right) \cdot \vec{v} = -2k\,\cos\left(\frac{\varphi}{2}\right)v.\cos\beta \tag{4}$$

with φ the bistatic angle and β the target heading relative to the bisector axis; hence,

$$\frac{\partial \omega}{\partial v} = -2k\,\cos\left(\frac{\varphi}{2}\right).v\,\cos\beta \tag{5}$$

The Doppler sensitivity loss factor $\cos\left(\frac{\varphi}{2}\right)$ is one component of the price we pay in return for whatever advantages we can extract from employing a bistatic configuration.

3.2. Transmitting Facility

Central to the design of the transmitting facility is the orientation of the illumination pattern relative to that of the receiver. For any given location $\vec{r}$ in the common zone, the radiated signal amplitude is proportional to $\widetilde{M}_T^{\vec{r}}\,\widetilde{T}(\theta,\varphi)w$, or simply $\widetilde{M}_T^{\vec{r}}\,\widetilde{T}(\varphi)w$ for HFSWR. The radar designer has the option to orient the maximum directive gain of the transmitting array towards that region in the receiving facility's field of view, which has been accorded the highest priority. More generally, for signal-to-noise dominated missions, we can formulate the HFSWR orientation problem as one of maximizing the figure of merit (FOM) of the priority-weighted pattern,

$$FOM = max_{\varphi_0} \iint_R P\left(\vec{r}\right) \widetilde{M}_{\vec{r}}^R \widetilde{M}_T^{\vec{r}}\,\widetilde{T}(\varphi,\varphi_0)w\,d\vec{r} \tag{6}$$

where φ_0 is the nominal boresight orientation of the transmit array and $P\left(\vec{r}\right)$ represents the priority weighting over the receiver processing zone R.

A complication that arises with clutter-related missions of HFSWR is the phenomenon of multiple scattering [21,22]. This can corrupt the received echoes when the sea state is significant, so in addition to providing sufficient incident power density, a sophisticated transmit antenna design would attempt to minimize the associated contributions, relative to the echoes received via the primary propagation path. To do this requires a regional wave climatology but is otherwise straightforward.

3.3. Propagation

The involvement of distinct outbound and inbound propagation paths has major ramifications for HF skywave radar, with a lesser, though still observable, impact on HFSWR. For monostatic skywave radars, frequency management systems probe the ionosphere and determine (i) the frequency band providing adequate power density in the target zone, and (ii) some measure of the quality of the propagation channel [23]. With bistatic configurations, the frequency that works best for propagation from transmitter to target zone will often be poor for propagation from target zone to receiver; in this case, the optimum frequency will effect a compromise, and may, on occasion, take a highly non-intuitive value.

In order to quantify the impact on performance, we can exploit the geometrical congruence of a single bistatic signal path and two monostatic paths [24]. Figure 3 shows the concept underlying this technique.

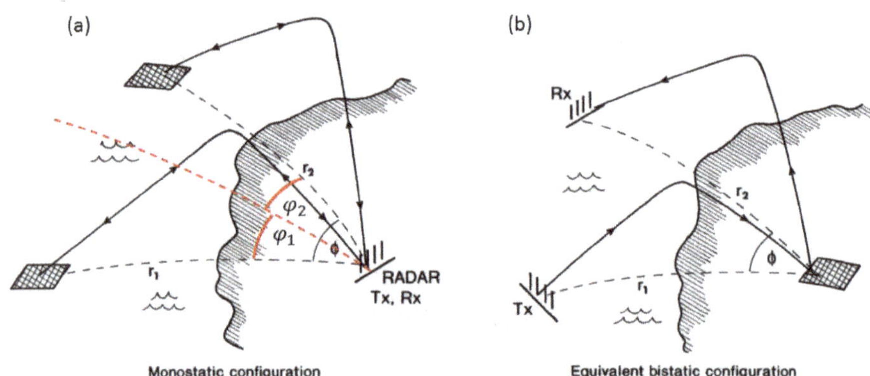

Figure 3. The geometrical congruence of (**a**) a pair of monostatic radar observations, and (**b**) a single bistatic radar observation. To see the equivalence, simply imagine that the area shown as land is actually sea and the area shown as sea is actually land, whereby Figure 3b appears as a land-based bistatic radar.

The first emission travels from the monostatic radar at location X to the target zone at relative coordinates (r_1, φ_1), scatters, and returns to the radar. Using a scalar form of (1) for notational simplicity, the complex amplitude of the received signal is given by

$$s_1 = R(\varphi_1)M^X_{\vec{r}_1}S(\vec{r}_1)M^{\vec{r}_1}_X T(\varphi_1)w \tag{7}$$

so the received power is $|s_1|^2$. Now, write $M^{\vec{r}_1}_X = a^T_1 e^{i\psi^T_1}$ and $M^X_{\vec{r}_1} = a^R_1 e^{i\psi^R_1}$. We can identify a^T_1 as the one-way propagation amplitude loss factor for the outbound signal and a^R_1 as the corresponding amplitude loss factor for the inbound signal. Power loss factors are then simply $\left|a^T_1\right|^2$ and $\left|a^R_1\right|^2$, and the propagation power loss for the two-way process is $\left|a^T_1\right|^2 \cdot \left|a^R_1\right|^2$.

A second observation is then made in a different direction, to a target zone at coordinates (r_2, φ_2),

$$s_2 = R(\varphi_2)M^X_{\vec{r}_2}S(\vec{r}_2)M^{\vec{r}_2}_X T(\varphi_2)w \tag{8}$$

with two-way propagation loss $\left|a^T_2\right|^2 \cdot \left|a^R_2\right|^2$, as shown in Figure 3a. Now imagine that there is a transmitter at location (r_1, φ_1) and a receiver at (r_2, φ_2) as shown in Figure 3b; that is, a bistatic radar configuration interrogating the region previously occupied by the monostatic radar. The complex amplitude for this case is given by

$$s_3 = R(\varphi_2)M^{\vec{r}_2}_X S(X)M^X_{\vec{r}_1} T(\varphi_1)w \tag{9}$$

where we have taken the orientation of the imagined arrays to be parallel to those of the monostatic system. Now, the propagation paths satisfy reciprocity, $M^X_{\vec{r}_1} = M^{\vec{r}_1}_X$ and $M^{\vec{r}_2}_X = M^X_{\vec{r}_2}$. Further, the gain patterns of the transmit and receive arrays are strongly determined by the array apertures but vary only weakly with steer angle over moderate departures from boresight. Thus, we can write

$T(\varphi_2) = \alpha\, T(\varphi_1)$ and $R(\varphi_2) = \beta R(\varphi_1)$ where $0.87 < \alpha, \beta < 1$ for a radar whose arrays each steer over a 60° arc. Substituting in (9), and invoking (7) and (8),

$$
\begin{aligned}
s_3 &= R(\varphi_2)M_X^{\vec{r}_2}S(X)M_{\vec{r}_1}^X\, T(\varphi_1)w = \sqrt{s_3^2} \\[4pt]
&= \sqrt{R(\varphi_2)M_X^{\vec{r}_2}S(X)M_{\vec{r}_1}^X\, T(\varphi_1)w . R(\varphi_2)M_X^{\vec{r}_2}S(X)M_{\vec{r}_1}^X\, T(\varphi_1)w} \\[4pt]
&= S(X)\sqrt{R(\varphi_2)M_{\vec{r}_2}^X M_{\vec{r}_1}^X \tfrac{T(\varphi_2)}{\alpha}w . R(\varphi_1)M_X^{\vec{r}_2}M_X^{\vec{r}_1}T(\varphi_1)w} \\[4pt]
&= S(X)\sqrt{R(\varphi_2)M_{\vec{r}_2}^X M_X^{\vec{r}_2}\tfrac{T(\varphi_2)}{\alpha}w . \beta R(\varphi_1)M_{\vec{r}_1}^X M_X^{\vec{r}_1}T(\varphi_1)w} \\[4pt]
&= \frac{S(X)}{\sqrt{S(\vec{r}_1)S(\vec{r}_2)}} . \sqrt{\tfrac{\beta}{\alpha}R(\varphi_1)M_{\vec{r}_1}^X S(\vec{r}_1)M_X^{\vec{r}_1}T(\varphi_1)w . R(\varphi_2)M_{\vec{r}_2}^X S(\vec{r}_2)M_X^{\vec{r}_2}T(\varphi_2)w} \\[4pt]
&= \frac{S(X)}{\sqrt{S(\vec{r}_1)S(\vec{r}_2)}}\sqrt{\tfrac{\beta}{\alpha}} . \sqrt{s_1 s_2} \approx \frac{S(X)}{\sqrt{S(\vec{r}_1)S(\vec{r}_2)}} . \sqrt{s_1 s_2}
\end{aligned}
\tag{10}
$$

The magnitudes of $S(\vec{r}_1)$ and $S(\vec{r}_2)$ can be estimated by inversion of the respective Doppler spectra, or even approximated at zero cost by assuming fully developed seas—typically valid for HF frequencies above 15 MHz. The steer directivity loss factor $\sqrt{\tfrac{\beta}{\alpha}} \approx 1$ so its effect is insignificant compared with the variability of the other terms. Thus, from measurements of the returned clutter power from monostatic observations s_1 and s_2, we can predict the echo power for the bistatic configuration observation s_3 for an arbitrary specified scattering coefficient $S(X)$. One point to note here is that we have simplified the discussion by ignoring the polarization domain; this is not a significant issue for HFSWR and can be avoided in the skywave radar case by a combination of spatial and temporal averaging.

HF skywave radars routinely collect backscatter ionograms (BSI) over the arc of coverage, typically out to a range of 5000–6000 km, so there is a wealth of propagation data available from which to derive statistical predictions that can be used for bistatic system design. A representative BSI is shown in Figure 4, with the instantaneous range depth marked for a nominal radar frequency of 15 MHz. Assuming a slow variation with azimuth, both r_1 and r_2 need to lie between 1400 km and 2300 km. Figure 5 shows an instance of an inferred sub-clutter visibility (SCV) map computed for a representative radar network (the monostatic input data are real but not obtained from these radars).

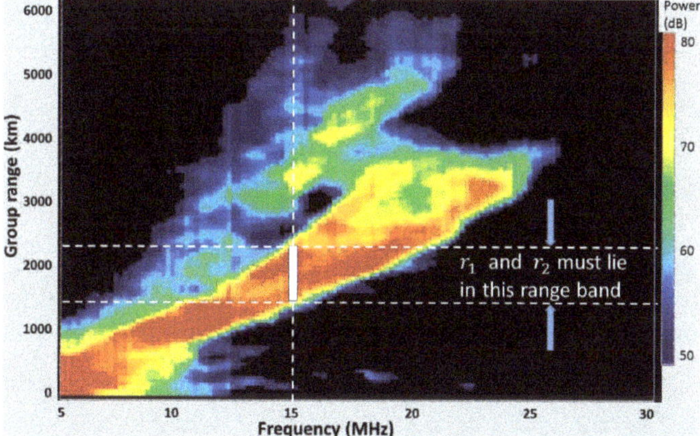

Figure 4. A backscatter ionogram—a map of echo strength as a function of (group) range and radar frequency. The dashed lines show, for a representative frequency, how the outbound and inbound group ranges must both lie in the band indicated for the system to operate successfully.

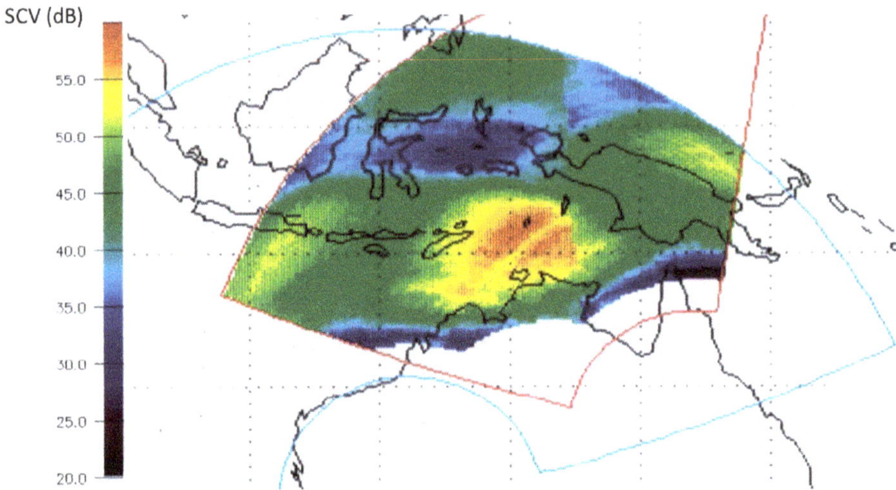

Figure 5. A map showing a single instance of the predicted bistatic sub-clutter visibility (clutter-to-noise ratio) in the overlap region of two skywave radars, as inferred using the geometrical congruence technique from a single azimuthal scan recorded with a separate monostatic radar.

While it is necessary to exceed some target-specific threshold of power density in order to achieve detection, for slow-moving targets, such as ships, that may not be sufficient. The presence of multimode propagation and phase path fluctuations associated with field line resonances and other ionospheric disturbances can blur the Doppler spectrum of the radar returns and thereby obscure the desired echoes. This raises the question: Can we extend the analysis discussed in the preceding paragraphs so as to obtain statistical information on the phase path modulation spectrum over bistatic paths?

The answer is a qualified 'yes'. Techniques to estimate and then correct for phase path variations have been developed and installed in operational systems since the 1980s [12,13] so the individual phase path modulation time series are available for each leg of the synthesized bistatic path. A rudimentary synthesis approach would simply concatenate the phase modulation histories, then halve them, but that could introduce Doppler spreading due to phase discontinuity at the junction point. A superior method involves first phase-shifting the second half to ensure phase continuity and then applying a conjugate taper weighting around the junction to affect a smooth first derivative.

This approach works for the most important class of fluctuations, where the spatial scale is of the order of 10^2 km, and latitude-dependent, being linked to the geomagnetic field line resonances (FLR) that are observed as micro-pulsations at ground level. At times, other dynamical processes cause fluctuations over much smaller spatial scales. Figure 6 illustrates these two types of modulation: Each frame shows the measured phase fluctuation time series over a two-way skywave channel. In Figure 6a, the modulation estimated from the echoes originating in four individual range cells spaced over a range depth of about 150 km shows a high degree of spatial correlation, suggesting that the outbound and inbound legs of a bistatic skywave radar observation would experience related modulation sequences. In contrast, when other types of modulation prevail, the paths can experience uncorrelated and often more erratic modulations. In Figure 6b, the cells shown are spaced over a total of only 20 km, yet the modulation patterns are quite distinct. In both cases, we can construct a simulated bistatic path resultant modulation sequence, using the ideas of the previous paragraph, but only for the former type can we hope to associate the observed modulation with the known properties of geophysical wave processes in the ionosphere.

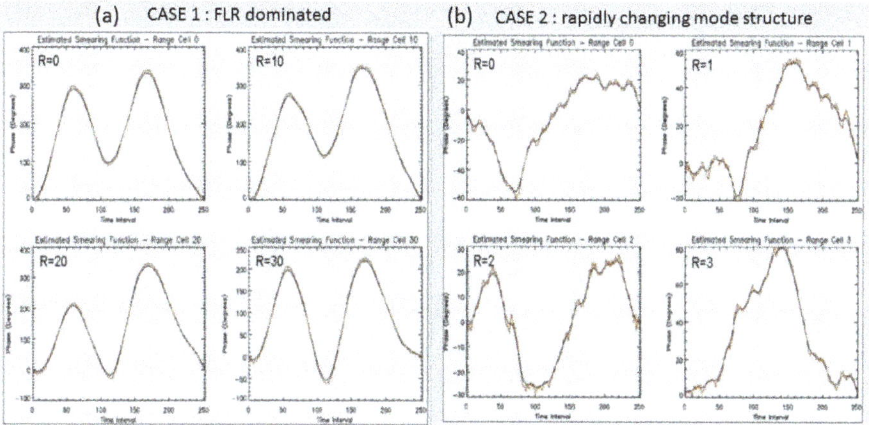

Figure 6. Phase modulation sequences measured over skywave propagation paths; (**a**) an example of a field line resonance modulation, with slow spatial variation over a distance of 150 km, and (**b**) an example where the modulation arises from other geophysical mechanisms, with spatial decorrelation occurring within 20 km.

It is perhaps apposite to note here that an operationally significant relative of the problem of joint path optimization is the converse—the selection of frequencies that guarantee strong propagation over one path and little over the other for the same radar frequency. Such a bistatic configuration has direct relevance to the detection of nonlinear target echoes and the ability to suppress sea clutter by many tens of dB. A description of this scheme can be found in [25].

3.4. Scattering

The ability of HF radar to address a wide range of missions brings with it the need for mathematical techniques for computing the radar signatures of the diverse phenomena involved. Bistatic HF radar has been implemented in the form of operational systems since the 1960s but, for most of its history, practice has dominated theory. We can perceive four main lines of development in HF scattering theory: One for the ocean surface, one for plasma formations in the ionosphere, one for land surfaces, and one for discrete targets such as ships, aircraft, and missiles.

3.4.1. Scattering from the Ocean Surface

The perturbation theoretic approach of Barrick [26], building on the Rice theory for scattering from static rough surfaces [27], has served as the cornerstone of HF radar oceanography for the past five decades. Quite a few generalizations of the Barrick theory have appeared over the years (e.g., [28–30]), as well as a different approach [31] based on the Walsh theory for scattering from static surfaces [32] and extended by Gill and co-workers (e.g., [33–36]). As ocean applications of HF radars dominate, and as bistatic configurations become more widespread, it is hardly surprising to find an emerging literature of papers that apply the fundamental theories to particular circumstances. As a guide, we have tabulated some of these bistatic scatter papers against key parameters: (i) The perturbation order of the approximation, (ii) the scattering geometry, (iii) whether platform motions were taken into account, (iv) the polarization states addressed, and (v) the hydrodynamic dispersion relation employed. This file is available from the author.

The general expression for the scattered field in $\left(\vec{k}, \omega\right)$ space, to second order, has the form

$$
\begin{aligned}
S\left(\vec{k}, \omega\right) = \int \, d\vec{k}_{scat}\widetilde{R}\left(\vec{k}_{inc}\right)\delta\left(\vec{k}_{scat} - \vec{k}_{inc} + 2\vec{k}_{inc}\cdot\hat{n}\,\hat{n}\right)\delta(\omega - \omega_0) \\
+ \int \, d\vec{\kappa}_1 \, F_1\left(\vec{k}_{scat}, \vec{k}_{inc}, \vec{\kappa}_1\right)\delta\left(\vec{k}_{scat} - \vec{k}_{inc} + \vec{\kappa}_1\right)\delta\left(\omega - \omega_0 + \Omega\left(\vec{\kappa}_1\right)\right) \\
+ \int\int d\vec{\kappa}_1 \, d\vec{\kappa}_2 \, F_2\left(\vec{k}_{scat}, \vec{k}_{inc}, \vec{\kappa}_1, \vec{\kappa}_2\right)\delta\left(\vec{k}_{scat} - \vec{k}_{inc} + \vec{\kappa}_1 + \vec{\kappa}_2\right) \times \delta\left(\omega - \omega_0 + \Omega\left(\vec{\kappa}_2\right) + \Omega\left(\vec{\kappa}_1\right)\right)
\end{aligned}
\tag{11}
$$

In this equation, the dispersion relation appears as $\Omega\left(\vec{\kappa}_i\right)$; it is this function that determines the contours of integration that yield the Doppler power spectral density. Recent investigations of HF scatter from sea ice motivated the development of a computational model able to solve for any explicit dispersion relation [11,37]. Expressions for the kernel functions F_1 and F_2 can be found in the cited literature, while the Fresnel reflection coefficient $\widetilde{R}$ is a function of the water temperature and salinity.

Viewed diagrammatically in the spatial frequency domain, as shown in Figure 7, an important feature of bistatic Bragg scattering becomes evident, namely the smaller modulus of the resultant wave vector increment, shown in black. At first order, this means that longer ocean waves can be measured directly, while at second order, it means that a broad angular spread of the shorter waves in the participating wave field favors the scattering contributions.

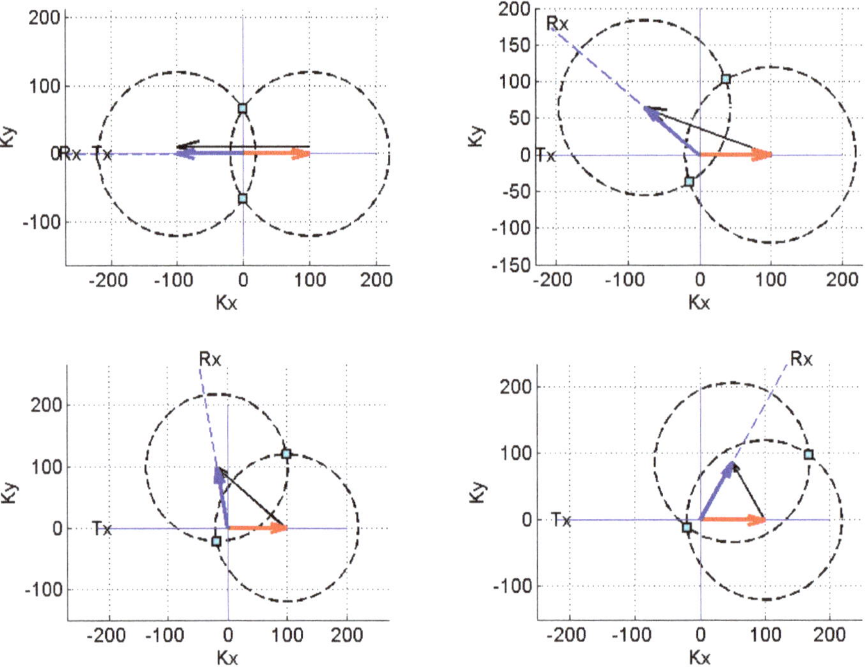

Figure 7. Double Bragg scattering processes in $\vec{\kappa}$-space. The red vector represents the incident radiowave, the blue vector the scattered radiowave, and the black vector, the required change in wave vector to be delivered by pairs of ocean waves whose $\vec{\kappa}$-vectors meet at the intersection of the circles of given wavenumbers. Here they are drawn for the case of equal wavenumbers. Tx and Rx indicate the directions of the transmitter and receiver.

It is instructive to see the form of the Doppler spectrum as a function of the bistatic geometry for particular situations and as a function of various parameters. First, we present Figure 8, taken from [18], which shows, at a most basic level, how a particular seastate modulates the frequency of the scattered signal depending on the scattering geometry: Back scatter, forward scatter, side scatter, or up scatter. In this example, only the VV (vertical in, vertical out) element of the polarization power scattering matrix is presented.

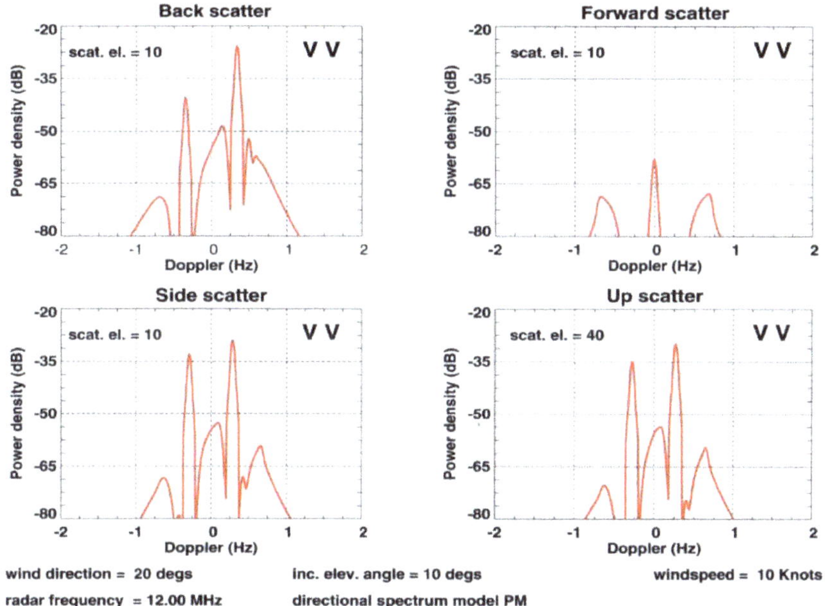

Figure 8. An example of the variation of the Doppler spectrum for representative bistatic scattering geometries incident on the same sea state: Backscatter, forward scatter, side scatter, and up scatter, as applicable to different HF radar configurations (reproduced from [18]).

Next, in Figures 9–11, we show the full polarization power scattering matrix for three different radar frequencies, in each case plotting the spectrum for four different bistatic skywave scattering geometries, at a fixed sea state. Here, 180° corresponds to backscatter, i.e., monostatic geometry. The vertical angles of incidence and reflection are 40° in all cases.

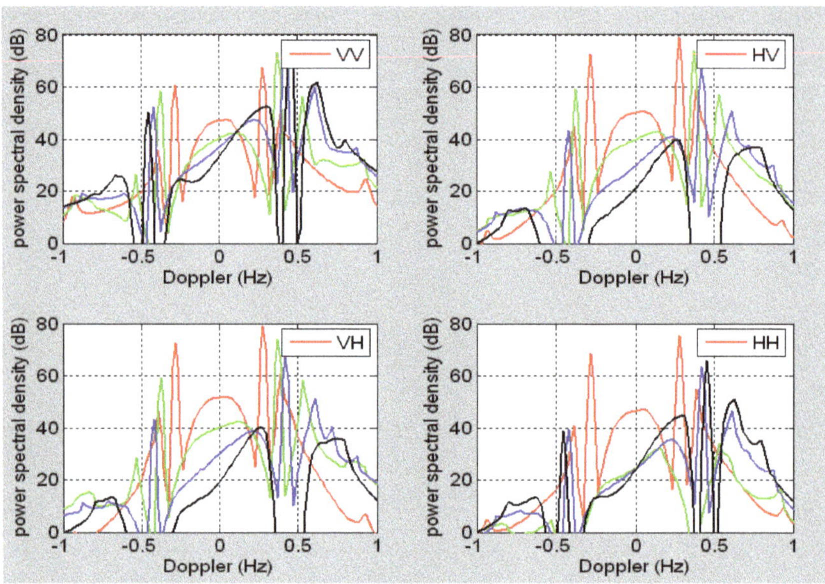

Figure 9. Sea clutter Doppler spectrum: $\varphi = 45°$(red), $90°$(green), $135°$(blue), $180°$(black); F = 25 MHz.

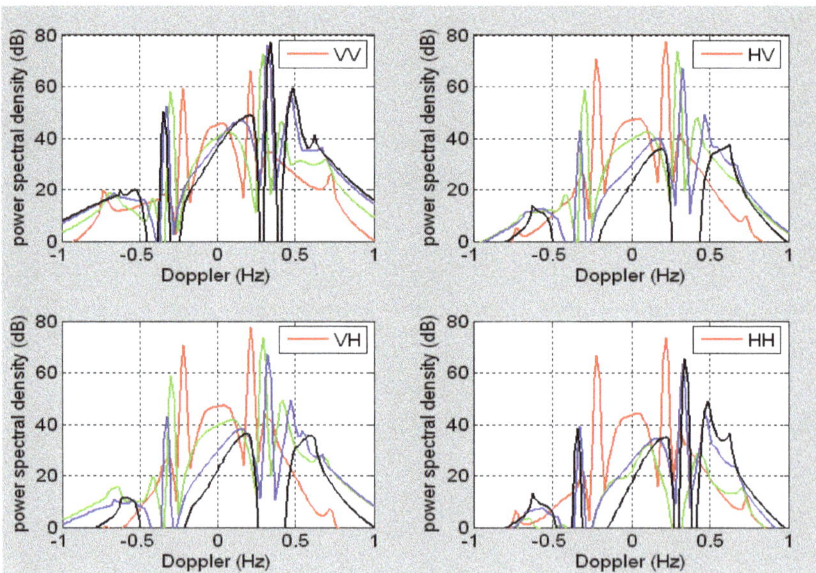

Figure 10. Sea clutter Doppler spectrum: $\varphi = 45°$(red), $90°$(green), $135°$(blue), $180°$(black); F = 15 MHz.

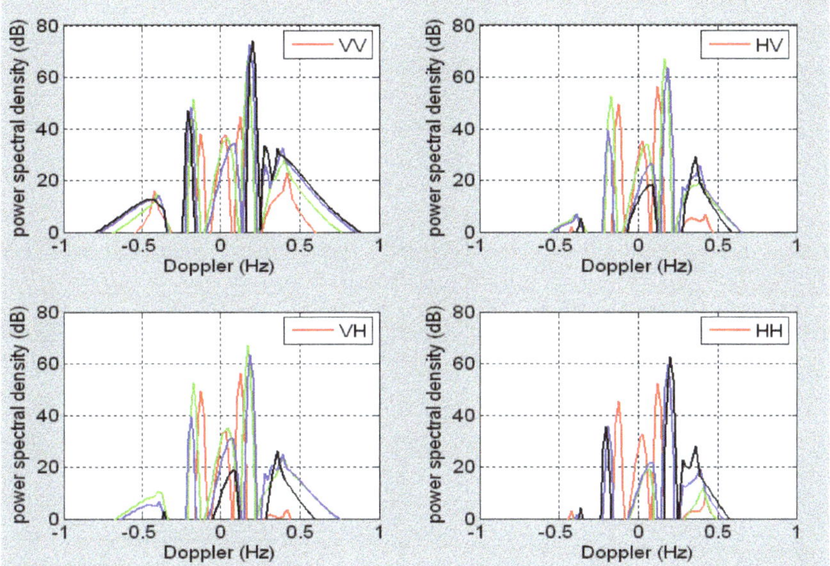

Figure 11. Sea clutter Doppler spectrum: $\varphi = 45°$ (red), $90°$ (green), $135°$ (blue), $180°$ (black); F = 5 MHz.

Finally, in Figure 12, we present a range-Doppler map measured with a bistatic HFSWR, along with two modelled range-Doppler maps computed for scattering geometries close to that used for the measurement.

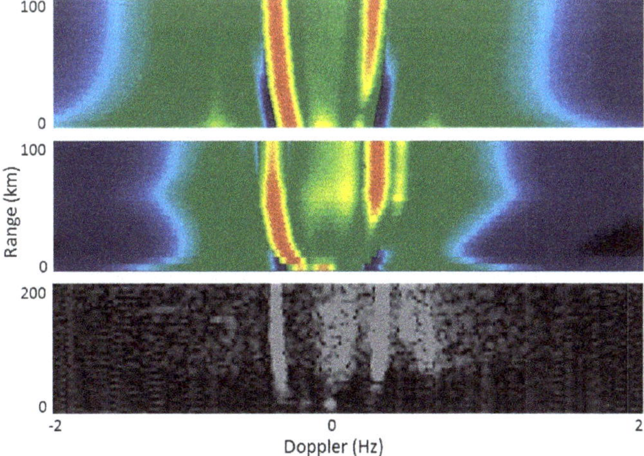

Figure 12. Modelled and measured Doppler spectra presented as range-Doppler maps. The variation of Bragg frequency with bistatic angle is clearly seen, especially at low ranges as the bistatic angle approaches its maximum. At near ranges, a decrease in signal power is evident in the measurement—this is due to the transmit antenna gain pattern falling off at angles close to the bearing to the receiver.

3.4.2. Scattering from Ships and Aircraft

Although target detection was once the province of HF radars deployed only for national defense, many low-power remote-sensing radars are now addressing this mission, primarily focusing on ship detection. Modelling HF radiowave scattering from ships and aircraft has been carried out for many decades by means of computational electromagnetic codes. Early work was almost exclusively carried out using the method-of-moments code NEC, but nowadays, other software packages such as FEKO, CST Studio, and HFSS are widely used, along with more advanced in-house codes. Here, we illustrate the general characteristics of bistatic HF radar cross sections (RCS) of ships and aircraft with results computed for the Fremantle-Class patrol boat 42 m in length, and the Aermacchi MB 326H aircraft, wingspan 10.6 m. These platforms are pictured in Figure 13.

Figure 13. The Fremantle Class patrol boat and the Aermacchi MB 326H trainer aircraft.

A format developed to present bistatic HF radar cross section is shown in Figure 14. The figure shows the predicted bistatic RCS of the Fremantle Class patrol boat, evaluated at four different radar frequencies. More precisely, it shows the squared magnitude of the VV component of the polarization scattering matrix, but we will use the term RCS where no confusion is likely. This format uses columns to index the azimuthal angle at which the radar signal is incident on the target and rows to index the azimuthal angle of departure or scattered angle. In this example, the elevation angle is set to $0°$ (vertical angle of incidence=$90°$), which closely approximates the field structure for HFSWR. Below each panel showing the bistatic RCS is the monostatic RCS, i.e., the trailing diagonal of the matrix.

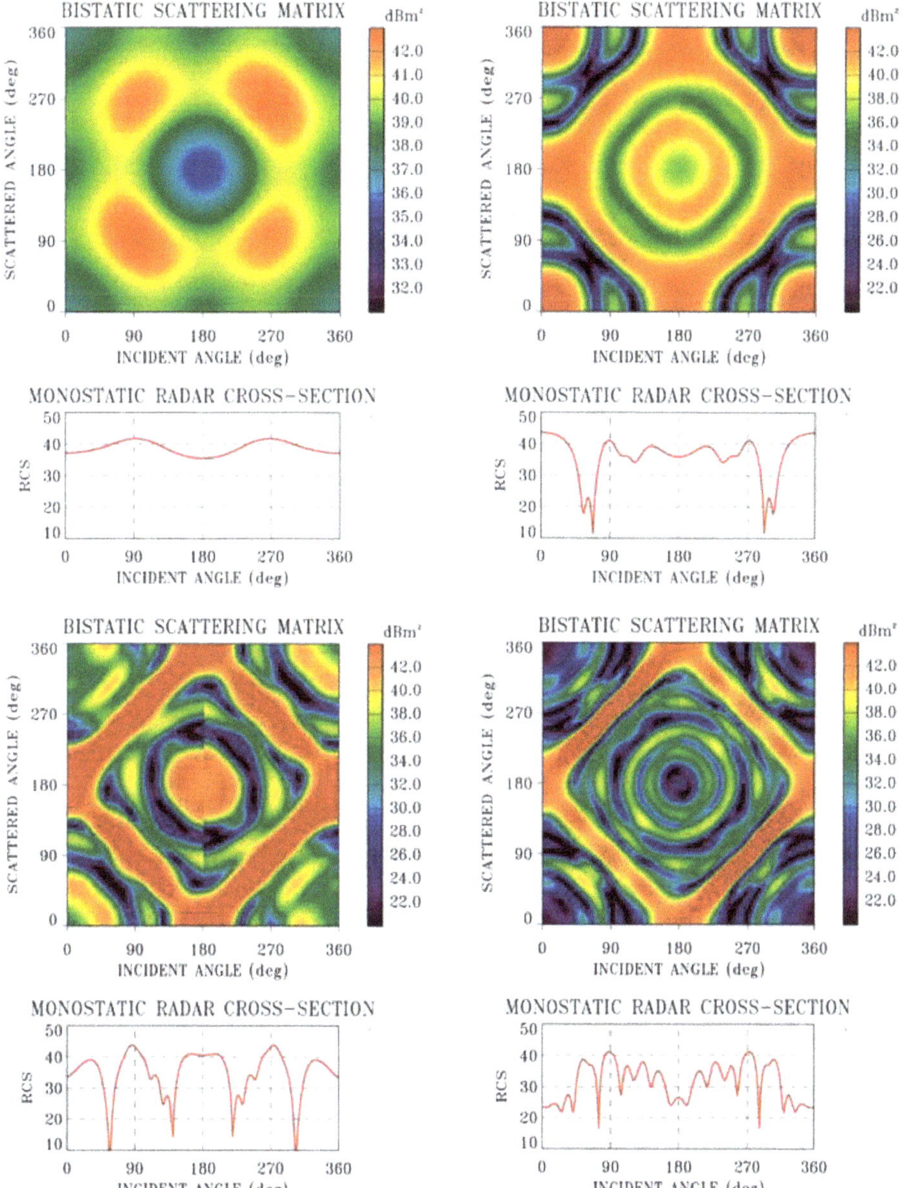

Figure 14. The bistatic RCS of the Fremantle Class patrol boat, evaluated at HF radar frequencies 5, 10, 15, and 20 MHz for HFSWR configurations. The lower panel in each case shows the monostatic RCS which is just the cut along the trailing diagonal.

For detection, we require that a ship echo exceed the clutter and noise power in the same Doppler bin by some margin ε; that is, there exists $\omega \in [-\Omega, \Omega]$ such that

$$s(\omega) > c(\omega) + n(\omega) + \varepsilon \tag{12}$$

where $s(\omega)$, $c(\omega)$, and $n(\omega)$ are the target, clutter, and noise power spectral densities. This situation is illustrated in Figure 15, where a nominal target echo is superimposed on Doppler spectra evaluated for four different sea states.

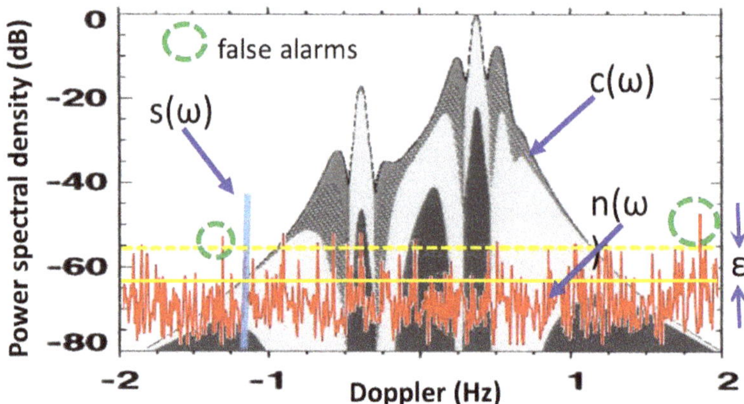

Figure 15. Ship detection in clutter and noise, showing the importance of thresholding.

The rich structure of the RCS patterns reveals the potential for exploitation by both the radar designer selecting sites for his transmit and receive systems, and the vessel seeking to minimize its detectability, in concert with other strategies that a clandestine mission might exploit. As we have seen in the previous section, the option of utilizing a bistatic configuration gives us an extra degree of freedom for 'controlling' the clutter spectrum.

While aircraft detection has seldom been a priority for HFSWR, it is certainly an established capability. Figure 16 uses a different format to show the VV RCS of the Aermacchi MB 326H as a function of azimuth (i.e., aspect) and radar frequency. The five panels correspond to different bistatic angles, namely, 0° (i.e., monostatic), 20°, 40°, 60°, and 80°. As aircraft move at speeds that carry them rapidly across typical HFSWR coverage, presenting a changing aspect, a distinctive temporal variation of echo strength can be observed, potentially useful for target classification.

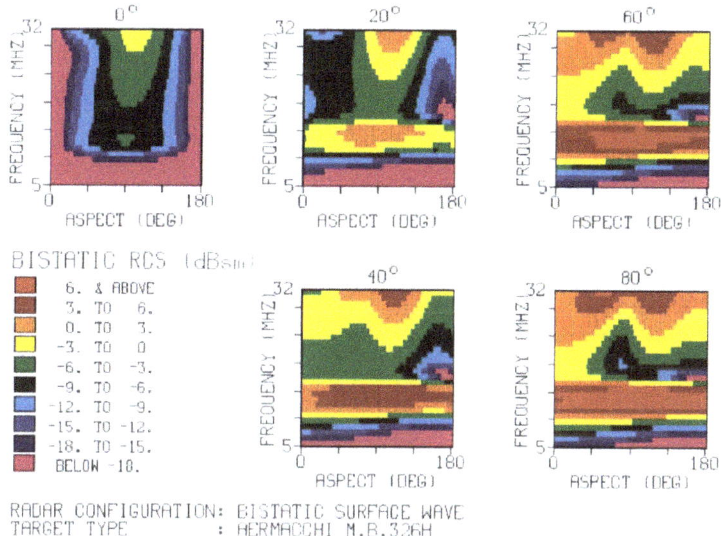

Figure 16. The HFSWR RCS of the Aermacchi MB 326H aircraft as a function of aspect and radar frequency. The five panels correspond to different bistatic angles as indicated.

3.4.3. Scattering from Slow-Moving Vehicles

Another example of the benefits of bistatic scattering at HF is the improved skywave detectability of slow land-based vehicles. Measurements of the spatial spectra of two types of terrain are presented in Figure 17a,b. These are based on very high-resolution digital elevation maps, covering thousands of square kilometers, with postings at 5 m spacing.

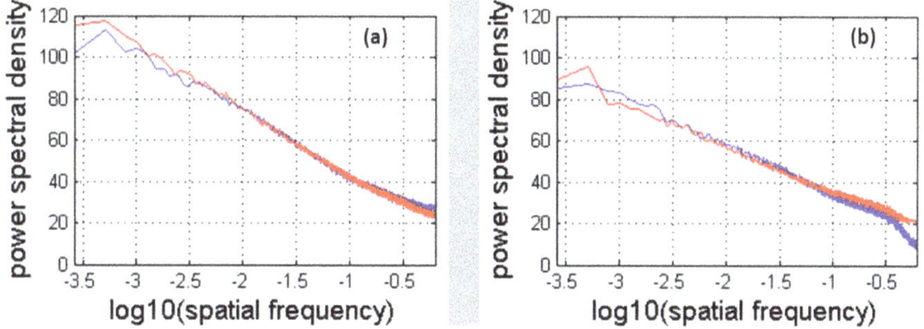

Figure 17. Power spectra of terrain elevation for (**a**) a hilly region, and (**b**) a relatively flat region.

From Figure 17a, corresponding to a hilly region, the surface elevation η follows a one-dimensional power law with $S(\kappa) \propto \kappa^{-3}$ while, for the relatively flat region treated in Figure 17b, the power law behaves as $S(\kappa) \propto \kappa^{-2.4}$. It is not unreasonable to regard these exponents as rough limits on typical terrestrial surfaces when the vegetation density is light. Now, the local HF bistatic scattering coefficient for natural land surfaces can be modelled with first-order small perturbation theory, taking account of the dependence of the scattering coefficient on the local angle of incidence, weighted by the medium-scale slope probability density function. To proceed, we facet the resolution cell into n patches that are planar in the mean, but with small scale roughness,

$$\sigma_{cell}(\theta_{scat}, \varphi_{scat}, \theta_{inc}, \varphi_{inc}) = \int \sigma\left(\theta'_{scat}, \varphi'_{scat}, \theta'_{inc}, \varphi'_{inc}; \vec{r}\right) P\left(\theta'_{inc}, \varphi'_{inc}\right) d\theta'_{inc} d\varphi'_{inc} d\vec{r}$$
$$= \sum_{m=1}^{n} \sigma\left(\theta'_{scat}(m), \varphi'_{scat}(m), \theta'_{inc}(m), \varphi'_{inc}(m)\right) \mu(\Delta_m) \tag{13}$$

where the patches over each of which a local normal vector is defined are obtained by faceting the surface using a local flatness criterion.

The general expression for the first-order bistatic scattering coefficient takes the form

$$\sigma_{pq}^{(1)}(\theta_{scat}, \varphi_{scat}, \theta_{inc}, \varphi_{inc}) = 2^6 \pi k_0^4 P_{pq} \sum_{m=1,2} S(k_0 \sin\theta_{scat} \cos(\varphi_{scat} - \varphi_{inc})$$
$$-k_0 \sin\theta_{inc}, \; k_0 \sin\theta_{scat} \sin(\varphi_{scat} - \varphi_{inc})) \tag{14}$$

where q and p index the incident and scattered polarisation states and the function P_{pq} accounts for the polarisation-dependence. To apply this to the tilted facets, we need only to transform to local coordinates via the appropriate rotation matrix, apply the scattering formula, then perform the inverse transformation.

There are two important considerations here that relate to the merits of bistatic scattering geometry. First, substituting the measured power laws of the land surfaces in the scattering formula, and writing the expression for the case of $\theta_{scat} = \theta_{inc} = \frac{\pi}{2}$ for clarity, we find

$$\sigma_{pq}^{(1)}\left(\frac{\pi}{2}, \varphi_{scat}, \frac{\pi}{2}, \varphi_{inc}\right) \propto k_0^4 . P_{pq} \left| k_0 \cos(\varphi_{scat} - \varphi_{inc}) - k_0 \right|^{-\alpha} \equiv k_0^{4-\alpha}. P_{pq} \left| \cos(\varphi_{scat} - \varphi_{inc}) - 1 \right|^{-\alpha} \tag{15}$$

where $\alpha \in [2.4, 3,]$. There are two features of interest here. The first is the slow fall-off of the power spectrum, which implies that the scattering coefficient will increase with radar frequency, so clutter can be reduced by operating at lower frequencies. The other feature is the variation of the scattering coefficient with bistatic angle. The case of the VV element of the polarization scattering matrix is plotted in Figure 18 over a typical range of bistatic angles. It is evident that by employing a moderately bistatic scattering geometry, the land clutter can be reduced by a factor of 2–3. As spectral leakage is a key limiting factor in slow land target detection, this is a significant gain in detectability.

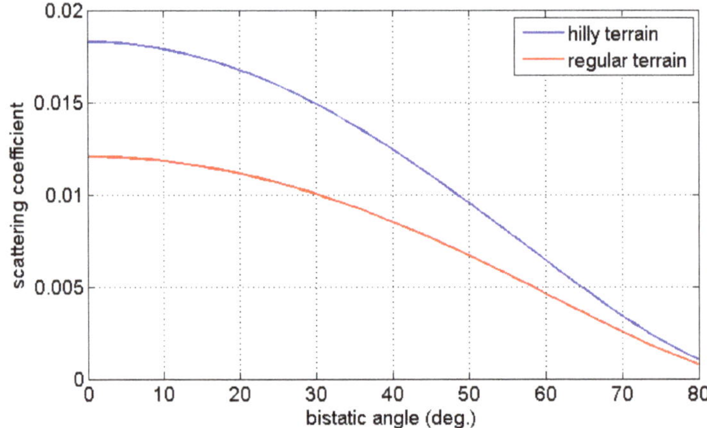

Figure 18. The variation of the VV scattering coefficient as a function of bistatic angle.

3.4.4. Scattering from Ship Wakes

The possibility of detecting ship wakes was recognized in the 1970s, but early studies focused on first-order scattering, which is critically dependent on radar geometry and frequency and hence not a practical solution. The development of a second-order theory [38–40] has opened the way to a meaningful capability, but here too, careful consideration of the radar geometry and frequency is required to approach optimum sensitivity.

According to this theory, the wake signature depends not only on the vessel, it is also a function of the ambient sea state, so no simple assertion of optimality of one or other configuration can be made. Nevertheless, extensive modelling has shown that bistatic geometries routinely emerge as optimal solutions.

One indication of the possible advantage of bistatic geometry can be seen from Figure 19, showing a measured omni-directional wave spectrum from the open sea, on which is superimposed the corresponding spectrum for a ship wake. As the annotations reveal, the wake components are concentrated at low frequencies (and hence wavenumbers) for which monostatic first-order Bragg scatter at HF is not available, but at a bistatic angle φ the resonant wavenumber is decreased by a factor $\sin(\varphi/2)$.

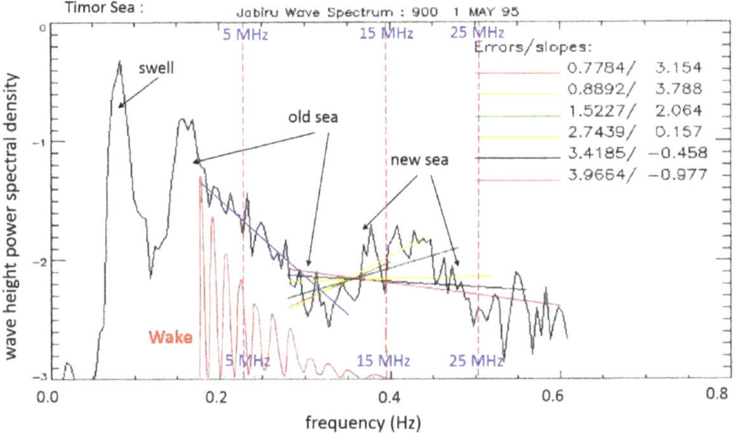

Figure 19. The omnidirectional spectrum of an ambient sea, and, superimposed (in red), the wake of a merchant ship at normal sailing speed; note that the abscissa is frequency, not wavenumber, connected here by the familiar deep-water dispersion relation. The radar frequencies corresponding to monostatic Bragg scatter are marked by the dashed lines.

3.4.5. Scattering from Ice-Covered Seas

There have been many oceanographic applications of HF radar but with a sole exception, they share the attribute that the dispersion relation assumed to govern the sea surface dynamics is that corresponding to a free water surface. In almost all applications, the general form of the dispersion relation,

$$\omega^2 = \left(g\kappa + \frac{\tau}{\rho}\kappa^3\right)\tanh\kappa H \tag{16}$$

is approximated by the inviscid, deep-water limit, $\tau \to 0$, $H \to \infty$

$$\omega^2 = g\kappa \tag{17}$$

though several studies have considered the shallow water case [29,41,42].

In the polar regions, the sea surface freezes and a complex ice structure forms over millions of square kilometers, varying dramatically with the season. This 'skin' is not completely rigid; it possesses mechanical properties that allow waves from the open sea to penetrate the ice zone, causing the surface to undulate as they propagate. In order to measure the surface motions and, from the Doppler signature, infer the structural and mechanical properties of the ice, the free surface dispersion relation has to be replaced by a new model that takes flexural and viscoelastic characteristics into account. Moreover, the HF scattering theory that is the basis of radar echo interpretation has to be reformulated with the appropriate ice dispersion relation. Different forms of ice have different dispersion relations, so a prerequisite to ice field mapping is a computational scattering theory that can handle any situation. Such a theory has been developed and reported in [11,37]; it solves the forward problem for HF radar, treating the most general case of bistatic geometry and polarimetric dependence. Figure 20 compares the Doppler spectra for a particular sea state and radar configuration, with and without ice present.

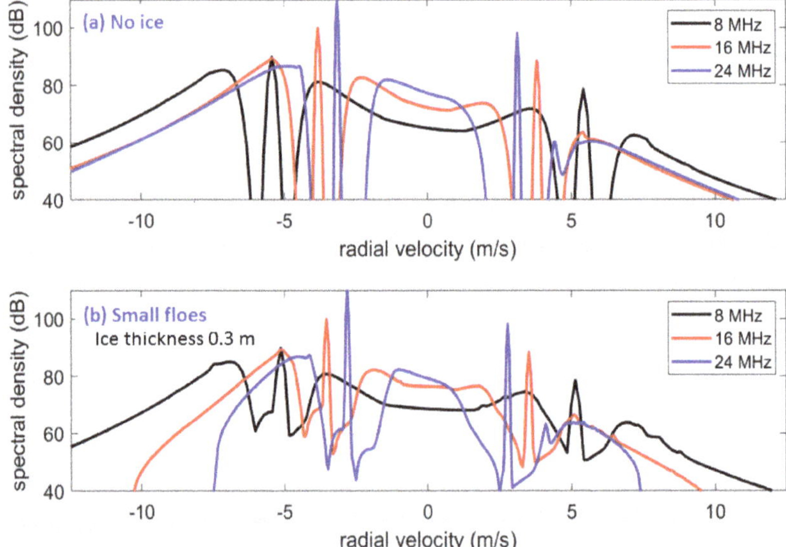

Figure 20. Doppler spectra computed for two situations: (**a**) A free ocean surface, and (**b**) a sea covered with small ice floes ('pancake' ice). Results for three radar frequencies are superimposed.

There are two key features of waves in the ice field that immediately draw our attention to bistatic HF radar geometries. First, the short waves tend to be attenuated far more quickly than long waves, in line with physical intuition. To give an idea of scale, waves in the swell frequency band can penetrate well over 100 km into the ice zone. Second, the most accessible properties of the ice that we seek to measure are encoded in the first-order Bragg scatter components. Together these imply is that we need to be able to sample the first-order Bragg scatter from the longer waves with our radars. As we noted earlier, bistatic scattering gives us greater access to the long waves.

Of course, the radar signal needs to propagate to areas of interest within the ice field, and this is another component of the remote sensing problem, but the local properties of the ice are reflected in the local ice dynamics and hence the local intrinsic Doppler spectrum.

3.4.6. Scattering from the Ionosphere

In addition to providing the propagation channel for HF skywave radar, the ionosphere is relevant to a wide range of human activities, so its physical properties are of keen interest. To HFSWR users, it poses a hazard in the form of field-aligned irregularities that result in ionospheric clutter, a potent source of obscuration of desired echoes. Techniques aimed at mitigating ionospheric clutter are employed in some systems but seldom achieve the goal of peeling away the elevated clutter to reveal the echoes from the sea. The most successful methods exploit detailed knowledge of the physics involved, and this applies to other technologies that rely on the ionosphere, such as communication with spacecraft, and to climate studies.

To determine the properties of the ionosphere, various sounding systems are used. Vertical incidence sounders provide useful point information but to sample a wide area, we need to employ oblique illumination. Line-of-sight HF radars, included in the table of Figure 1, support some types of observation, but to obtain some more subtle properties, a bistatic sounding technique is required; that is, a form of bistatic radar. There is a vast literature on this, but for now it suffices to point out that the demands of modern HF radar systems go far beyond the 'traditional' channel transfer function parameterization—the distribution of energy over group range and Doppler [23]. More advanced applications of HF skywave radar and HF communications demand information on the structure and dynamics of the ionospheric plasma as manifested in wavefront geometry [14], repolarization and depolarization [15], wideband phase path modulation [13], nonlinear effects [25], and a variety of higher-order parameters. Bistatic skywave radar is a powerful tool for exploring these phenomena, as well as being a beneficiary of the derived knowledge.

3.4.7. Polarization Considerations in Bistatic Scattering Configurations

The scattering theories that are most widely employed to model HF radar scattering phenomena and hence to solve the associated inverse problems—the small perturbation method, the Born approximation, and physical optics—share the attribute that, at first order, they predict zero cross-polarized return for backscatter. If we wish to extract information from the cross-polarized components of the scattered field by applying the inverse scattering operator to the measurements, and staying with a monostatic radar configuration, we need to extend the theories to second order, thereby complicating the inversion procedures. Moreover, the cross-polarized elements of the monostatic polarization scattering matrix are frequently small, though this certainly may not be the case for the value of the information they contain. Not only can bistatic scattering geometries provide access to the cross-polarized echo information content at first order, it is often the case that cross-polarization becomes more significant as the bistatic angle increases. Thus, even at the quite fundamental level, there can be strong reasons for adopting a bistatic configuration even when a monostatic configuration is simpler to engineer and install.

That said, we need to bear in mind that the surface wave propagation mode heavily favours transverse magnetic field propagation (ie, approximately vertical, but with a forward tilt), the line-of-sight mode departs only slightly from being polarization-blind, and skywave propagation introduces complex field transformations, with both repolarization and depolarization [15] entering the picture. It follows that the practicalities of antenna design and installation are heavily dependent on the propagation modalities involved. Bistatic surface wave radars, like their monostatic counterparts, rely predominantly on relatively simple, vertically polarized antenna elements, though auxiliary horizontally polarized elements can be used to reduce external noise and interference very effectively. Purely line-of-sight radars in any configuration can measure the full scattering matrix, even through an intervening ionosphere [43], if they are equipped with appropriate dual-polarized elements. For skywave radars, it is not yet clear whether the cost and complexity of deploying a polarimetric capability can achieve, in practice, the benefits that theory suggests might be accessible. In particular, to design polarimetric antennas able to radiate a controlled polarization state over a substantial range of azimuths and elevations is a formidable challenge.

4. Bistatic Configurations that Are Currently of Particular Interest

The great majority of HF radar systems in operation today employ a single propagation mechanism, be it the skywave or surface wave mode. Recently, though, there has been a surge of interest in 'hybrid-mode' mode radars, so here we shall remark very briefly on three of these special configurations.

First consider these two configurations:

- The Tx→[skywave]→target→[surface-wave]→Rx configuration,

- The Tx→[skywave]→target→[line-of-sight]→Rx configuration.

These configurations are not new: Designs for both types were proposed in the 1980s for land-based transmitters and shore-based or shipborne receivers. Associated experiments carried out, but several studies concluded that these configurations were optimum only for niche applications. Nevertheless, the concept of augmenting skywave radars with forward-based receiving facilities was resurrected by several groups in the mid-2000s [44,45] and by others more recently [46–51]. To date, the modelling studies reported in the open literature have ignored many of the complexities of the skywave leg of the propagation path, though we may anticipate improvements in this area. The other area where endless complications enter the picture is platform dynamics and the associated impact on the Doppler spectrum of received signals, as noted in some recent publications [35,36,52–54].

We should also note the emergence of another, potentially potent configuration:

- Tx→[surface-wave]→target→[line-of-sight]→Rx, where the reception takes place in space.

Of the many non-standard configurations listed in Figure 1, this one has only recently been explored in the form where the receiver is mounted on an orbiting spacecraft; this is illustrated in the schematic in Figure 21a. As there are hundreds of HFSWR systems in operation, a single satellite might be able to collect information from a large number of radars as it travels around the Earth. Bernhardt et al. [55] conducted an experiment to test the closely related concept Tx→[skywave]→target→[line-of-sight]→Rx using the Canadian ePOP/CASSIOPE satellite and were able to identify land and sea features in the received echoes.

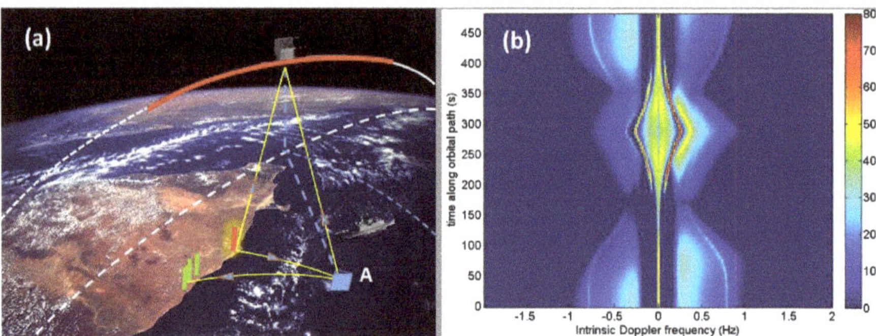

Figure 21. (**a**) The bistatic configuration of the satellite-borne receiver acquiring sea clutter echoes produced by a shore-based HFSWR, and (**b**) the computed Doppler spectrum of the signal arriving at points along the satellite orbital path.

An independent investigation [56] modelled the second-order intrinsic Doppler spectrum of the upwards-propagating signals as they reach orbital heights and demonstrated the viability of the concept by measuring the Doppler spectra of sea clutter collected over a Tx→[skywave]→target→[surface-wave]→Rx path. As shown in Figure 21b, the Doppler spectrum from a fixed patch on the sea varies in a complex and potentially highly informative way along the satellite's orbital path, though of course it must be acquired and processed by the spacecraft. Figure 22 depicts the geometry of an experiment that confirmed the viability of a signal path that includes mode (iii) as a subset in a time-reversed sense, i.e., retracing the signal path from receiver to sea to ionosphere.

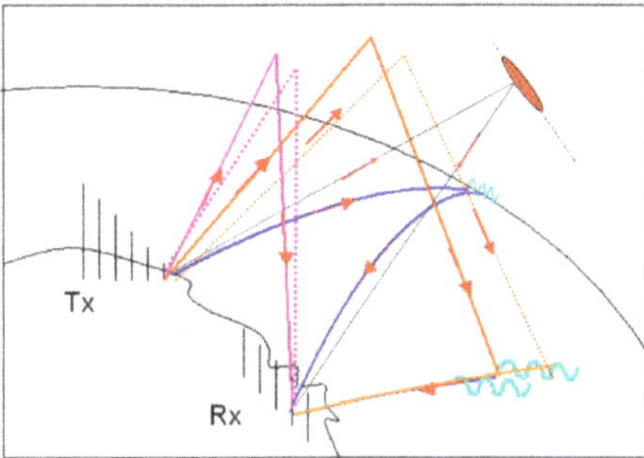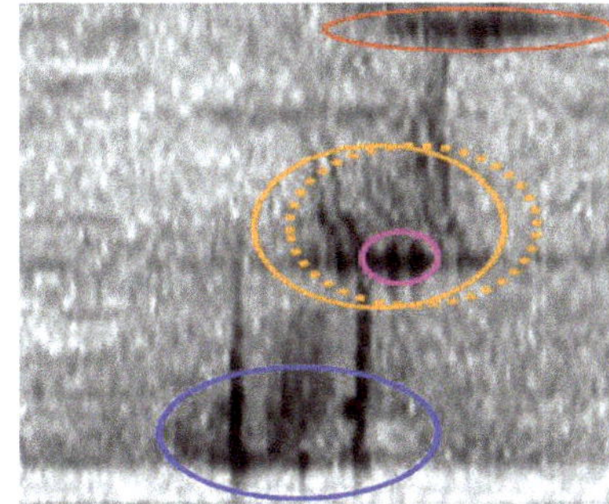

Figure 22. Experimental data from a nominal HF surface wave radar, which by happenstance recorded signals on Tx -> skywave -> sea scatter -> surface wave -> receiver. The magenta traces indicate the direct overhead reflection, the orange traces show the path of interest. Note that dual reflection points in the corrugated ionosphere provided a pair of echo traces in each case, slightly displaced in Doppler.

5. Site Selection for Bistatic HF Radars

If the challenge of optimum site selection for a monostatic radar may be considered difficult, then the equivalent task for a bistatic radar system is formidable. All the complexities of propagation and scattering that we have reviewed in the preceding pages must be taken into account and their impact assessed, carefully weighed against the missions to be addressed and the statistics of environmental parameters [57].

5.1. The Orthogonality Criterion

Intuitively, one might suspect that the benefits of bistatic configurations should be maximized when the bistatic angle is close to 90°, just as it is for stereoscopic radar configurations measuring currents or tracking targets—minimizing the well-known geometric dilution of precision (GDOP). This reasoning does not lead to practical or even good solutions for many applications. First, there is the obvious limitation of site availability—suitable locations offering near-orthogonality over the priority part of the coverage criterion may not be exist, or if they do, may be inaccessible for some reason. Second, near-orthogonality is no guarantee of increased radar cross section. Indeed, it is often the case that the bistatic RCS of targets of interest is consistently low for such scattering geometries; examining the trailing diagonal offset by 90° in either row or column space in Figure 9 makes this very clear.

Even for stereoscopic configurations, the issue of optimum radar disposition is nontrivial. As an example, in the case of ship detection against the sea clutter background, the signal-to-clutter ratio is not a monotonic function of bistatic angle, as shown schematically in Figure 23. While stereoscopic viewing clearly unmasks air targets, where the clutter may be taken as a narrow band centered on zero Doppler, the case for ship detection in clutter is complicated by the existence of Bragg line pairs and the associated second-order clutter. As shown in the figure, for some course-speed combinations, the ship can remain close to Bragg lines for both radars simultaneously.

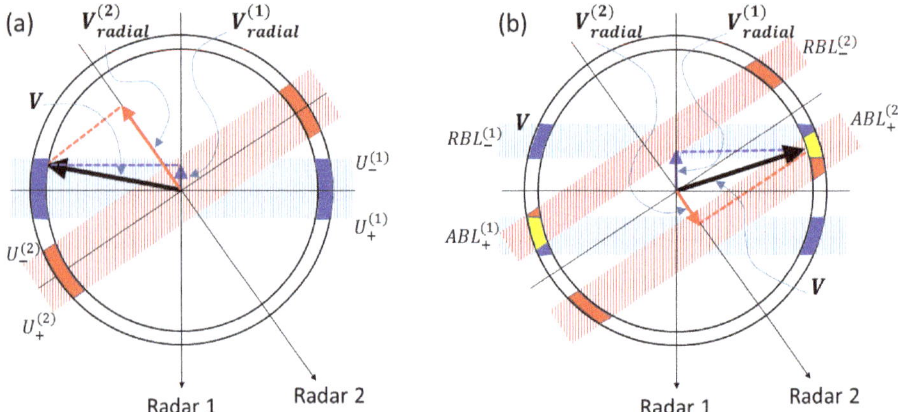

Figure 23. Blind speed diagrams for a bistatic radar system: (**a**) Aircraft detection, (**b**) ship detection. The blue arrows show the projection of the target velocity vector (black) onto the axis of Radar 1; the red arrows show the projection onto the axis of Radar 2. The blind speeds are shaded in the velocity annulus: Blue for Radar 1, red for Radar 2, yellow for doubly blind azimuths that can arise in the case of sea clutter.

Thus, an appreciation of the spatial distribution of the geometric attributes of a radar configuration plays a significant role in system design, and it has been found helpful to present this information in a graphical format. For the case of stereoscopic radar system operation, where the bistatic signal path may not be exploited, and focusing on current vector measurement and aircraft detection, where GDOP is important, it suffices to map the degree of orthogonality at each cell in the surveillance zone, as illustrated in Figure 24. It presents the information in two ways. Figure 24a plots the sine of the included angle in areas of common illumination, while Figure 24b converts this to yield the minimum radial speed that a target can present to one or other radar in its monostatic mode, expressed as a percentage of the target's actual speed. This kind of display facilitates site selection that optimizes performance over priority areas for the specified missions. It is a straightforward matter to generate equivalent displays for the case where bistatic modes of operation are available.

One must also bear in mind that potentially damaging crosstalk can occur when multiple radars in a network, perhaps stereoscopic, are obliged to share a common frequency band. In this circumstance, special signal processing is required to cancel the interfering transmissions [10].

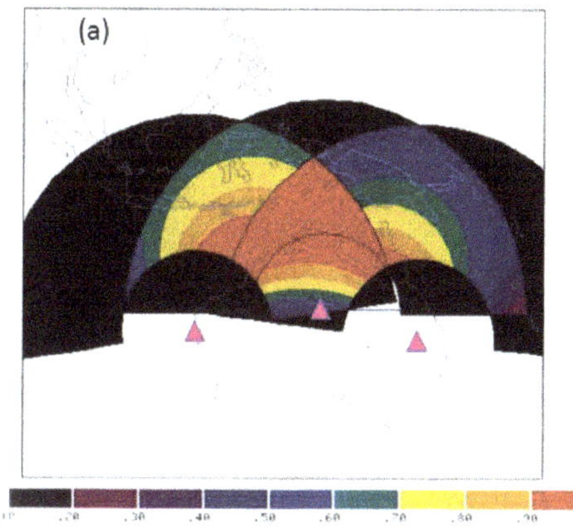
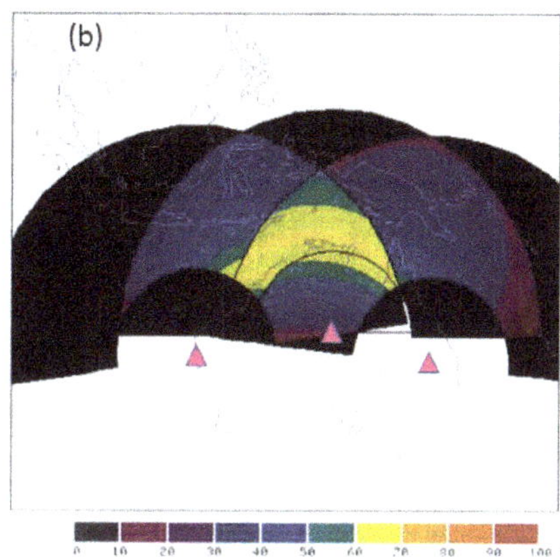

Orthogonality of illumination (sine of bistatic angle)

Minimum radial speed as a percentage of target speed

Figure 24. Trigonometric properties of overlapped radar coverage from the viewpoint of stereoscopic modes of operation. Figure 24a shows the degree of orthogonality as measured by the sine of the bistatic angle, while Figure 24b expresses it in terms of the minimum component of radial velocity that a target can present. The figure is for illustrative purposes only: The radar sites shown do not correspond to existing radars.

5.2. Site Selection via Multi-Objective Optimization

In order to take account of a multiplicity of mission types, each with its own performance criteria and dependence on prevailing environmental parameters, a far more rigorous approach to radar siting is needed. One methodology that has been used with success [58] is based on the concept of Pareto dominance. In this approach, a number of figures of merit are defined and then the site parameter space is searched to locate those solutions that possess the following property: That they are at least as good as their competitors against every FOM and better than any competitor in at least one FOM. The set of solutions that emerge from this search define the so-called Pareto front, and site selection can be carried out far more easily when one need only deal with this greatly reduced number of possibilities. Attributes such as clustering can be exploited to assess robustness of the solutions, and additional factors may then be taken into account, including less quantifiable criteria such as visual impact or even personal taste.

An important practical aspect of this methodology is its numerical implementation when the search space is large, as happens when one is designing a network of radars. Typically, one is confronted with need to select m sites out of n possibilities, where m is modest, but n may be large. One study of the South China Sea [59] encountered a sample space of $\sim 10^{23}$ solutions. To handle realistic cases successfully, an efficient nonlinear optimization algorithm is essential. One such technique has been reported in [60]; it employs a genetic algorithm fitted with a special acceleration routine that enables it to address highly demanding radar network design problems.

6. Conclusions

It may seem surprising that we have not partitioned this paper into distinct sections dealing with individual radar configurations, such as skywave and surface wave radars. Our decision was based on two considerations. First, bistatic configurations such as the hybrid sky–surface mode obviously straddle the disciplines that may be relevant to either skywave or surface wave monostatic radars, but not both. Second, and just as importantly, during half a century of first-hand experience with several forms of HF radar, we frequently encountered circumstances where cross-fertilization of ideas provided significant benefits.

Accordingly, in this paper, we have set out with four goals:

(i)   To present a taxonomy that encompasses essentially all possible HF radar configurations;

(ii) To illustrate how bistatic geometries impact on the various components of the radar process;

(iii) To provide some examples of missions for which bistatic geometries offer distinct advantages;

(iv) To describe some practical techniques that have been found efficacious in the design and siting of bistatic HF radar systems

The taxonomy speaks for itself; it may serve to provoke thought on the possible merits of alternative bistatic configurations in particular applications. The consequences of bistatism for waveforms, propagation, scattering, and so on extend beyond the brief treatment we have provided here, but hopefully the underlying message is clear: Some of the basic radar principles that we take for granted and seldom bother to reflect upon need careful re-examination when we leave the monostatic domain. We have described several radar missions where clear benefits of bistatic configurations are evident; some of these missions are familiar ones that have long been addressed with monostatic configurations, but others, such as ice monitoring and characterization, ship wake detection and analysis, and space-borne interception of HFSWR clutter for global scale radio oceanography are only now entering the HF radar user's lexicon. The list is hardly exhaustive, and no doubt there are surprises in store for us all. Finally, the practical techniques that are mentioned in the text have all been used successfully in serious applications.

Funding: This research received no external funding.

Conflicts of Interest: The author declares no conflict of interest.

References

1.   Headrick, J.M.; Anderson, S.J. HF Over-the-Horizon Radar, Chapter 20. In *Radar Handbook*, 3rd ed.; Skolnik, M., Ed.; McGraw-Hill: New York, NY, USA, 2008.

2.   Roarty, H.; Cook, T.; Hazard, L.; Harlan, J.; Cosoli, S.; Wyatt, L.; Alvarez Fanjul, E.; Terrill, E.; Otero, M.; Largier, J.; et al. The global high frequency radar network. *Front. Mar. Sci.* **2019**. [CrossRef]

3.   Lyon, E. Missile Attack Warning. In *Advances in Bistatic Radar*; Willis, N.J., Griffiths, H.D., Eds.; SCITECH Publishing: Rayleight, NC, USA, 2007; pp. 47–55.

4.   CODAR Ocean Sensors. Available online: http://www.codar.com/SeaSonde.shtml (accessed on 17 January 2020).

5.   WERA Ocean Radar. Available online: https://helzel-messtechnik.de/de/6035-WERA-Remote-Ocean-Sensing (accessed on 17 January 2020).

6.   Cosoli, S.; de Vos, S. Interoperability of direction-finding and beam-forming high-frequency radar systems: An example from the Australian high-frequency ocean radar network. *Remote Sens.* **2019**, *11*, 291. [CrossRef]

7.   Australian Defence Business Review. June 2019. Available online: https://adbr.com.au/over-the-horizon/ (accessed on 17 January 2020).

8.   Raytheon Relocatable Over-the-Horizon Radar (ROTHR) for Homeland Security. Available online: http://www.mobileradar.org/Documents/ROTHR.pdf (accessed on 17 January 2020).

9.   Francis, D.B.; Cervera, M.A.; Frazer, G.J. Performance prediction for design of a network of skywave over-the-horizon radars. *IEEE Aerosp. Electron. Syst. Mag.* **2017**, *32*, 18–28. [CrossRef]

10. Zhang, Y.; Frazer, G.J.; Amin, M.G. Simultaneous operation of two over-the-horizon radars. In Proceedings of the SPIE 5559, Advanced Signal Processing Algorithms, Architectures and Implementations XIV, Denver, CO, USA, 26 October 2004.

11. Anderson, S.J. Monitoring the marginal ice zone with HF radar. In Proceedings of the IEEE Radar Conference, Seattle, WA, USA, 8–12 May 2017.

12. Parent, J.; Bourdillon, A. A method to correct HF skywave backscattered signals for ionospheric frequency modulation. *IEEE Trans. Antennas Propag.* **1988**, *36*, 127–135. [CrossRef]

13. Anderson, S.J.; Abramovich, Y.I. A unified approach to detection, classification and correction of ionospheric distortion in HF skywave radar systems. *Radio Sci.* **1998**, *33*, 1055–1067. [CrossRef]

14. Anderson, S.J. Multiple scattering of HF skywave radar signals—Physics, interpretation and exploitation. In Proceedings of the IEEE Radar Conference, Rome, Italy, 26–30 May 2008.

15. Anderson, S.J. Skywave channel characterisation for polarimetric OTH radar. In Proceedings of the IEEE US Radar Conference, Philadelphia, PA, USA, 2–6 May 2016.

16. Anderson, S.J. The Challenge of Signal Processing for HF Over-the-Horizon Radar. In Proceedings of the Workshop on Signal Processing and Applications (WOSPA'93), Queensland University of Technology, Brisbane, Australia, 13–15 December 1993.

17. Anderson, S.J. OTH Radar Phenomenology: Signal Interpretation and Target Characterization at HF. *IEEE Aerosp. Electron. Syst. Mag.* **2017**, *32*, 4–16. [CrossRef]

18. Anderson, S.J. Limitations to the extraction of information from multihop skywave radar signals. In Proceedings of the IEEE International Conference on Radar, Adelaide, Australia, 3–5 September 2003.

19. Purdy, D.S. Receiver antenna scan rate requirements needed to implement pulse chasing in a bistatic radar receiver. *IEEE Trans. Aerosp. Electron. Syst.* **2001**, *37*, 285–288. [CrossRef]

20. Willis, N. *Bistatic Radar*; SCITECH Publishing: Rayleigh, NC, USA, 2005.

21. Anderson, S.J. Multiple scattering of HF surface waves: Implications for radar design and sea clutter interpretation. *IET Radar Sonar Navig.* **2010**, *4*, 195–208. [CrossRef]

22. Anderson, S.J. Scattering effects on DOA estimation and geolocation from HF surface wave intercepts. In Proceedings of the HF/DF Symposium, South-West Research Institute, San Antonio, TX, USA, 10–12 May 2008.

23. Earl, G.F.; Ward, B.D. The frequency management system of the Jindalee over-the-horizon backscatter HF radar. *Radio Sci.* **1987**, *22*, 275–291. [CrossRef]

24. Anderson, S.J. Stereoscopic and Bistatic Skywave Radars: Assessment of Capabilities and Limitations. In Proceedings of the Radarcon-90, Adelaide, Australia, 18–20 April 1990; pp. 305–313.

25. Anderson, S.J. Nonlinear scattering at HF: Prospects for exploitation in OTH radar systems. *Turk. J. Electron. Eng. Comput. Sci.* **2010**, *18*, 439–456.

26. Barrick, D.E. Remote Sensing of sea state by radar. In *Remote Sensing of the Troposphere*; Derr, V.E., Ed.; NOAA/Environmental Research Laboratories: Boulder, CO, USA, 1972; Chapter 12.

27. Rice, S.O. Reflection of EM waves from slightly rough surfaces. Comm. *Pure Appl. Math.* **1951**, *4*, 351–378. [CrossRef]

28. Johnstone, D. *Second-Order Electromagnetic and Hydrodynamic Effects in High-Frequency Radio Wave Scattering from the Sea*; Technical Report 3615-3; Stanford Electronics Laboratories, Stanford University: Stanford, CA, USA, 1975.

29. Barrick, D.E.; Lipa, B.J. The second-order shallow water hydrodynamic coupling coefficient in interpretation of HF radar sea echo. *IEEE J. Ocean. Eng.* **1986**, *11*, 310–315. [CrossRef]

30. Anderson, S.J.; Anderson, W.C. Bistatic scattering from the ocean surface and its application to remote sensing of seastate. In Proceedings of the IEEE APS International Symposium, Blacksburg, VA, USA, 15–19 June 1987.

31. Srivastava, S.K.; Walsh, J. An analysis of the second-order Doppler return from the ocean surface. *IEEE J. Oceanic Eng.* **1985**, *10*, 443–445. [CrossRef]

32. Walsh, J. *On the Theory of Electromagnetic Propagation across a Rough Surface and Calculations in the VHF Region*; Technical Report N00232; DREA: Ottawa, ON, Canada, 1980.

33. Gill, E.W.; Walsh, J. High-frequency bistatic cross sections of the ocean surface. *Radio Sci.* **2002**, *36*, 1459–1475. [CrossRef]

34. Gill, E.W.; Huang, W.; Walsh, J. On the development of a second-order bistatic radar cross section of the ocean surface: A high frequency results for a finite scattering patch. *IEEE J. Oceanic Eng.* **2004**, *31*, 740–750. [CrossRef]

35. Walsh, J.; Huang, W.; Gill, E.W. The first-order high frequency radar ocean surface cross section for an antenna on a floating platform. *IEEE Trans. Antennas Propag.* **2010**, *58*, 2994–3003. [CrossRef]

36. Walsh, J.; Huang, W.; Gill, E.W. The second-order high frequency radar ocean surface cross section for an antenna on a floating platform. *IEEE Trans. Antennas Propag.* **2012**, *60*, 4804–4813. [CrossRef]

37. Anderson, S.J. Prospects for the inversion of HF radar echoes from sea ice to recover structural and viscoelastic parameters. In Proceedings of the 3rd Australasian Conference on Wave Science, Auckland, New Zealand, 12–14 February 2018.

38. Anderson, S.J. HF radar signatures of ship and submarine wakes. *J. Eng.* **2019**, *2019*, 7512–7520. [CrossRef]

39. Anderson, S.J. Sensitivity of HF radar signatures of ship wakes to hull geometry parameters. In Proceedings of the 21st Australasian Fluid Mechanics Conference, Adelaide, Australia, 10–13 December 2018.

40. Anderson, S.J. An OTH Plimsoll line? Remote measurement of ship loading with HF Radar. In Proceedings of the IEEE International Conference Radar 2019, Toulon, France, 23–27 September 2019.
41. Holden, G.J.; Wyatt, L.R. Extraction of sea state in shallow water using HF radar. *IET Proc. F (Radar Signal Process.)* **1992**, *139*, 175–181. [CrossRef]
42. Lipa, B.; Nyden, B.; Barrick, D.E.; Kohut, J. HF radar sea-echo from shallow water. *Sensors* **2008**, *8*, 4611–4635. [CrossRef]
43. Anderson, S.J.; Abramovich, Y.I.; Boerner, W.-M. On the solvability of some inverse problems in radar polarimetry. In Proceedings of the SPIE Conference on Wideband Interferometric Sensing and Imaging Polarimetry, San Diego, CA, USA, 23 December 1997; Volume 3120.
44. Frazer, G.J. Forward based receiver augmentation for OTHR. In Proceedings of the IEEE Radar Conference, Boston, MA, USA, 17–20 May 2007; pp. 373–378.
45. Riddolls, R. *Ship Detection Performance of a High Frequency Hybrid Sky-Surface Wave Radar*; Ottawa Technical Memorandum TM 2007-327; Defence R&D Canada: Ottawa, ON, Canada, 2007.
46. Li, Y.J.; Wei, Y.S.; Xu, R.Q.; Shang, C. Simulation analysis and experimentation study on sea clutter spectrum for high-frequency hybrid sky-surface wave propagation mode. *IET Radar Sonar Navig.* **2014**, *8*, 917–930.
47. Zhu, Y.P.; Wei, Y.S.; Li, Y.J. First order sea clutter model for HF hybrid sky-surface wave radar. *Radioengineering* **2014**, *23*, 1180–1191.
48. Walsh, J.; Gill, E.W.; Huang, W.M.; Chen, S. On the development of a high-frequency radar cross section model for mixed path ionosphere–ocean propagation. *IEEE Trans. Antennas Prop.* **2015**, *63*, 2655–2664. [CrossRef]
49. Chen, S.; Gill, E.W.; Huang, W.M. A first-order HF radar cross-section model for mixed-path ionosphere–ocean propagation with an FMCW source. *IEEE J. Ocean Eng.* **2016**, *41*, 982–992. [CrossRef]
50. Wang, S.W.; Li, Y.J.; Xu, R.Q.; Wei, Y.S.; Wang, Z.Q.; Sheng, S. Research on characteristics of first-order sea clutter for HF sky-surface wave radar. *IET Microw. Antennas Propag.* **2016**, *10*, 1124–1134.
51. Yang, L.W.; Fan, J.M.; Guo, L.X.; Liu, W.; Li, X.; Feng, J.; Lou, P.; Lu, Z.X. Simulation analysis and experimental study on the echo characteristics of high-frequency hybrid sky–surface wave propagation mode. *IEEE Trans. Antennas Propag.* **2018**, *66*, 4821–4831. [CrossRef]
52. Ma, Y.; Gill, E.W.; Huang, W.M. First-order bistatic high-frequency radar ocean surface cross-section for an antenna on a floating platform. *IET Radar Sonar Navig.* **2016**, *10*, 1136–1144. [CrossRef]
53. Ma, Y.; Huang, W.M.; Gill, E.W. The Second-Order Bistatic High-Frequency Radar Cross Section of Ocean Surface for an Antenna on a Floating Platform. *Can. J. Remote Sens.* **2016**, *42*. [CrossRef]
54. Ma, Y.; Huang, W.M.; Gill, E.W. Bistatic high-frequency radar ocean surface cross section for an FMCW source with an antenna on a floating platform. *Int. J. Antennas Propag.* **2016**. [CrossRef]
55. Bernhardt, P.A.; Siefring, C.L.; Briczinski, S.C.; Vierinen, J.; Miller, E.; Howarth, A.; James, H.G.; Blincoe, E. Bistatic observations of the ocean surface with HF radar, satellite and airborne receivers. In Proceedings of the IEEE OCEANS 2017-Anchorage, Anchorage, AK, USA, 18–21 September 2017; pp. 1–5.
56. Anderson, S.J. Space-borne interception of HF radar echoes for environmental intelligence. In Proceedings of the 16th Australian Space Research Conference, Melbourne, Australia, 26–28 September 2016.
57. Anderson, S.J. Optimising bistatic HF radar configurations for target and environmental signature discrimination. In Proceedings of the IEEE Conference on Information, Decision and Control, Adelaide, Australia, 12–14 February 2007.
58. Anderson, S.J. Optimizing HF radar siting for surveillance and remote sensing in the Strait of Malacca. *IEEE Trans. Geosci. Remote Sens.* **2013**, *51*, 1805–1816. [CrossRef]
59. Anderson, S.J. HF radar network design for remote sensing of the South China Sea. In *Geoscience and Remote Sensing*; Margheny, M., Ed.; InTech: London, UK, 2014.
60. Anderson, S.J.; Darces, M.; Helier, M.; Payet, N. Accelerated convergence of genetic algorithms for application to real-time problems. In Proceedings of the 4th International Symposium on Inverse Problems, Design and Optimization, Albi, France, 26–28 June 2013.

Article

The Scattering Coefficient for Shore-to-Air Bistatic High Frequency (HF) Radar Configurations as Applied to Ocean Observations

Zezong Chen [1,2,*], **Jian Li** [1], **Chen Zhao** [1], **Fan Ding** [1] and **Xi Chen** [1]

[1] School of Electronic Information, Wuhan University, Wuhan 430072, China; ivanlijian@whu.edu.cn (J.L.); zhaoc@whu.edu.cn (C.Z.); dingf242@whu.edu.cn (F.D.); ivanchenxi@whu.edu.cn (X.C.)

[2] School of Electronic Information and the Collaborative Innovation Center for Geospatial Technology, Wuhan University, Wuhan 430072, China

* Correspondence:chenzz@whu.edu.cn; Tel.: +86-133-0711-8527

Received: 28 October 2019; Accepted: 9 December 2019; Published: 11 December 2019

Abstract: To extend the scope of high frequency (HF) radio oceanography, a new HF radar model, named shore-to-air bistatic HF radar, has been proposed for ocean observations. To explore this model, the first-order scattering coefficient and the second-order electromagnetic scattering coefficient for shore-to-air bistatic HF radar are derived using the perturbation method. In conjunction with the contribution of the hydrodynamic component, the second-order scattering coefficient is derived. Based on the derived scattering coefficients, we analyzed the simulated echo Doppler spectra for various scattering angles and azimuthal angles, operation frequencies, wind speeds, and directions of wind, which may provide the guideline on the extraction of sea state information for shore-to-air bistatic HF radar. The singularities in the simulated echo Doppler spectra are discussed using the normalized constant Doppler frequency contours. In addition, the scattering coefficients of shore-to-air bistatic HF radar are compared with that of monostatic HF radar and land-based bistatic HF radar. The results verify the correctness of the proposed scattering coefficients. The model of shore-to-air bistatic HF radar is effective for ocean observations.

Keywords: shore-to-air bistatic HF radar; scattering coefficient; Doppler spectra

---

1. Introduction

High frequency (HF) radars have been efficient tools for ocean current, wave, and wind measurement, as well as target detection in the past four decades [1–4]. The interpretation of HF radio scattering from the ocean surface in monostatic and bistatic mode has been developed for several decades. A monostatic HF radar system consisting of a colocated transmitter and receiver operates in backscattering case. The first-order and second-order scattering coefficients for monostatic HF radar, derived by Barrick [5,6] based on Rice's work [7], have been widely accepted [8,9]. Subsequently, new monostatic HF electromagnetic scattering coefficients were proposed by Walsh using generalized functions from rough surfaces [10]. Hisaki and Tokuda also presented the monostatic results using the perturbation method when the illuminated area is finite [11,12]. In contrast to monostatic mode, bistatic HF radar system operates in the non-backscattering case in general. Johnstone presented the scattering coefficients of bistatic HF radar [13]. Anderson et al. [14] obtained a general solution to the bistatic scattering problem and have published a number of papers applying the formulae to various configurations and presenting computed spectra (e.g., [15]); however, they did not publish the details of their derivation. Other theoretical results were proposed by Anderson et al. and validated by field experiments [16–18]. Gill and Walsh developed the first-order and second-order scattering coefficients for land-based bistatic HF radar based on a generalized function [19–21]. Some theoretical results were validated by Huang et al. using the wind direction measurements from the land-based bistatic HF radar [22]. Recently, Bernhardt gave an incoherent scattering coefficient related to the wave-height spectra for HF Ground-Ionosphere-Ocean-Space (GIOS) system [23].

Air-borne radars have the ability to detect large areas of the sea, which is meaningful to extend coverage for ocean radars [23–25]. To extend the scope of HF radio oceanography and meet the demands for large-area ocean observation, a new bistatic radar model so-called shore-to-air bistatic HF radar is designed for ocean observation [26]. The configuration is shown in Figure 1: The transmitter installed on the coast emits vertically polarized and narrow-beam electromagnetic waves to illuminate the ocean patch in a grazing incidence; the electromagnetic waves are scattered to a radar receiver deployed at an air platform (airplane or airship) owing to the rough sea surface; then the power spectra are estimated from radar echoes to extract the sea state information.

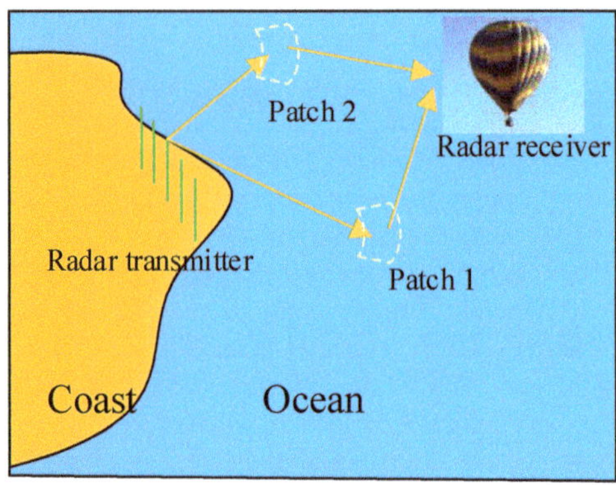

Figure 1. Model of shore-to-air bistatic high frequency (HF) radar.

In this paper, beginning with the establishment of the geometry of shore-to-air bistatic HF radar, the electric field intensity is obtained for the perfectly conducting rough ocean surface. Then the first-order and second-order electric field intensities are derived using Rice's method. The electric field near the observation point is obtained based on Kirchhoff theory [27]. The first-order and second-order electromagnetic scattering coefficients are given. Finally, the second-order scattering coefficient is obtained in conjunction with the contribution of hydrodynamic coupling. In order to validate the proposed scattering coefficients, the Doppler spectra are simulated in various scattering and azimuthal angles, operating frequencies, wind speeds, and wind directions.

This paper is organized as follows. The derivation of the first-order scattering coefficient and second-order scattering coefficient is given in Section 2. In Section 3, the simulated Doppler spectra in various operating modes and sea states are presented and analyzed. In Section 4, the singularities that occur in the simulated Doppler spectra are discussed. Conclusions are drawn in Section 5.

2. Model and Scattering Coefficients

2.1. The Geometry of Shore-To-Air Bistatic Hf Radar

The geometry of shore-to-air bistatic HF radar is shown in Figure 2. The x axis is assumed as the direction of the radar beam and the y axis is perpendicular to the x axis. The z axis is vertical to the sea surface. The incident electromagnetic wave (the wavenumber vector is $\vec{k_0}$) lies on x-z plane and the incident angle θ_i is the angle of the incident radar wave from the z axis. For the near-grazing incident wave, $\theta_i \approx \frac{\pi}{2}$. θ_s is the angle of the scattered radar wave (the wavenumber vector is $\vec{k_{sc}}$) from the z axis. φ_s is the azimuthal angle of the scattered radar wave from the incidence plane. The half angle between the transmitter and the projection in the x-y plane of the receiver as viewed from the scatter patch is φ_0, which satisfies the equation $\varphi_s = 180° - 2\varphi_0$. The scattering coefficients of bistatic HF radar can be approximated as the sum of the first-order and second-order scattering coefficients

$$\sigma(\omega) = \sigma^{(1)}(\omega) + \sigma^{(2)}(\omega). \tag{1}$$

The equation of the perfectly conducting rough time-varying surface z can be expressed as a Fourier series:

$$z = f(x,y,t) = \sum_{mnl} P(m,n,l)exp\left\{(-j(\frac{2\pi m}{L}x + \frac{2\pi n}{L}y) - j\omega lt)\right\} \tag{2}$$

where the triple summation extends from $-\infty$ to $+\infty$ for l, m, and n, $P(m,n,l)$ are the Fourier expansion coefficients, and $T = 2\pi/\omega$ is the time period of the Fourier expansion and corresponds to the spatial period L (assumed to be large).

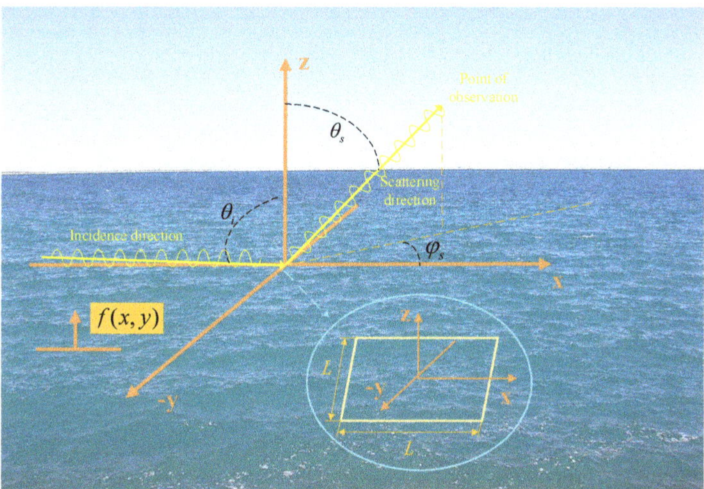

Figure 2. Geometry of shore-to-air bistatic HF radar.

2.2. Electromagnetic Scattering on the Rough Sea Surface

As shown in Figure 3, the magnitude of the vertical polarized incident wave is assumed to be unity. The electric-field vector of the incident wave can be expressed as:

$$\overrightarrow{E_i}(x,z,t) = (cos\theta_i\hat{x} + sin\theta_i\hat{z})exp(-jk_0 sin\theta_i x + jk_0 \cos\theta_i z - j\omega_0 t) \tag{3}$$

where $\hat{x}$ and $\hat{z}$ are unit vectors along with the x axis and z axis, respectively, k_0 is the magnitude of radar radio wavenumber vector $\overrightarrow{k_0}$, which is defined by the equation $k_0 = 2\pi/\lambda$ (λ is the wavelength of incident wave), ω_0 is the circular frequency, and t is the time.

The total scattering field is the sum of reflected fields for lack of surface roughness (specular scattering), and scattered fields due to the roughness of sea surface (nonspecular scattering). The specular scattering fields can be expressed as:

$$\overrightarrow{E_1}(x,z,t) = (-cos\theta_i\hat{x} + sin\theta_i\hat{z})exp(-jk_0 sin\theta_i x - jk_0 \cos\theta_i z - j\omega_0 t). \tag{4}$$

The components of nonspecular scattering fields $\overrightarrow{E_2}$ in the x, y, and z directions can be expressed as:

$$
\begin{cases}
E_{2x} = \sum_{mnl} A_{mnl} E(m,n,z,l) \\
E_{2y} = \sum_{mnl} B_{mnl} E(m,n,z,l) \\
E_{2z} = \sum_{mnl} C_{mnl} E(m,n,z,l)
\end{cases}
\tag{5}
$$

where $A_{mnl}, B_{mnl}, C_{mnl}$ are constants, and

$$
E(m,n,z,l) = \exp\left(-j\frac{2\pi m}{L}x - j\frac{2\pi n}{L}y - jb(m,n)z - j\omega lt\right).
\tag{6}
$$

The reflected wave field should satisfy the wave equation

$$
b^2(m,n) = k_0{}^2 - \left(\frac{2\pi m}{L}\right)^2 - \left(\frac{2\pi n}{L}\right)^2
\tag{7}
$$

and the divergence of the reflected field should be zero. Therefore, the coefficients A_{mnl}, B_{mnl}, and C_{mnl} are determined by the relation

$$
\frac{2\pi m}{L} A_{mnl} + \frac{2\pi n}{L} B_{mnl} + b(m,n) C_{mnl} = 0.
\tag{8}
$$

$E(m,n,f,l)$ can be expanded as exponential series and A_{mnl} can be expressed using the perturbation method

$$
E(m,n,f,l) = E(m,n,0,l)[1 - jb(m,n)f + \ldots]
\tag{9}
$$

$$
A_{mnl} = A_{mnl}^{(1)} + A_{mnl}^{(2)} + \ldots
\tag{10}
$$

where $f = z$, $A_{mnl}^{(1)}$ denotes $o(f)$, and $A_{mnl}^{(2)}$ denotes $o(f^2)$. B_{mnl}, and C_{mnl} can be expressed in a similar way.

The total electromagnetic field above the surface is expressed as a sum of the incident field and scattered field (including the specular scattered field and nonspecular scattered field). For the vertical polarized incident wave illuminating the ocean surface, the components of total electromagnetic fields $E(x,y,z,l)$ in the x, y, and z directions can be expressed as

$$
\begin{aligned}
E_x = {} & 2j\cos\theta_i[\sin(k_0\cos\theta_i z)]e^{-jk_0\sin\theta_i x - j\omega_0 t} + \\
& \sum_{mnl}(A_{mnl}^{(1)} + A_{mnl}^{(2)} + \ldots)[1 - jb(m,n)f + \ldots]E(m,n,0,l)
\end{aligned}
\tag{11}
$$

$$
E_y = \sum_{mnl}(B_{mnl}^{(1)} + B_{mnl}^{(2)} + \ldots)[1 - jb(m,n)f + \ldots]E(m,n,0,l)
\tag{12}
$$

$$
\begin{aligned}
E_z = {} & 2\sin\theta_i[\cos(k_0\cos\theta_i z)]e^{-jk_0\sin\theta_i x - j\omega_0 t} + \\
& \sum_{mnl}(C_{mnl}^{(1)} + C_{mnl}^{(2)} + \ldots)[1 - jb(m,n)f + \ldots]E(m,n,0,l).
\end{aligned}
\tag{13}
$$

Substituting Formulas (11)–(13) into Rice boundary conditions [7] and separating the first-order and the second-order terms in these formulas, $A_{mnl}^{(1)}, B_{mnl}^{(1)}, C_{mnl}^{(1)}$ and $A_{mnl}^{(2)}, B_{mnl}^{(2)}, C_{mnl}^{(2)}$ can be solved. Then the electric components of the reflected electromagnetic waves can be obtained. According to the relationship between the electric field and the magnetic field of Maxwell's equation $\nabla \times \overrightarrow{E} = -j\omega\mu\overrightarrow{H}$, where ∇ denotes the Hamilton operator, and μ is the permeability to the volume material, the magnetic components of the reflected electromagnetic field can be obtained.

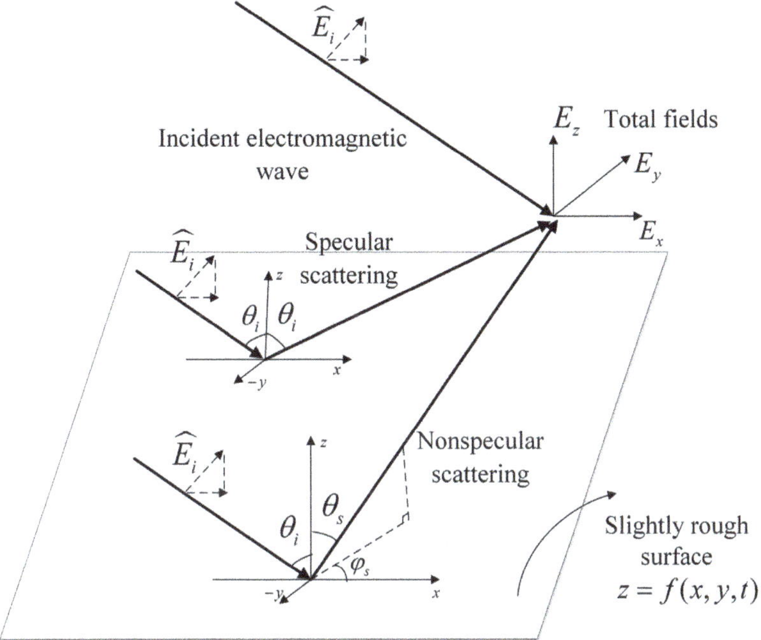

Figure 3. Fields above the sea surface. The total field is the sum of incident field, specular scattering field, and nonspecular scattering field. $\widehat{E}_i$ denotes the magnitude of vertical polarized incident wave, which is assumed to be unity.

The electric field of the observation point can be derived based on Kirchhoff theory [27]

$$\vec{E}(\vec{r}) = \nabla \times \int_{S'} \vec{N} \times \vec{E}(\vec{r'}) G_0(\vec{r'}, \vec{r}) dS' - \frac{j}{\omega \varepsilon} \nabla \times \nabla \times \int_{S'} \vec{N} \times \vec{H}(\vec{r'}) G_0(\vec{r'}, \vec{r}) dS' \qquad (14)$$

where S' is the patch of ocean surface, $\vec{N}$ is the unit normal to the surface S', $\vec{r}$ is the position vector designating point of observation, $\vec{r'}$ is the position vector designating source field, ε is the permittivity of the volume material, and $G_0(r', r)$ is the Green function:

$$G_0(\vec{r'}, \vec{r}) = \frac{\exp(-jk|\vec{r} - \vec{r'}|)}{4\pi|\vec{r} - \vec{r'}|}. \qquad (15)$$

The scattering coefficient of the vertical polarized waves can be defined by the equation in [13]

$$\sigma_v = \frac{4\pi|E_s{}^2|R^2}{|E_i{}^2|L^2} \qquad (16)$$

where E_s is the vertically polarized components of scattered electric field in the position of observation, and R is the distance from scattering patch to the receiver.

2.3. The First-Order Scattering Coefficient

The first-order electric field and magnetic field can be calculated using $A_{mnl}^{(1)}, B_{mnl}^{(1)}, C_{mnl}^{(1)}$. Then the first-order scattering coefficient derived from Formulas (14) and (16) can be expressed as

$$
\begin{aligned}
\sigma^{(1)}(\omega,\theta_s,\varphi_s) = {}& 2^4\pi k_0{}^4 \times (sin\theta_s - cos\varphi_s)^2 \\
& \times \sum_{m=\pm 1} S[k_0(sin\theta_s cos\varphi_s - 1), k_0 sin\theta_s sin\varphi_s]\delta(\omega - m\omega_B)
\end{aligned}
\tag{17}
$$

where $S(\cdot)$ denotes the ocean directional wavenumber spectrum. The delta-function $\delta(\cdot)$ represents the condition in which Bragg resonance occurs. Thus, ideally, the first-order Bragg peaks located at the the Bragg frequencies of $\pm\omega_B$, are defined by the dispersion equation for deep water

$$
\omega_B = \sqrt{gk_B} = \sqrt{g}\left(k_x{}^2 + k_y{}^2\right)^{1/4}
\tag{18}
$$

where k_B denotes the magnitude of Bragg wavenumber vector $\overrightarrow{k_B}$, g is the gravitational acceleration, and k_x, k_y are the component of $\overrightarrow{k_B}$ in the x and y directions, respectively:

$$
\begin{cases}
k_x = k_0(sin\theta_s cos\varphi_s - 1) \\
k_y = k_0 sin\theta_s sin\varphi_s
\end{cases}
\tag{19}
$$

Assuming that the angle of Bragg wave from x axis is β, the tan β can be expressed as

$$
\tan\beta = \frac{k_y}{k_x} = \frac{sin\theta_s sin\varphi_s}{sin\theta_s cos\varphi_s - 1}.
\tag{20}
$$

For the shore-to-air bistatic HF radar, the magnitude of the Bragg wavenumber vector is determined by the radar wavenumber, the scattering angle and the azimuth angle. The direction of the Bragg wavenumber vector is determined by the scattering angle and the azimuth angle.

2.4. The Second-Order Scattering Coefficient

Similar to the derivation of the first-order scattering coefficient, the second-order electromagnetic scattering coefficient can be derived using $A_{mnl}^{(2)}, B_{mnl}^{(2)}, C_{mnl}^{(2)}$

$$
\begin{aligned}
\sigma^{(2)}{}_{EM}(\omega,\theta_s,\varphi_s) = {}& 2^4\pi k_0{}^4(sin\theta_s - cos\varphi_s)^2 \\
& \times \sum_{m,m'=\pm 1} \int_{-\infty}^{+\infty}\int_{-\infty}^{+\infty} \left|\frac{A+B}{2}\right|^2 S(m\overrightarrow{k_1})S(m'\overrightarrow{k_2})\delta(\omega - m\sqrt{gk_1} - m'\sqrt{gk_2})dxdy
\end{aligned}
\tag{21}
$$

where A and B are

$$
A = \frac{-\dfrac{(\overrightarrow{k_0}\cdot\overrightarrow{k_1})(\overrightarrow{k_2}\cdot\overrightarrow{k_s})}{(sin\theta_s - cos\varphi_s)sin\theta_s\cdot k_0{}^2} - (k_0{}^2 - (\overrightarrow{k_0} + \overrightarrow{k_1})(\overrightarrow{k_s} - \overrightarrow{k_2}))}{\sqrt{k_0{}^2 - (\overrightarrow{k_0} + \overrightarrow{k_1})(\overrightarrow{k_s} - \overrightarrow{k_2})}}
\tag{22}
$$

$$
B = \frac{-\dfrac{(\overrightarrow{k_0}\cdot\overrightarrow{k_2})(\overrightarrow{k_1}\cdot\overrightarrow{k_s})}{(sin\theta_s - cos\varphi_s)sin\theta_s\cdot k_0{}^2} - (k_0{}^2 - (\overrightarrow{k_0} + \overrightarrow{k_2})(\overrightarrow{k_s} - \overrightarrow{k_1}))}{\sqrt{k_0{}^2 - (\overrightarrow{k_0} + \overrightarrow{k_2})(\overrightarrow{k_s} - \overrightarrow{k_1})}}.
\tag{23}
$$

In practice, it is impossible for a vertically polarized wave to propagate exactly parallel to any surface that is imperfect, and satisfy the required boundary conditions at the interface. Both roughness and finite conductivity of the medium below the surface force an effective boundary condition at the mean interface that gives an apparent vertical wave vector component $-k_0\Delta$. For the rough and imperfect sea at HF, a typical value of normalized surface impedance Δ is $\Delta \approx 0.011 - i0.012$ [28,29]. The second-order electromagnetic scattering coefficient involving $-k_0\Delta$ can be expressed as

$$\sigma^{(2)}{}_{EM}(\omega, \theta_s, \varphi_s) = 2^4 \pi k_0{}^4 (\sin \theta_s - \cos \varphi_s)^2$$
$$\times \sum_{m,m'=\pm 1} \int_{-\infty}^{+\infty} \int_{-\infty}^{+\infty} \left| \frac{A_1 + B_1}{2} \right|^2 S(m\vec{k_1}) S(m'\vec{k_2}) \delta(\omega - m\sqrt{gk_1} - m'\sqrt{gk_2}) dx dy \tag{24}$$

where A_1 and B_1 are

$$A_1 = \frac{-\dfrac{(\vec{k_0} \cdot \vec{k_1})(\vec{k_2} \cdot \vec{k_s})}{(\sin \theta_s - \cos \varphi_s) \sin \theta_s \cdot k_0{}^2} - (k_0{}^2 - (\vec{k_0} + \vec{k_1})(\vec{k_s} - \vec{k_2}))}{\sqrt{k_0{}^2 - (\vec{k_0} + \vec{k_1})(\vec{k_s} - \vec{k_2})} - k_0\Delta} \tag{25}$$

$$B_1 = \frac{-\dfrac{(\vec{k_0} \cdot \vec{k_2})(\vec{k_1} \cdot \vec{k_s})}{(\sin \theta_s - \cos \varphi_s) \sin \theta_s \cdot k_0{}^2} - (k_0{}^2 - (\vec{k_0} + \vec{k_2})(\vec{k_s} - \vec{k_1}))}{\sqrt{k_0{}^2 - (\vec{k_0} + \vec{k_2})(\vec{k_s} - \vec{k_1})} - k_0\Delta}. \tag{26}$$

Electromagnetic coupling coefficient Γ_{EM} can be defined as

$$\Gamma_{EM} = \frac{A_1 + B_1}{2}. \tag{27}$$

The second-order electromagnetic scattering process can be illustrated as in Figure 4. Figure 4 is a view of Figure 2 in the x-y plane. The direction of the radar beam $\vec{k_0}$ is in the x direction. $\vec{k_s}$ is the projection vector of scattering wave vector $\vec{k_{sc}}$ in the x-y plane. The incident radar wave (wavenumber vector is $\vec{k_0}$) interacts with the first ocean wave $\vec{k_1}$, to produce an intermediate scattered wave $\vec{k}$. The interactions between an intermediate scattered wave and a second ocean wave $\vec{k_2}$, produce a scattered wave $\vec{k_s}$. These waves obey the constraints

$$\vec{k_s} = \vec{k_0} + \vec{k_B} \tag{28}$$

and

$$\vec{k_B} = \vec{k_1} + \vec{k_2}. \tag{29}$$

In addition to the contribution of the second-order electromagnetic scattering, the second-order scattering also contains the contribution of hydrodynamic coupling. The process of hydrodynamic coupling arises from the combination of two ocean waves to produce a second-order ocean wave that generates Bragg scattering. The hydrodynamic coupling coefficient Γ_H in deep water can be found in [30,31]

$$\Gamma_H = \frac{-i}{2} \left[\frac{(\vec{k_1} \cdot \vec{k_2} - k_1 k_2)(\omega^2 + \omega_B^2)}{mm' \sqrt{k_1 k_2}(\omega^2 - \omega_B^2)} + k_1 + k_2 \right]. \tag{30}$$

Therefore, the second-order scattering coefficient can be written as

$$\sigma^{(2)}(\omega, \theta_s, \varphi_s) = 2^4 \pi k_0{}^4 (\sin \theta_s - \cos \varphi_s)^2$$
$$\times \sum_{m,m'=\pm 1} \int_{-\infty}^{+\infty} \int_{-\infty}^{+\infty} |\Gamma|^2 S(m\vec{k_1}) S(m'\vec{k_2}) \delta(\omega - m\sqrt{gk_1} - m'\sqrt{gk_2}) dx dy \tag{31}$$

where the value of m and m' denotes the four cases of how the two ocean waves are combined. Γ is the sum of the electromagnetic coupling coefficient and the hydrodynamic coupling coefficient

$$\Gamma = \Gamma_H + \Gamma_{EM}. \tag{32}$$

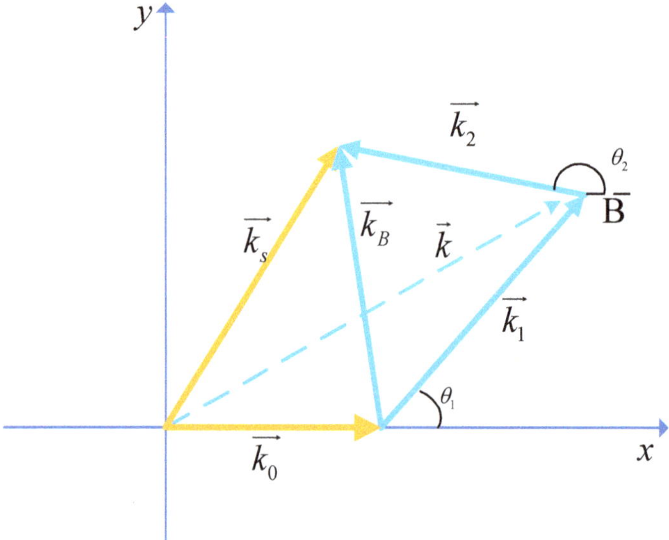

Figure 4. Illustration of the second-order electromagnetic interaction process.

3. Simulation Results and Analysis of the Echo Spectrum

To explore the dependence of the scattering coefficient of a single ocean patch on the environmental parameters, we have employed the Pierson–Moskowiz wave spectrum model [32] with the Longuet–Higgins directional distribution [33] for a wind-driven sea. Many factors, including scattering and azimuthal angles, operating frequencies, and wind speeds and wind directions, are input to the model to examine the effects on the Doppler spectrum.

Figure 5 shows four simulated Doppler spectra for different scattering angles and azimuth angles. The radar operating frequency and wind speed are set to 18 MHz and 12 m/s, respectively. The wind direction is 90°, which is referenced to the direction of the Bragg wave. The first-order Bragg peaks, which have the maxima amplitude, can be seen from the figure at the normalized Bragg frequency $F_B=1$. Figure 5a shows the Doppler spectrum when the scattering angle and azimuth angle are $\theta_s = 90°$, $\varphi_s = 180°$, which is in a monostatic case. Figure 5b shows the Doppler spectrum when the scattering angle and azimuth angle are $\theta_s = 90°$, $\varphi_s = 120°$, respectively, which is in the land-based bistatic case. The singularities at $\pm\sqrt{2}F_B = 1.414$ result from the second-order electromagnetic coupling and the hydrodynamic coupling discussed in [29]. Other singularities at the normalized frequency of $\pm 2^{3/4}F_B$ for monostatic HF radar and at $f_d = \pm 2^{3/4}\sqrt{\frac{(1\pm sin\phi_0)^{1/2}}{\cos\phi_0}}F_B$ for land-based bistatic radar, resulting from "corner reflection" condition of second-order electromagnetic scattering [21,29]. Figure 5c,d show the Doppler spectrum in the shore-to-air bistatic radar configuration. Except for the singularities at the normalized frequency of $\pm\sqrt{2}F_B$, there are other singularities resulting from "corner reflection" in the Doppler spectrum. They will be discussed in next section. In addition, the Doppler spectrum is asymmetric about the zero frequency when the wind direction is perpendicular to the direction of reference. This asymmetry is caused by second-order electromagnetic scattering, which is different from monostatic and land-based bistatic cases.

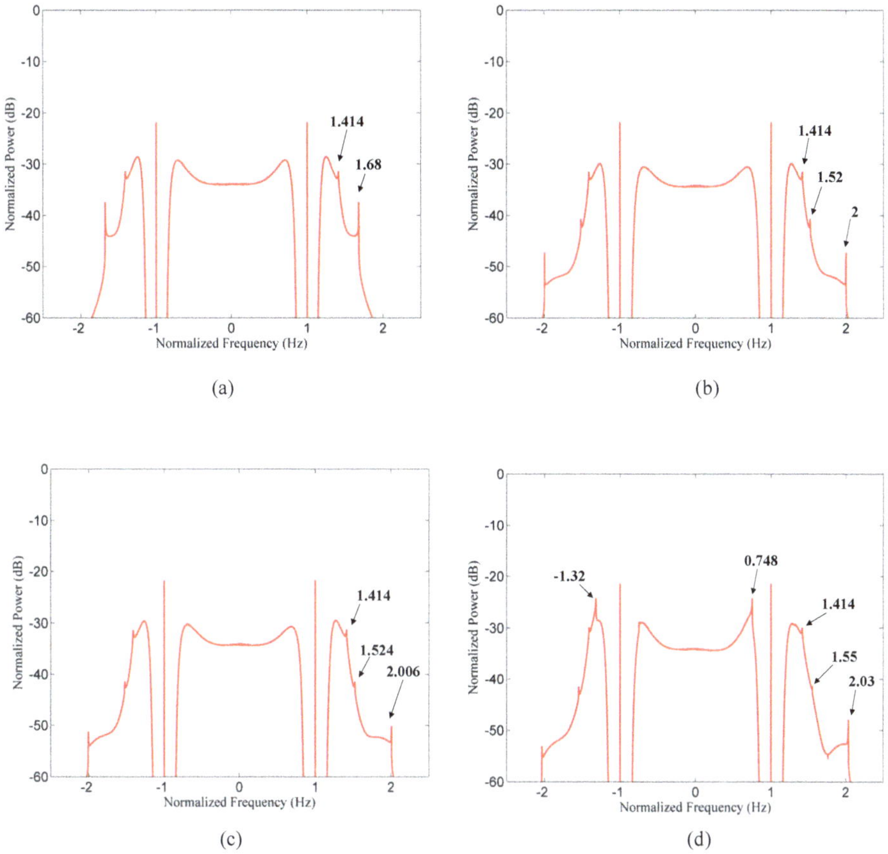

Figure 5. Simulated Doppler spectra for different scattering angles and azimuth angles.The scattering angle and azimuth angle are (**a**) $\theta_s = 90°$, $\varphi_s = 180°$, (**b**) $\theta_s = 90°$, $\varphi_s = 120°$, (**c**) $\theta_s = 75°$, $\varphi_s = 120°$, and (**d**) $\theta_s = 60°$, $\varphi_s = 120°$.

Figure 6 shows the Doppler spectra at different operating frequencies. The wind speed, wind direction, scattering angle, and azimuth angle are set to 12 m/s, 90°, 60°, and 120°, respectively. It can be noted that the magnitude of the Bragg peaks does not dramatically vary with the change of operating frequency since the Bragg wave is in the saturated zone of the wave height spectrum. The Bragg frequency increases with the operating frequency. The relation between the Bragg frequency and operating frequency can be given by

$$f_B = \left(\frac{g f_0}{2\pi c} \phi \right)^{1/2} \tag{33}$$

where f_0 is the frequency of the electromagnetic wave emitted by radar, c is the speed of light, and $\phi = \sqrt{\sin^2\theta_s + 1 - 2\sin\theta_s\cos\varphi_s}$. In addition, the magnitude of the Doppler spectra near the first-order peaks increases with the operating frequency as well.

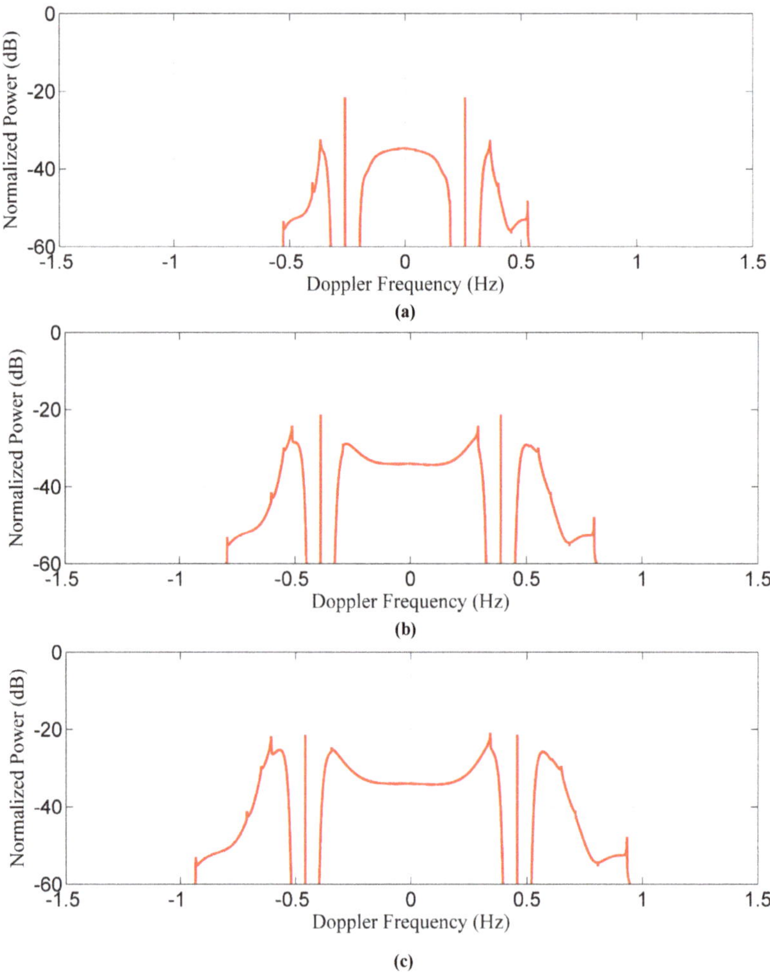

Figure 6. Simulated Doppler spectra at different operating frequencies. The operating frequencies are
(**a**) 8 MHz, (**b**) 18 MHz, and (**c**) 25 MHz.

Figure 7 shows the results for different wind speeds. The radar operating frequency, wind
direction, scattering angle, and azimuth angle are set to 18 MHz, 90°, 60°, and 120°, respectively.
It can be found that the magnitude of the Bragg peaks does not significantly change when the wind
speed varies. As aforementioned, this is because the circular frequency of the Bragg wave is about
2.44 rad/s in the above operating status, which indicates that the Bragg wave is fully developed.
However, the magnitude of Doppler spectrum near the first-order peaks is sensitive to the wind
speed. It indicates that the long ocean waves corresponding to this part of the Doppler spectrum
have more energy when wind speed above the ocean surface becomes higher. Just as for monostatic
radar, the second-order spectrum can be applied to extract the information of ocean waves [34–37].
Additionally, the magnitude of the second-order spectra far away from Bragg peaks (e.g., at 0.2 Hz
and 1.8 Hz) is hardly influenced by the wind speed, since the ocean waves responsible for this portion
of the Doppler spectra stay in the saturated region of the ocean spectrum.

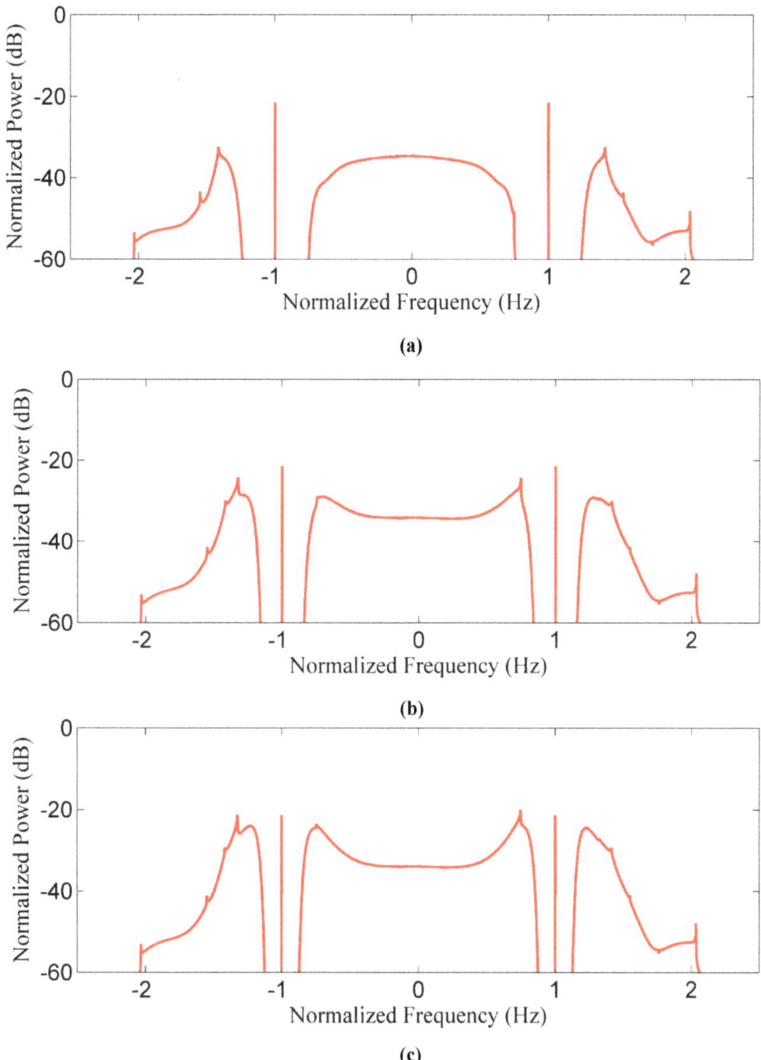

Figure 7. Simulated Doppler spectra with different wind speeds. The wind speed is (**a**) 8 m/s, (**b**) 12 m/s, and (**c**) 15 m/s.

Figure 8 shows the simulated Doppler spectra for different wind directions when the wind speed, scattering angle, and azimuth angle are 12 m/s, 60°, and 120°, respectively. It can be noted that the ratio of the magnitude of the left and right Bragg peaks in the Doppler spectrum varies with the change of wind direction. In the monostatic case, the energy of left and right Bragg peaks is equivalent when the wind direction is perpendicular to the direction of radar beam. In other cases, the energy of one Bragg peak is enhanced and the other will be weakened. For shore-to-air bistatic HF radar, a similar phenomenon occurs, in which the two Bragg peaks do not carry similar amounts of energy when wind direction is not perpendicular to the direction of the Bragg wave. Some researchers have used the ratio of the left and right first-order Bragg peaks to estimate wind direction [38–40].

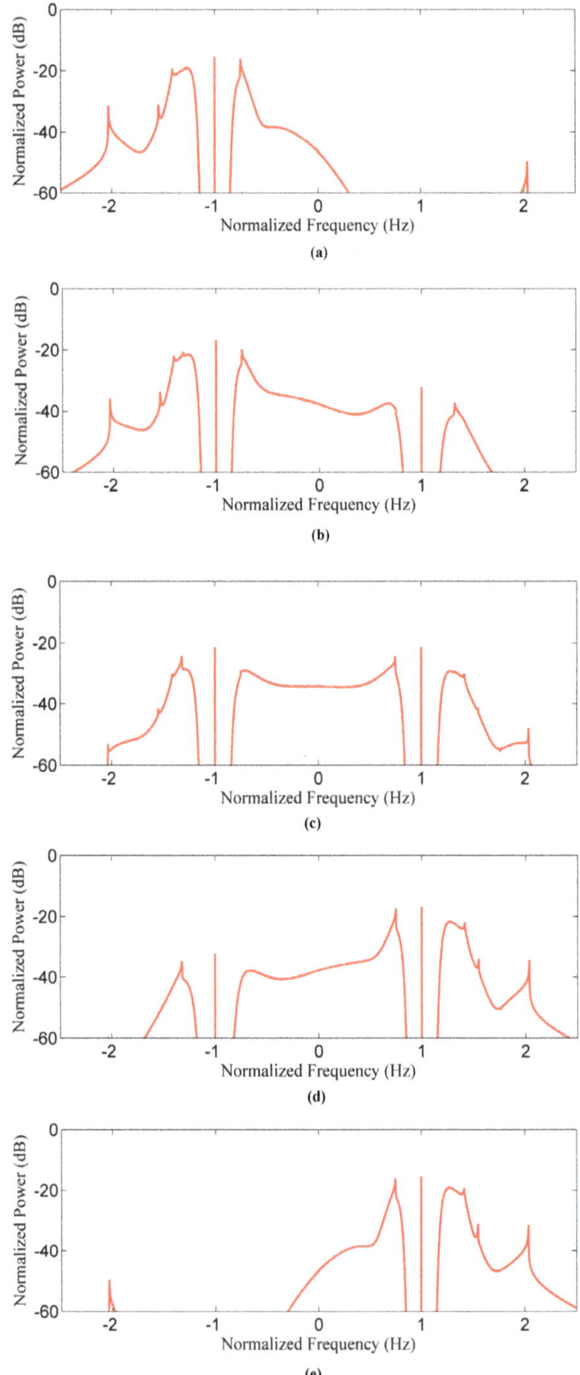

Figure 8. Simulated Doppler spectra with different wind directions. The wind direction is (**a**) 0°, (**b**) 45°, (**c**) 90°, (**d**) 135°, and (**e**) 180°.

4. Discussion

The singularities results from "corner reflection" [41], except for the singularity at $\pm\sqrt{2}F_B$ in Figure 5c,d. Taking Figure 5d for example, where the scattering angle and the azimuth angle are $\theta_s = 60°$ and $\varphi_s = 120°$ respectively, additional singularities appear in the simulated Doppler spectrum. The normalized constant Doppler frequency contours are determined by formula (29) and the delta function in (31). Each point on the constant Doppler frequency contour gives a pair of $\vec{k_1}$ and $\vec{k_2}$, as shown in Figure 9. The p axis is parallel to the direction of the Bragg wave, and the contours of constant normalized frequency ω for $\vec{k_1}$ and $\vec{k_2}$ satisfy the formulas (29) and (31). The case of "corner reflection" is presented by the black dashed curve, where $\vec{k_1}$ and $\vec{k_2}$ satisfy the relation

$$k_0{}^2 - (\vec{k_0} + \vec{k_2})(\vec{k_s} - \vec{k_1}) = 0 \tag{34}$$

or

$$k_0{}^2 - (\vec{k_0} + \vec{k_1})(\vec{k_s} - \vec{k_2}) = 0. \tag{35}$$

The singularities occur where the frequency contour is tangential to the circle of dash curve. Figure 9a shows the normalized constant Doppler frequency contours for $|\omega| > \omega_B$. The frequency contour whose normalized frequency equals 1.32, 1.55, and 2.03 is tangential to the circle of the dash curve. In addition, the singularity at $|\omega| = \sqrt{2}\omega_B$ occurs when the contours separate. Figure 9b shows the normalized constant Doppler frequency contours for $|\omega| < \omega_B$. The normalized frequency equals 0.748 when the frequency contour is tangential to the circle of the dash curve. Therefore, there are ten singularities in the simulated Doppler spectrum at the normalized frequency of ±0.748, ±1.32, ±1.44, ±1.55, and ±2.03, which is the same in Figure 5d.

When the scattering angle and the azimuth angle are $\theta_s = 90°$ and $\varphi_s = 180°$, the first-order and second-order scattering coefficients for shore-to-air bistatic HF radar using the perturbation method will be reduced to the scattering coefficients for monostatic HF radar, which are identical to the result in [29]. The characteristic of the simulated echo spectrum (see Figure 5a) is similar to the characteristic of the simulated spectrum in [29]. Meanwhile, the scattering coefficient for shore-to-air bistatic radar will be reduced to the scattering coefficient of land-based bistatic radar when the scattering angle is $\theta_s = 90°$. While the scattering coefficient for land-based bistatic radar in [21] differs from this work, they have the same forms. The reason is that additional contributions have been incorporated in [21]. The characteristic of the simulated echo spectrum (see Figure 5b) is similar to the characteristic of the simulated spectrum in [21]. It indicates that the scattering coefficient for shore-to-air bistatic radar incorporates the cases of monostatic operation and land-based bistatic operation.

The first-order and second-order scattering coefficients of shore-to-air bistatic HF radar are derived based on the perturbation method. The sea surface needs to satisfy the perturbation condition: $0.2 < \frac{H_s k_0}{2} < 1$, where H_s is the significant wave height, and k_0 is the wavenumber of the radar electromagnetic wave. Therefore, the radar operating frequency determines the limitations of the wave height measurement [42]. For instance, when the operating frequency is 8 MHz, the upper and lower limits of the corresponding significant wave height measurement are about 12 m and 2.4 m, respectively. Whereas at 25 MHz, the upper and lower limits of significant wave height measurement are about 0.7 m and 3.8 m, respectively.

For the bistatic HF radar system where the receiver is deployed on the land or a buoy, the coverage of radar will not be significantly improved comparing with monostatic HF radar since radio waves scattered from the ocean surface to the radar receiver propagate along the sea surface and the attenuation of radio in the receiving path is the same as monostatic HF radar. However, for the shore-to-air bistatic radar, the attenuation of radio in the receiving path is quite small since the radio waves scattered from the ocean surface to the radar receiver propagate in the free space. This advantage will significantly increase the maximum detection range of the radar. In addition, the transmitting and receiving antenna of shore-to-air bistatic radar can be placed in different areas, and this configuration is flexible, which will reduce the space requirements for radar deployment. Nonetheless, since the receiver is placed on the airborne platform, the size of the receiving antenna will be greatly limited. Therefore, higher requirements are placed on the design of the receiving antenna. A smaller antenna is needed to meet the requirements of size.

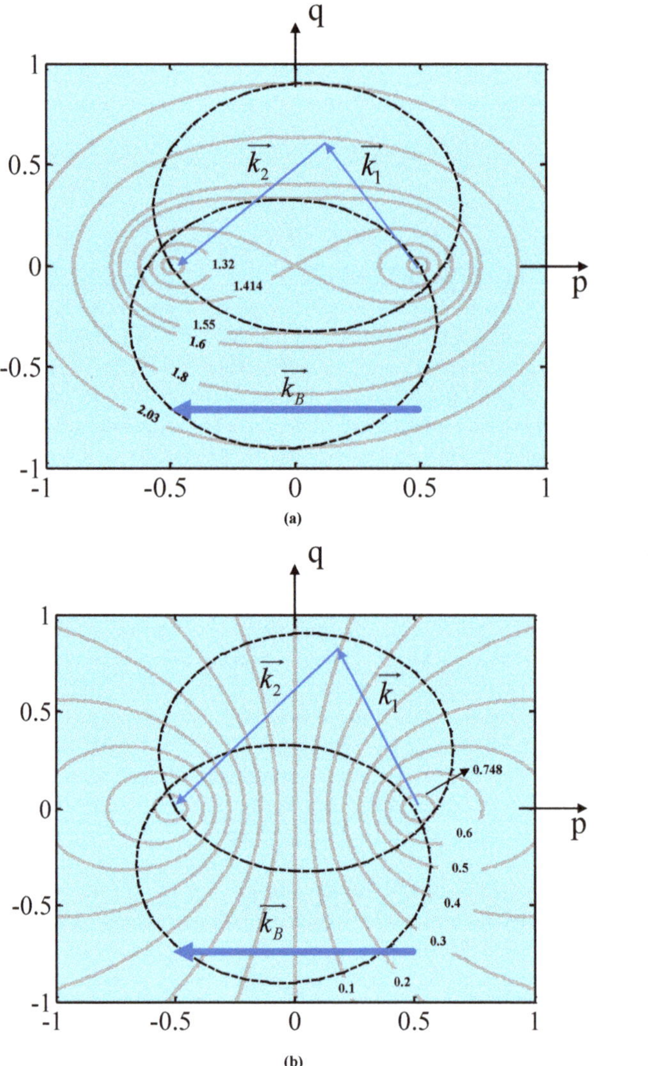

Figure 9. The normalized constant Doppler frequency contours when the scattering angle and the azimuth angle are $\theta_s = 60°$ and $\varphi_s = 120°$, for (**a**) $|\omega| > \omega_B$ $(m = m')$ and (**b**)$|\omega| < \omega_B$ $(m \neq m')$. The blue bold arrow denotes the vector of Bragg wave. The normalized constant Doppler frequency contours of ω for $\vec{k}_1$ and $\vec{k}_2$ are determined by Formulas (29) and (31) as shown in thin arrows.

5. Conclusions

A new model we have named "shore-to-air bistatic HF radar" has been proposed for ocean observation. The first-order and second-order scattering coefficients for the shore-to-air bistatic radar are derived using the perturbation method. The scattering coefficient of shore-to-air bistatic HF radar can be reduced to the case of monostatic radar when the scattering angle and the azimuth angle are 90° and 180° respectively, which is identical to the result in [29].

The scattering mechanism presented herein between the radar electromagnetic waves and the sea surface may be the foundation for shore-to-air bistatic HF radar development and validation. The Doppler spectra are simulated for various operating conditions and sea states based on the proposed scattering coefficients, which may provide a guideline on the extraction of sea state information using shore-to-air bistatic HF radar. The characteristic of the simulated echo spectrum is similar to the characteristic of the Doppler spectrum for the monostatic and land-based bistatic radar, which verifies the correctness of the developed scattering coefficients. In addition, the singularities in the Doppler spectrum for shore-to-air bistatic radar are analyzed using the normalized constant Doppler frequency contours. Further research will be conducted to apply the proposed scattering coefficients to practical situations.

Author Contributions: Conceptualization, Z.C. and C.Z.; funding acquisition, Z.C. and C.Z.; investigation, J.L., F.D.; Methodology, C.Z., J.L. and F.D.; project daministration, Z.C. and C.Z.; software, J.L.; supervision, Z.C. and C.Z.; Validation, J.L., F.D. and X.C.; visualization, J.L.; writing-Original draft, J.L.; writing-review&editing C.Z.

Funding: This research was supported in part by the National Natural Science Foundation of China under Grant 61871296, Grant 41506201, and Grant 41376182, in part by the National Key Research and Development Program of China under Grant 2017YFF0206404 and Grant 2016YFC1400504, in part by the Public Science and Technology Research Funds Projects of Ocean under Grant No. 201205032 in part by Fundamental Research Funds for the Central Universities under Grant 2042018kf1009.

Acknowledgments: The authors would like to thank anonymous reviewers for their valuable comments in improving the quality of this paper.

Conflicts of Interest: The authors declare no conflicts of interest.

Abbreviations

The following abbreviations are used in this manuscript:

HF   High frequency
GIOS   Ground-Ionosphere-Ocean-Space

References

1. Emery, B.M.; Washburn, L.; Harlan, J.A. Evaluating radial current measurements from CODAR high-frequency radars with moored current meters. *J. Atmos. Ocean. Technol.* **2004**, *21*, 1259–1271. doi:10.1175/1520-0426(2004)021<1259:ERCMFC>2.0.CO;2. [CrossRef]
2. Maresca, S.; Braca, P.; Horstmann, J.; Grasso, R. Maritime surveillance using multiple high-frequency surface-wave radars. *IEEE Trans. Geosci. Remote Sens.* **2013**, *52*, 5056–5071. doi:10.1109/TGRS.2013.2286741. [CrossRef]
3. Shen, W.; Gurgel, K.W. Wind direction inversion from narrow-beam HF radar backscatter signals in low and high wind conditions at different radar frequencies. *Remote Sens.* **2018**, *10*, 1480. doi:10.3390/rs10091480. [CrossRef]
4. Wyatt, L.R.; Green, J.J.; Middleditch, A.; Moorhead, M.D.; Howarth, J.; Holt, M.; Keogh, S. Operational wave, current, and wind measurements with the Pisces HF radar. *IEEE J. Ocean. Eng.* **2006**, *31*, 819–834. doi:10.1109/JOE.2006.888378. [CrossRef]
5. Barrick, D. First-order theory and analysis of MF/HF/VHF scatter from the sea. *IEEE Trans. Antennas Propag.* **1972**, *20*, 2–10. doi:10.1109/TAP.1972.1140123. [CrossRef]

6. Barrick, D. Remote sensing of sea state by radar. In Proceedings of the Ocean 72-IEEE International Conference on Engineering in the Ocean Environment, Newport, RI, USA, 13–15 September 1972; pp. 186–192. doi:10.1109/OCEANS.1972.1161190. [CrossRef]

7. Rice, S.O. Reflection of electromagnetic waves from slightly rough surfaces. *Commun. Pure Appl. Math.* **1951**, *4*, 351–378. doi:10.1002/cpa.3160040206. [CrossRef]

8. Chen, Z.; Zhao, C.; Jiang, Y.; Hu, W. Wave measurements with multi-frequency HF radar in the East China Sea. *J. Electromag. Waves Appl.* **2011**, *25*, 1031–1043. doi:10.1163/156939311795253902. [CrossRef]

9. Green, J.; Wyatt, L. Row-action inversion of the Barrick–Weber equations. *J. Atmos. Ocean. Technol.* **2006**, *23*, 501–510. doi:10.1175/JTECH1854.1. [CrossRef]

10. Walsh, J.; Howell, R.; Dawe, B. *Model Development for Evaluation Studies of Ground Wave Radar*; Contract Report; Center Cold Ocean Resources Eng., Mem. Univ. Newfoundland: St. John's, NL, Canada, 1990; pp. 90–C14.

11. Hisaki, Y.; Tokuda, M. VHF and HF sea echo Doppler spectrum for a finite illuminated area. *Radio Sci.* **2001**, *36*, 425–440. doi:10.1029/2000RS002343. [CrossRef]

12. Hisaki, Y. Doppler spectrum of radio wave scattering from ocean-like moving surfaces for a finite illuminated area. *Int. J. Remote Sens.* **2003**, *24*, 3075–3091. doi:10.1080/01431160210153057. [CrossRef]

13. Johnstone, D. Second-Order Electromagnetic and Hydromagnetic Effects In High-Frequency Radio-Wave Scattering From the Sea. PhD Thesis, Stanford University, Stanford, CA, USA, 1975.

14. Anderson, S.; Anderson, W. Bistatic HF scattering from the ocean surface and its application to remote sensing of seastate. In Proceedings of the 1987 IEEE APS International Symposium, Blacksburg, VA, USA, 15–19 June 1987.

15. Anderson, S. Limitations to the extraction of information from multi-hop skywave radar signals. In Proceedings of the RADAR, Adelaide, Australia, 3–5 September 2003.

16. Anderson, S.; Fuks, I.; Praschifka, J. Multiple scattering of HF radio waves propagating across the sea surface. *Waves Random Media* **1998**, *8*, 283–302. doi:10.1088/0959-7174/8/3/002. [CrossRef]

17. Anderson, S. Directional wave spectrum measurement with multistatic HF surface wave radar. In Proceedings of the IGARSS 2000. IEEE 2000 International Geoscience and Remote Sensing Symposium. Taking the Pulse of the Planet: The Role of Remote Sensing in Managing the Environment. Proceedings (Cat. No. 00CH37120), Honolulu, HI, USA, 24–28 July 2000; Volume 7, pp. 2946–2948. doi:10.1109/IGARSS.2000.860299. [CrossRef]

18. Anderson, S.; Edwards, P.; Marrone, P.; Abramovich, Y. Investigations with SECAR-a bistatic HF surface wave radar. In Proceedings of the International Conference on Radar (IEEE Cat. No. 03EX695), Adelaide, Australia, 3–5 September 2003; pp. 717–722. doi:10.1109/RADAR.2003.1278831. [CrossRef]

19. Walsh, J.; Gill, E. An analysis of the scattering of high-frequency electromagnetic radiation from rough surfaces with application to pulse radar operating in backscatter mode. *Radio Sci.* **2000**, *35*, 1337–1359. doi:10.1029/2000RS002532. [CrossRef]

20. Gill, E.; Walsh, J. High-frequency bistatic cross sections of the ocean surface. *Radio Sci.* **2001**, *36*, 1459–1475. doi:10.1029/2000RS002525. [CrossRef]

21. Gill, E.; Huang, W.; Walsh, J. On the development of a second-order bistatic radar cross section of the ocean surface: a high-frequency result for a finite scattering patch. *IEEE J. Ocean. Eng.* **2006**, *31*, 740–750. doi:10.1109/JOE.2006.886228. [CrossRef]

22. Huang, W.; Gill, E.; Wu, X.; Li, L. Measurement of sea surface wind direction using bistatic high-frequency radar. *IEEE Trans. Geosci. Remote Sens.* **2012**, *50*, 4117–4122. doi:10.1109/tgrs.2012.2188298. [CrossRef]

23. Bernhardt, P.A.; Briczinski, S.J.; Siefring, C.L.; Barrick, D.E.; Bryant, J.; Howarth, A.; James, G.; Enno, G.; Yau, A. Large area sea mapping with Ground-Ionosphere-Ocean-Space (GIOS). In Proceedings of the OCEANS 2016 MTS/IEEE Monterey, Monterey, CA, USA, 19–23 September 2016; pp. 1–10. doi:10.1109/OCEANS.2016.7761503. [CrossRef]

24. Bernhardt, P.A.; Siefring, C.L.; Briczinski, S.C.; Vierinen, J.; Miller, E.; Howarth, A.; James, H.G.; Blincoe, E. Bistatic observations of the ocean surface with HF radar, satellite and airborne receivers. In Proceedings of the IEEE OCEANS 2017-Anchorage, Anchorage, AK, USA, 18–21 September 2017; pp. 1–5.

25. Voronovich, A.G.; Zavorotny, V.U. Measurement of ocean wave directional spectra using airborne HF/VHF synthetic aperture radar: A theoretical evaluation. *IEEE Trans. Geosci. Remote Sens.* **2017**, *55*, 3169–3176. doi:10.1109/TGRS.2017.2663378. [CrossRef]

26. Zhao, C.; Chen, Z. Model of shore-to-air bistatic HF radar for ocean observation. In Proceedings of the OCEANS 2018 MTS/IEEE Charleston, Charleston, SC, USA, 22–25 October 2018; pp. 1–4. doi:10.1109/OCEANS.2018.8604581. [CrossRef]

27. Tai, C.T. Kirchhoff theory: scalar, vector, or dyadic? *IEEE Trans. Antennas Propag.* **1972**, *20*, 114–115. doi:10.1109/TAP.1972.1140146. [CrossRef]

28. Barrick, D.E. Theory of HF and VHF propagation across the rough sea, 1, The effective surface impedance for a slightly rough highly conducting medium at grazing incidence. *Radio Sci.* **1971**, *6*, 517–526. doi:10.1029/RS006i005p00517. [CrossRef]

29. Lipa, B.J.; Barrick, D.E. Extraction of sea state from HF radar sea echo: Mathematical theory and modeling. *Radio Sci.* **1986**, *21*, 81–100. doi:10.1029/RS021i001p00081. [CrossRef]

30. Barrick, D.; Weber, B. On the nonlinear theory for gravity waves on the ocean's surface. Part II: Interpretation and applications. *J. Phys. Oceanogr.* **1977**, *7*, 11–21. doi:10.1175/1520-0485(1977)007<0011:OTNTFG>2.0.CO;2. [CrossRef]

31. Weber, B.; Barrick, D. On the nonlinear theory for gravity waves on the ocean's surface. Part I: Derivations. *J. Phys. Oceanogr.* **1977**, *7*, 3–10. doi:10.1175/1520-0485(1977)007<0003:OTNTFG>2.0.CO;2. [CrossRef]

32. Moskowitz, L. Estimates of the power spectrums for fully developed seas for wind speeds of 20 to 40 knots. *J. Geophys. Res.* **1964**, *69*, 5161–5179. doi:10.1029/JZ069i024p05161. [CrossRef]

33. Longuet-Higgins, M.S. Observations of the directional spectrum of sea waves using the motions of a floating buoy. *Ocean Wave Spectra* **1961**. doi:10.1007/BF02861508. [CrossRef]

34. Howell, R.; Walsh, J. Measurement of ocean wave spectra using narrow-beam HE radar. *IEEE J. Ocean. Eng.* **1993**, *18*, 296–305. doi:10.1109/JOE.1993.236368. [CrossRef]

35. Gill, E.; Khandekar, M.; Howell, R.; Walsh, J. Ocean surface wave measurement using a steerable high-frequency narrow-beam ground wave radar. *J. Atmos. Ocean. Technol.* **1996**, *13*, 703–713. doi:10.1175/1520-0426(1996)013<0703:OSWMUA>2.0.CO;2. [CrossRef]

36. Huang, W.; Wu, S.; Gill, E.; Wen, B.; Hou, J. HF radar wave and wind measurement over the Eastern China Sea. *IEEE Trans. Geosci. Remote Sens.* **2002**, *40*, 1950–1955. doi:10.1109/TGRS.2002.803718. [CrossRef]

37. Barrick, D.E. Extraction of wave parameters from measured HF radar sea-echo Doppler spectra. *Radio Sci.* **1977**, *12*, 415–424. doi:10.1029/RS012i003p00415. [CrossRef]

38. Heron, M.L.; Prytz, A. Wave height and wind direction from the HF coastal ocean surface radar. *Can. J. Remote Sens.* **2002**, *28*, 385–393. doi:10.5589/m02-031. [CrossRef]

39. Wyatt, L.R. An evaluation of wave parameters measured using a single HF radar system. *Can. J. Remote Sens.* **2002**, *28*, 205–218. doi:10.5589/m02-018. [CrossRef]

40. Zeng, Y.; Zhou, H.; Lai, Y.; Wen, B. Wind-Direction Mapping With a Modified Wind Spreading Function by Broad-Beam High-Frequency Radar. *IEEE Geosci. Remote Sens. Lett.* **2018**, *15*, 679–683. doi:10.1109/LGRS.2018.2809558. [CrossRef]

41. Ivonin, D.V.; Shrira, V.I.; Broche, P. On the singular nature of the second-order peaks in HF radar sea echo. *IEEE J. Ocean. Eng.* **2006**, *31*, 751–767. doi:10.1109/joe.2006.886080. [CrossRef]

42. Wyatt, L.R.; Green, J.J.; Middleditch, A. HF radar data quality requirements for wave measurement. *Coastal Eng.* **2011**, *58*, 327–336. doi:10.1016/j.coastaleng.2010.11.005. [CrossRef]

 *remote sensing*

Article

Ocean Surface Cross Section for Bistatic HF Radar Incorporating a Six DOF Oscillation Motion Model

Guowei Yao [1], Junhao Xie [1,*] and Weimin Huang [2]

[1] Department of Electronic Engineering, Harbin Institute of Technology, Harbin 150001, China; yaoguoweiygw@hit.edu.cn

[2] Faculty of Engineering and Applied Science, Memorial University of Newfoundland, St. Johns, NL A1B 3X5, Canada; weimin@mun.ca

* Correspondence: xj@hit.edu.cn

Received: 29 September 2019; Accepted: 19 November 2019; Published: 21 November 2019

Abstract: To investigate the characteristics of sea clutter, based on ocean surface electromagnetic scattering theory, the first- and second-order ocean surface scattering cross sections for bistatic high-frequency (HF) radar incorporating a multi-frequency six degree-of-freedom (DOF) oscillation motion model are mathematically derived. The derived radar cross sections (RCSs) can be reduced to the floating platform based monostatic case or onshore bistatic case for corresponding geometry setting. Simulation results show that the six DOF oscillation motion will result in more additional peaks in the radar Doppler spectra and the amplitudes and frequencies of these motion-induced peaks are decided by the amplitudes and frequencies of the oscillation motion. The effect of the platform motion on the first-order radar spectrum is greater than that of the second-order, and the motion-induced peaks in the first-order spectrum may overlap with the second-order spectrum. Furthermore, yaw is the dominant factor affecting the radar spectra, especially the second-order. Moreover, the effect of platform motion on radar spectra and the amplitudes of the second-order spectrum decreases as the bistatic angle increases. In addition, it should be noted that the amplitudes of the Bragg peaks may be lower than those of the motion-induced peaks due to the low frequency (LF) oscillation motion of the floating platform, which is an important finding for the applications of the floating platform based bistatic HF radar in moving target detection and ocean surface dynamics parameter estimation.

Keywords: bistatic HF radar; radar cross section (RCS); sea clutter

---

1. Introduction

High-frequency (HF) radar has been successfully deployed to detect ocean surface moving target and remote sensing of ocean surface dynamics such as wind direction and speed, current and wave parameters in many countries for decades [1–13] because it can provide real-time, all-weather surveillance beyond the horizon. Based on the geometry, HF radar can be generally divided into monostatic (transmitter and receiver are collocated) and bistatic (transmitter and receiver are separated) types. Bistatic HF radar possesses some inherent advantages over the monostatic case: (1) bistatic HF radar can improve the detection capability of stealth targets; (2) bistatic HF radar can address the ambiguous problem of sea state information extraction; (3) bistatic HF radar can suppress mutual interference between the transmitting and receiving antennas. Therefore, bistatic HF radar has attracted increasing attention internationally. For example, based on the ocean surface electromagnetic scattering theory presented by Walsh [14], Gill et al. [15] developed the first- and second-order ocean surface scattering cross sections for bistatic HF radar, which provides an important theoretical basis for the applications of bistatic HF radar. Subsequently, they investigated the effect of bistatic angle on radar cross sections (RCSs) in detail [16], which is helpful to determine a suitable geometry for the deployment of bistatic HF radar. Based on the characteristics of sea clutter for bistatic HF radar, Lipa et al. [17] extracted ocean surface current. Trizna [18] successfully implemented an experiment to map ocean surface current and track the ship target using bistatic HF radar. Grosdidier et al. [19] validated the simulated bistatic HF radar Doppler spectra with the experimental data. Based on the RCS model developed by Gill et al. [15], Huang et al. [20] successfully obtained the unambiguous wind direction on the Southern China coast and the directional ocean wave spectra are respectively extracted from simulated noisy bistatic HF radar data [21] and synthetic bistatic HF radar data [22]. As the application of onshore bistatic HF radar matures, the floating platform based bistatic HF radar (the transmitter is deployed on a floating ocean platform and the receiver is installed on shore) gradually becomes a deployment trend. However, the platform motion may have an important effect on the application of bistatic HF radars.

In the floating platform based monostatic HF radar experiment, some researchers have observed that the platform motion can be viewed as phase modulation of radar Doppler spectra [23–26]. Theoretically, Walsh et al. [27,28] developed the first- and second-order ocean surface scattering cross section models for the case of a transmitter being installed on a floating platform with sway. They pointed out sway can induce additional peaks in radar Doppler spectra. Subsequently, Sun et al. [29] and Ma et al. [30,31] derived corresponding RCSs for shipborne and bistatic cases, respectively. However, based on the seakeeping theory [32,33], the deep-water floating platform generally has six degree-of-freedom (DOF) oscillation motion with multi-frequency due to the interaction between the complex ocean environment and floating platform. More recently, Ma et al. [34] extended the floating platform based bistatic RCSs to a dual-frequency platform motion case incorporating sway and surge. They presented that more additional peaks caused by the combined motion will symmetrically appear in radar Doppler spectra. Yao et al. [35,36] extended the first-order shipborne and bistatic RCSs to a horizontal oscillation motion case with a single-frequency and pointed out yaw may have a more important effect on radar Doppler spectra. Therefore, only considering two-dimensional platform motion with a dual-frequency model for the floating platform based bistatic HF radar is not realistic in practice. In this paper, on the basis of previous works [30,31,34,36], the first- and second-order ocean surface scattering cross sections for the floating platform based bistatic HF radar incorporating a more realistic multi-frequency six DOF oscillation motion model are presented. The results may have significant implications in future investigations for the application of floating-based bistatic HF radars.

The rest of this paper is organized as follows. In Section 2, a multi-frequency six DOF oscillation motion physical model is first developed. Subsequently, the first- and second-order ocean surface scattering cross sections for the floating platform based bistatic HF radar incorporating a single-frequency six DOF oscillation motion model are derived and, then, the results are extended to the multi-frequency case. Section 3 presents the simulation results and comparative analyses with different oscillation motion models and bistatic angles. Section 4 discusses the effect of six DOF oscillation motion and bistatic angle on the application of the floating-based bistatic HF radar in moving target detection and ocean surface dynamics parameter estimation. Conclusions are provided in Section 5.

2. Derivation

2.1. Physical Model

Due to the interaction between the ocean floating platform and complex ocean environment, based on the seakeeping theory of deep-water floating platform, the motion of the ocean floating platform can be viewed as the superposition of sway, surge, heave, yaw, pitch, and roll with a multi-frequency model [32,33]. Figure 1 shows the diagram of six DOF motion for a transmitting sensor on a floating platform. It is assumed that the source is at (a, b, h). According to the work of Walsh et al. in [27], the motion components in vertical direction will not result in additional Doppler effect. Thus, heave will not be considered in the physical model as well as the components of pitch and roll in vertical direction.

The displacement vectors in horizontal direction caused by sway, surge, yaw, pitch, and roll can be respectively expressed as

$$\delta\vec{\rho}_{01}(t) = \sum_{j=1}^{N_1} a_{1,j}\sin\left(\omega_{1,j} + \phi_{1,j}\right)\delta\hat{\rho}_{01},$$ (1)

$$\delta\vec{\rho}_{02}(t) = \sum_{j=1}^{N_2} a_{2,j}\sin\left(\omega_{2,j} + \phi_{2,j}\right)\delta\hat{\rho}_{02},$$ (2)

$$\delta\vec{\rho}_{03}(t) = 2l_3\sin\left[\frac{\theta_3(t)}{2}\right]\delta\hat{\rho}_{03}(t),$$ (3)

$$\delta\vec{\rho}_{04}(t) = 2l_4\sin\left[\frac{\theta_4(t)}{2}\right]\sin\left[\frac{\pi}{2} + \frac{\theta_4(t)}{2} - \alpha_4\right]\delta\hat{\rho}_{04},$$ (4)

$$\delta\vec{\rho}_{05}(t) = 2l_5\sin\left[\frac{\theta_5(t)}{2}\right]\sin\left[\frac{\pi}{2} + \frac{\theta_5(t)}{2} - \alpha_5\right]\delta\hat{\rho}_{05},$$ (5)

where $a_{1,j}$ and $a_{2,j}$, $\omega_{1,j}$ and $\omega_{2,j}$, and $\phi_{1,j}$ and $\phi_{2,j}$ are the amplitudes, angular frequencies, and initial phases for each frequency component of sway and surge, respectively. $l_3 = \sqrt{a^2 + b^2}$, $l_4 = \sqrt{a^2 + h^2}$, $l_5 = \sqrt{b^2 + h^2}$, $\alpha_4 = \arctan(a/h)$, $\alpha_5 = \arctan(b/h)$. $\theta_3(t)$, $\theta_4(t)$, and $\theta_5(t)$ are the rotation angles of yaw, pitch, and roll, respectively, which can be written as

$$\theta_3(t) = \sum_{j=1}^{N_3} \theta_{m3,j}\sin\left(\omega_{3,j}t + \phi_{3,j}\right),$$ (6)

$$\theta_4(t) = \sum_{j=1}^{N_4} \theta_{m4,j}\sin\left(\omega_{4,j}t + \phi_{4,j}\right),$$ (7)

$$\theta_5(t) = \sum_{j=1}^{N_5} \theta_{m5,j}\sin\left(\omega_{5,j}t + \phi_{5,j}\right),$$ (8)

in which $\theta_{m3,j}$, $\theta_{m4,j}$ and $\theta_{m5,j}$, $\omega_{3,j}$, $\omega_{4,j}$ and $\omega_{5,j}$, and $\phi_{3,j}$, $\phi_{4,j}$ and $\phi_{5,j}$ are the amplitudes, angular frequencies, and initial phases of yaw, pitch, and roll, respectively. $j = 1, 2, \ldots, N_i$ $(i = 1, 2, \ldots, 5)$ indicates the number of frequency components associated with sway, surge, yaw, pitch, and roll, respectively. $\delta\hat{\rho}_{01}(t)$, $\delta\hat{\rho}_{02}(t)$, $\delta\hat{\rho}_{03}(t)$, $\delta\hat{\rho}_{04}(t)$, and $\delta\hat{\rho}_{05}(t)$ are the corresponding motion directions, respectively, which can be represented by angles $\theta_{01}(t)$, $\theta_{02}(t)$, $\theta_{03}(t)$, $\theta_{04}(t)$, and $\theta_{05}(t)$.

Therefore, the overall displacement vector caused by six DOF oscillation motion with a multi-frequency model can be expressed as

$$\delta\vec{\rho}_0(t) = \delta\vec{\rho}_{01}(t) + \delta\vec{\rho}_{02}(t) + \delta\vec{\rho}_{03}(t) + \delta\vec{\rho}_{04}(t) + \delta\vec{\rho}_{05}(t).$$ (9)

2.2. RCS Incorporating a Single-Frequency Six DOF Motion Model

In order to simplify the derivation, a single-frequency six DOF oscillation motion model is first considered. That is, $N_i = 1$ $(i = 1, 2, \ldots, 5)$ in Equation (9).

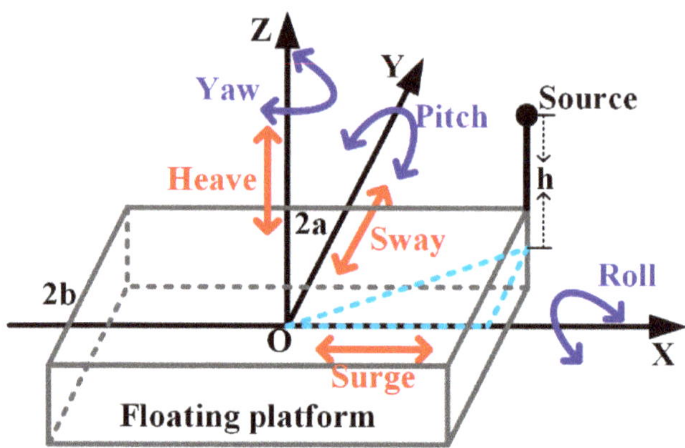

Figure 1. Diagram of six degree-of-freedom (DOF) motion of floating platform.

2.2.1. First-Order RCS

Figure 2 shows the first-order bistatic HF radar scatter geometry for the case of the source being installed on an ocean floating platform. In [30], Ma et al. derived the first-order bistatic HF RCS when the source is deployed on a floating platform with a single-frequency sway motion. Then, different platform motion models are introduced to derive corresponding RCS [29,34–36]. In this study, based on the ocean surface electromagnetic scattering theory, a more realistic six DOF oscillation motion model is established and the first-order bistatic HF RCS model can be modified to

$$\sigma_1(\omega_d) = 2^2 k_0^2 \Delta\rho \sum_{m=\pm 1} \int_K K^2 \cos\phi_0 S_1\left(m\vec{K}\right) Sa^2\left[\frac{\Delta\rho}{2}\left(\frac{K}{\cos\phi_0} - 2k_0\right)\right]$$
$$\cdot \int_\tau e^{-j\tau(m\sqrt{gK}+\omega_d)}\left\langle M\left(K, \theta_{\vec{K}}, \tau, t\right)\right\rangle d\tau dK \tag{10}$$

where ω_d is the Doppler frequency, k_0 is the radian wavenumber, $\Delta\rho$ is the patch width, $\vec{K} = \left(K, \theta_{\vec{K}}\right)$ is the ocean wave vector, $S_1(\cdot)$ indicates the directional ocean wave spectrum, $Sa(\cdot)$ represents the sinc function, τ is the interval between samples, g is the gravitational acceleration and

$$M\left(K, \theta_{\vec{K}}, \tau, t\right) = e^{-\frac{j}{2}\{K\delta\rho_0(t)\cos[\theta_{\vec{K}}-\theta_0(t)]\}} e^{\frac{j}{2}\{K\delta\rho_0(t+\tau)\sin[\theta_{\vec{K}}-\theta_0(t+\tau)]\}}$$
$$\cdot e^{-\frac{j\tan\phi_0}{2}\{K\delta\rho_0(t)\sin[\theta_{\vec{K}}-\theta_0(t)]\}} e^{\frac{j\tan\phi_0}{2}\{K\delta\rho_0(t+\tau)\sin[\theta_{\vec{K}}-\theta_0(t+\tau)]\}} \tag{11}$$

By substituting the displacement term in Equation (9) into Equation (11), the ensemble average of $M\left(K, \theta_{\vec{K}}, \tau, t\right)$ can be derived as

$$\left\langle M\left(K, \theta_{\vec{K}}, \tau, t\right)\right\rangle = \langle M_1 M_2 M_3 M_4 M_5\rangle, \tag{12}$$

where

$$M_1 = e^{jV_1\cos Q_1}, \tag{13}$$

$$M_2 = e^{jV_2\cos Q_2}, \tag{14}$$

$$M_3 = e^{jV_{31}\cos(2Q_3)} e^{jV_{32}\cos Q_3}, \tag{15}$$

$$M_4 = e^{jV_{41}\sin(2Q_4)} e^{jV_{42}\cos Q_4}, \tag{16}$$

and

$$M_5 = e^{jV_{51}\sin(2Q_5)} e^{jV_{52}\cos Q_5}, \tag{17}$$

in which

$$V_1 = a_1 K \left[\cos\left(\theta_{\vec{K}} - \theta_{01}\right) + \tan\phi_0 \sin\left(\theta_{\vec{K}} - \theta_{01}\right) \right] \sin \frac{\omega_1 \tau}{2}, \tag{18}$$

$$V_2 = a_2 K \left[\cos\left(\theta_{\vec{K}} - \theta_{02}\right) + \tan\phi_0 \sin\left(\theta_{\vec{K}} - \theta_{02}\right) \right] \sin \frac{\omega_2 \tau}{2}, \tag{19}$$

$$V_{31} = -\frac{K l_3 \theta_{m3}^2 \cos\left(\theta_{\vec{K}} - \theta'\right)}{4} (1 + \tan\phi_0) \sin \omega_3 \tau, \tag{20}$$

$$V_{32} = -K l_3 \theta_{m3} \cos\left(\theta_{\vec{K}} - \theta'\right)(-1 + \tan\phi_0) \sin \frac{\omega_3 \tau}{2}, \tag{21}$$

$$V_{41} = \frac{K l_4 \theta_{m4}^2 \cos\left(\theta_{\vec{K}} - \theta_{04}\right) \sin\alpha_4}{2} \left(1 + \frac{\tan\phi_0}{2}\right) \sin \omega_4 \tau, \tag{22}$$

$$V_{42} = 2 K l_4 \theta_{m4} \cos\left(\theta_{\vec{K}} - \theta_{04}\right) \cos\alpha_4 \left(1 + \frac{\tan\phi_0}{2}\right) \sin \frac{\omega_4 \tau}{2}, \tag{23}$$

$$V_{51} = \frac{K l_5 \theta_{m5}^2 \cos\left(\theta_{\vec{K}} - \theta_{05}\right) \sin\alpha_5}{2} \left(1 + \frac{\tan\phi_0}{2}\right) \sin \omega_5 \tau, \tag{24}$$

$$V_{52} = 2 K l_5 \theta_{m5} \cos\left(\theta_{\vec{K}} - \theta_{05}\right) \cos\alpha_5 \left(1 + \frac{\tan\phi_0}{2}\right) \sin \frac{\omega_5 \tau}{2}, \tag{25}$$

$$Q_i = \omega_i t + \frac{\omega_i \tau}{2} + \phi_i \ (i = 1, 2, \ldots, 5) \tag{26}$$

and $\theta' = \arctan(b/a)$.

Using the Euler equation and the property of the Bessel function

$$e^{jx} = \cos x + j \sin x, \tag{27}$$

$$\cos(V \sin Q) = J_0(V) + 2 \sum_{n=1}^{+\infty} J_{2n}(V) \cos(2nQ), \tag{28}$$

$$\sin(V \sin Q) = 2 \sum_{n=1}^{+\infty} J_{2n-1}(V) \cos[(2n-1)Q], \tag{29}$$

$$\cos(V \cos Q) = J_0(V) + 2 \sum_{n=1}^{+\infty} (-1)^n J_{2n}(V) \cos(2nQ), \tag{30}$$

and

$$\sin(V \cos Q) = -2 \sum_{n=1}^{+\infty} (-1)^n J_{2n-1}(V) \cos[(2n-1)Q], \tag{31}$$

where J_n is the n-th order Bessel function. Then, Equation (12) can be reduced to

$$\left\langle M\left(K, \theta_{\vec{K}}, \tau, t\right) \right\rangle = J_0(V_1) J_0(V_2) J_0(V_{31}) J_0(V_{32}) J_0(V_{41}) J_0(V_{42}) J_0(V_{51}) J_0(V_{52}). \tag{32}$$

By taking advantage of the relationship of Bessel function

$$J_0\left(2x \sin \frac{\Phi}{2}\right) = \sum_{n=-\infty}^{+\infty} J_n^2(x) \cos(n\Phi), \tag{33}$$

$$J_n(-x) = (-1)^n J_n(x), \tag{34}$$

and similar derivation in [34], Equation (32) can be further modified to

$$
\begin{aligned}
\left\langle M\!\left(K, \theta_{\vec{K}}, \tau, t\right)\right\rangle =\;& \sum_{n_1=-\infty}^{+\infty} J_{n_1}^2(X_1)\cos(n_1\omega_1\tau) \sum_{n_2=-\infty}^{+\infty} J_{n_2}^2(X_2)\cos(n_2\omega_2\tau) \\
& \cdot \sum_{n_{31}=-\infty}^{+\infty} J_{n_{31}}^2(X_{31})\cos(2n_{31}\omega_3\tau) \sum_{n_{32}=-\infty}^{+\infty} J_{n_{32}}^2(X_{32})\cos(n_{32}\omega_3\tau) \\
& \cdot \sum_{n_{41}=-\infty}^{+\infty} J_{n_{41}}^2(X_{41})\cos(2n_{41}\omega_4\tau) \sum_{n_{42}=-\infty}^{+\infty} J_{n_{42}}^2(X_{42})\cos(n_{42}\omega_4\tau) \\
& \cdot \sum_{n_{51}=-\infty}^{+\infty} J_{n_{51}}^2(X_{51})\cos(2n_{51}\omega_5\tau) \sum_{n_{52}=-\infty}^{+\infty} J_{n_{52}}^2(X_{52})\cos(n_{52}\omega_5\tau)
\end{aligned}
\tag{35}
$$

where

$$
X_1 = \frac{a_1 K\left[\cos\!\left(\theta_{\vec{K}} - \theta_{01}\right) + \tan\phi_0 \sin\!\left(\theta_{\vec{K}} - \theta_{01}\right)\right]}{2},
\tag{36}
$$

$$
X_2 = \frac{a_2 K\left[\cos\!\left(\theta_{\vec{K}} - \theta_{02}\right) + \tan\phi_0 \sin\!\left(\theta_{\vec{K}} - \theta_{02}\right)\right]}{2},
\tag{37}
$$

$$
X_{31} = -\frac{K l_3 \theta_{m3}^2 \cos\!\left(\theta_{\vec{K}} - \theta'\right)}{8}(1 + \tan\phi_0),
\tag{38}
$$

$$
X_{32} = -\frac{K l_3 \theta_{m3} \cos\!\left(\theta_{\vec{K}} - \theta'\right)}{2}(-1 + \tan\phi_0),
\tag{39}
$$

$$
X_{41} = \frac{K l_4 \theta_{m4}^2 \cos\!\left(\theta_{\vec{K}} - \theta_{04}\right)\sin\alpha_4}{4}\left(1 + \frac{\tan\phi_0}{2}\right),
\tag{40}
$$

$$
X_{42} = K l_4 \theta_{m4} \cos\!\left(\theta_{\vec{K}} - \theta_{04}\right)\cos\alpha_4\left(1 + \frac{\tan\phi_0}{2}\right),
\tag{41}
$$

$$
X_{51} = \frac{K l_5 \theta_{m5}^2 \cos\!\left(\theta_{\vec{K}} - \theta_{05}\right)\sin\alpha_5}{4}\left(1 + \frac{\tan\phi_0}{2}\right),
\tag{42}
$$

$$
X_{52} = K l_5 \theta_{m5} \cos\!\left(\theta_{\vec{K}} - \theta_{05}\right)\cos\alpha_5\left(1 + \frac{\tan\phi_0}{2}\right).
\tag{43}
$$

Substituting Equation (35) into Equation (10), using the relationship $\cos x = \frac{e^{jx}+e^{-jx}}{2}$ and then completing τ integration, the first-order ocean surface scattering cross section for bistatic HF radar incorporating a single-frequency six DOF oscillation motion model can be finally derived as

$$
\begin{aligned}
\sigma_1(\omega_d) =\;& 2^3 \pi k_0^2 \Delta\rho \sum_{m=\pm 1} \int_K K^2 \cos\phi_0 S_1\!\left(m\vec{K}\right) Sa^2\!\left[\tfrac{\Delta\rho}{2}\!\left(\tfrac{K}{\cos\phi_0} - 2k_0\right)\right] \\
& \cdot \sum_{n_1=-\infty}^{+\infty} J_{n_1}^2(X_1) \sum_{n_2=-\infty}^{+\infty} J_{n_2}^2(X_2) \sum_{n_{31}=-\infty}^{+\infty} J_{n_{31}}^2(X_{31}) \sum_{n_{32}=-\infty}^{+\infty} J_{n_{32}}^2(X_{32}) \\
& \cdot \sum_{n_{41}=-\infty}^{+\infty} J_{n_{41}}^2(X_{41}) \sum_{n_{42}=-\infty}^{+\infty} J_{n_{42}}^2(X_{42}) \sum_{n_{51}=-\infty}^{+\infty} J_{n_{51}}^2(X_{51}) \sum_{n_{52}=-\infty}^{+\infty} J_{n_{52}}^2(X_{52}) \\
& \cdot \delta\!\left(\omega_d + m\sqrt{gK} - \sum_{i=1}^{2} n_i\omega_i - \sum_{i=3}^{5}(2n_{i1} + n_{i2})\omega_i\right) dK
\end{aligned}
\tag{44}
$$

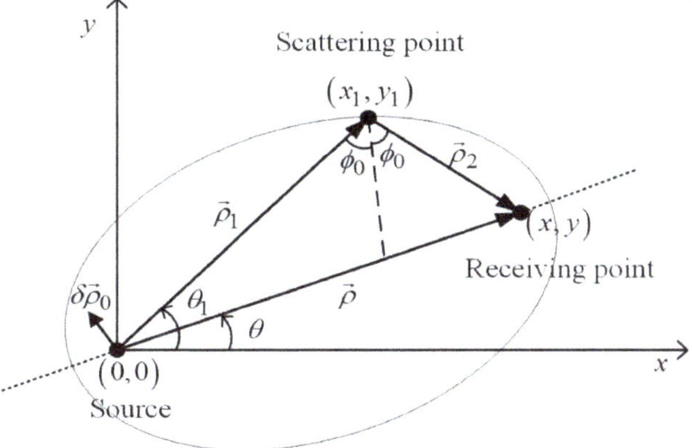

Figure 2. First-order bistatic high-frequency (HF) radar scatter geometry with antenna motion. $(0,0)$, (x_1, y_1), and (x, y) represent the source, the scattering point, and the receiving point, respectively. $\delta\vec{\rho}_0$ is the displacement vector caused by the source motion. θ is the angle between the x-axis and the direction of the receiving point. θ_1 is the angle between the x-axis and the direction of the scattering point. $\vec{\rho}$ is the displacement vector from the source to the receiving point. $\vec{\rho}_1$ is the displacement vector from the source to the scattering point. $\vec{\rho}_2$ is the displacement vector from the scattering point to the receiving point and ϕ_0 indicates the bistatic angle.

2.2.2. Second-Order RCS

In general, the second-order RCS is mainly composed of two parts. One is due to single ocean surface scatter from a second-order ocean wave and the scatter geometry is similar to Figure 2. Its difference from the first-order RCS is that the first-order ocean wave at the scattering point is replaced by a second-order ocean wave. The other is due to double scatters from two first-order ocean waves and the scatter geometry is shown in Figure 3. In [31], Ma et al. derived the second-order bistatic HF RCS when the source is deployed on a floating platform with a single-frequency sway motion. Then, different platform motion models are introduced to derive corresponding second-order RCS [29,34]. In this study, based on the scattering theory in [31], a more realistic six DOF oscillation motion model is established and the second-order bistatic HF RCS model can be modified to

$$
\begin{aligned}
\sigma_2(\omega_d) = 2^2 k_0^2 \Delta\rho \sum_{m_1=\pm1} \sum_{m_2=\pm1} \int_K \int_{\theta_{\vec{K}_1}} \int_{K_1} S_1\left(m_1\vec{K}_1\right) S_1\left(m_2\vec{K}_2\right) \\
\cdot |\Gamma|^2 K^2 K_1 \cos\phi_0 Sa^2\left[\frac{\Delta\rho}{2}\left(\frac{K}{\cos\phi_0} - 2k_0\right)\right] \\
\cdot \int_\tau e^{-j\tau(m_1\sqrt{gK_1}++m_2\sqrt{gK_2}+\omega_d)} \left\langle M\left(K, \theta_{\vec{K}}, \tau, t\right)\right\rangle d\tau dK_1 d\theta_{\vec{K}_1} dK
\end{aligned}
\tag{45}
$$

where $\vec{K} = \vec{K}_1 + \vec{K}_2$, $\vec{K}_1 = \left(K_1, \theta_{\vec{K}_1}\right)$, and $\vec{K}_2 = \left(K_2, \theta_{\vec{K}_2}\right)$ indicate the two first-order ocean waves, respectively. $|\Gamma| = |\Gamma_H| + |\Gamma_E|$ represents the total coupling coefficient, $|\Gamma_H|$ is the hydrodynamic coupling coefficient of two first-order ocean waves [37] and $|\Gamma_E|$ is the electromagnetic coupling coefficient [28].

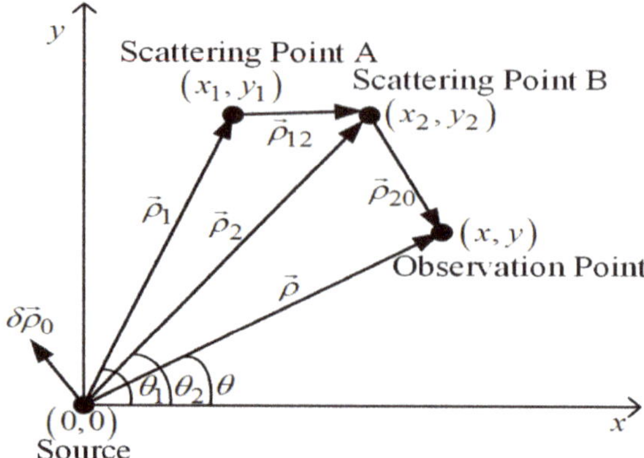

Figure 3. Second-order bistatic HF radar scatter geometry with antenna motion for the case of a double scatter from two first-order ocean waves. $(0,0)$, (x_1, y_1), (x_2, y_2), and (x, y) represent the source, the first scattering point A, the second scattering point B, and the receiving point, respectively. θ is the angle between the x-axis and the direction of the receiving point. θ_1 is the angle between x-axis and the direction of the scattering point A. θ_2 is the angle between x-axis and the direction of the scattering point B. $\delta\vec{\rho}_0$ is the displacement vector caused by the motion of the source. $\vec{\rho}$ is the displacement vector from the source to the receiving point. $\vec{\rho}_1$ is the displacement vector from the source to the scattering point A. $\vec{\rho}_2$ is the displacement vector from the source to the scattering point B. $\vec{\rho}_{12}$ is the displacement vector from the scattering point A to the scattering point B. $\vec{\rho}_{20}$ is the displacement vector from the scattering point B to the observation point.

Substituting a single-frequency six DOF oscillation motion model (Equation (9)) into the second-order bistatic HF RCS model (Equation (45)), similar to the analysis of the first-order RCS in Section 2.2.1, the second-order ocean surface scattering cross section for bistatic HF radar can be finally derived as

$$
\begin{aligned}
\sigma_2(\omega_d) = {} & 2^3 \pi k_0^2 \Delta\rho \sum_{m_1=\pm 1} \sum_{m_2=\pm 1} \int_K \int_{\theta_{\vec{K}_1}} \int_{K_1} S_1\!\left(m_1 \vec{K}_1\right) S_1\!\left(m_2 \vec{K}_2\right) \\
& \cdot |\Gamma|^2 K^2 K_1 \cos\phi_0 Sa^2\!\left[\tfrac{\Delta\rho}{2}\left(\tfrac{K}{\cos\phi_0} - 2k_0\right)\right] \\
& \cdot \sum_{n_1=-\infty}^{+\infty} J_{n_1}^2(X_1) \sum_{n_2=-\infty}^{+\infty} J_{n_2}^2(X_2) \sum_{n_{31}=-\infty}^{+\infty} J_{n_{31}}^2(X_{31}) \sum_{n_{32}=-\infty}^{+\infty} J_{n_{32}}^2(X_{32}) \\
& \cdot \sum_{n_{41}=-\infty}^{+\infty} J_{n_{41}}^2(X_{41}) \sum_{n_{42}=-\infty}^{+\infty} J_{n_{42}}^2(X_{42}) \sum_{n_{51}=-\infty}^{+\infty} J_{n_{51}}^2(X_{51}) \sum_{n_{52}=-\infty}^{+\infty} J_{n_{52}}^2(X_{52}) \\
& \cdot \delta\!\left(\omega_d + + m_1\sqrt{gK_1} + m_2\sqrt{gK_2} - \sum_{i=1}^{2} n_i\omega_i - \sum_{i=3}^{5}(2n_{i1} + n_{i2})\omega_i\right) dK_1 d\theta_{\vec{K}_1} dK
\end{aligned}
\tag{46}
$$

2.3. RCS Incorporating a Multi-Frequency Six DOF Motion Model

In this section, a general six DOF oscillation motion model incorporating multi-frequency components is considered. Combing the derivation in Section 2.2 and the analysis in [34], the first- and second-order ocean surface scattering cross sections for bistatic HF radar incorporating a multi-frequency six DOF oscillation motion model can be respectively expressed as

$$
\begin{aligned}
\sigma_1(\omega_d) = {}& 2^3 \pi k_0^2 \Delta\rho \sum_{m=\pm 1} \int_K K^2 \cos\phi_0 S_1\!\left(m\vec{K}\right) Sa^2\!\left[\tfrac{\Delta\rho}{2}\left(\tfrac{K}{\cos\phi_0} - 2k_0\right)\right] \\
& \cdot \sum_{j=1}^{N_1}\left[\sum_{n_{1,j}=-\infty}^{+\infty} J_{n_{1,j}}^2\!\left(X_{1,j}\right)\right] \sum_{j=1}^{N_2}\left[\sum_{n_{2,j}=-\infty}^{+\infty} J_{n_{2,j}}^2\!\left(X_{2,j}\right)\right. \\
& \cdot \sum_{j=1}^{N_3}\left[\sum_{n_{31,j}=-\infty}^{+\infty} J_{n_{31,j}}^2\!\left(X_{31,j}\right) \sum_{n_{32,j}=-\infty}^{+\infty} J_{n_{32,j}}^2\!\left(X_{32,j}\right)\right] \\
& \cdot \sum_{j=1}^{N_4}\left[\sum_{n_{41,j}=-\infty}^{+\infty} J_{n_{41,j}}^2\!\left(X_{41,j}\right) \sum_{n_{42,j}=-\infty}^{+\infty} J_{n_{42,j}}^2\!\left(X_{42,j}\right)\right] \\
& \cdot \sum_{j=1}^{N_5}\left[\sum_{n_{51,j}=-\infty}^{+\infty} J_{n_{51,j}}^2\!\left(X_{51,j}\right) \sum_{n_{52,j}=-\infty}^{+\infty} J_{n_{52,j}}^2\!\left(X_{52,j}\right)\right] \\
& \cdot \delta\!\left\{\omega_d + m\sqrt{gK} - \sum_{i=1}^{2}\left[\sum_{j=1}^{N_i} n_{i,j}\omega_{i,j}\right] - \sum_{i=3}^{5}\left[\sum_{j=1}^{N_i}\left(2n_{i1,j} + n_{i2,j}\right)\omega_{i,j}\right]\right\} dK
\end{aligned}
\tag{47}
$$

and

$$
\begin{aligned}
\sigma_2(\omega_d) = {}& 2^3 \pi k_0^2 \Delta\rho \sum_{m_1=\pm 1}\sum_{m_2=\pm 1} \int_K \int_{\theta_{\vec{K}_1}} \int_{K_1} S_1\!\left(m_1\vec{K}_1\right) S_1\!\left(m_2\vec{K}_2\right) \\
& \cdot |\Gamma|^2 K^2 K_1 \cos\phi_0 Sa^2\!\left[\tfrac{\Delta\rho}{2}\left(\tfrac{K}{\cos\phi_0} - 2k_0\right)\right] \\
& \cdot \sum_{j=1}^{N_1}\left[\sum_{n_{1,j}=-\infty}^{+\infty} J_{n_{1,j}}^2\!\left(X_{1,j}\right)\right] \sum_{j=1}^{N_2}\left[\sum_{n_{2,j}=-\infty}^{+\infty} J_{n_{2,j}}^2\!\left(X_{2,j}\right)\right. \\
& \cdot \sum_{j=1}^{N_3}\left[\sum_{n_{31,j}=-\infty}^{+\infty} J_{n_{31,j}}^2\!\left(X_{31,j}\right) \sum_{n_{32,j}=-\infty}^{+\infty} J_{n_{32,j}}^2\!\left(X_{32,j}\right)\right] \\
& \cdot \sum_{j=1}^{N_4}\left[\sum_{n_{41,j}=-\infty}^{+\infty} J_{n_{41,j}}^2\!\left(X_{41,j}\right) \sum_{n_{42,j}=-\infty}^{+\infty} J_{n_{42,j}}^2\!\left(X_{42,j}\right)\right] \\
& \cdot \sum_{j=1}^{N_5}\left[\sum_{n_{51,j}=-\infty}^{+\infty} J_{n_{51,j}}^2\!\left(X_{51,j}\right) \sum_{n_{52,j}=-\infty}^{+\infty} J_{n_{52,j}}^2\!\left(X_{52,j}\right)\right] \\
& \cdot \delta\!\left\{\omega_d + m_1\sqrt{gK_1} + m_2\sqrt{gK_2} - \sum_{i=1}^{2}\left[\sum_{j=1}^{N_i} n_{i,j}\omega_{i,j}\right] - \sum_{i=3}^{5}\left[\sum_{j=1}^{N_i}\left(2n_{i1,j} + n_{i2,j}\right)\omega_{i,j}\right]\right\} \\
& \cdot dK_1 d\theta_{\vec{K}_1} dK
\end{aligned}
\tag{48}
$$

where

$$
X_{1,j} = \frac{a_{1,j}K\left[\cos\left(\theta_{\vec{K}} - \theta_{01}\right) + \tan\phi_0 \sin\left(\theta_{\vec{K}} - \theta_{01}\right)\right]}{2},
\tag{49}
$$

$$
X_{2,j} = \frac{a_{2,j}K\left[\cos\left(\theta_{\vec{K}} - \theta_{02}\right) + \tan\phi_0 \sin\left(\theta_{\vec{K}} - \theta_{02}\right)\right]}{2},
\tag{50}
$$

$$
X_{31,j} = -\frac{Kl_3\theta_{m3,j}^2 \cos\left(\theta_{\vec{K}} - \theta'\right)}{8}(1 + \tan\phi_0),
\tag{51}
$$

$$
X_{32,j} = -\frac{Kl_3\theta_{m3,j} \cos\left(\theta_{\vec{K}} - \theta'\right)}{2}(-1 + \tan\phi_0),
\tag{52}
$$

$$
X_{41,j} = \frac{Kl_4\theta_{m4,j}^2 \cos\left(\theta_{\vec{K}} - \theta_{04}\right)\sin\alpha_4}{4}\left(1 + \frac{\tan\phi_0}{2}\right),
\tag{53}
$$

$$X_{42,j} = Kl_4\theta_{m4,j}\cos\left(\theta_{\vec{K}} - \theta_{04}\right)\cos\alpha_4\left(1 + \frac{\tan\phi_0}{2}\right),\tag{54}$$

$$X_{51,j} = \frac{Kl_5\theta_{m5,j}^2\cos\left(\theta_{\vec{K}} - \theta_{05}\right)\sin\alpha_5}{4}\left(1 + \frac{\tan\phi_0}{2}\right),\tag{55}$$

$$X_{52,j} = Kl_5\theta_{m5,j}\cos\left(\theta_{\vec{K}} - \theta_{05}\right)\cos\alpha_5\left(1 + \frac{\tan\phi_0}{2}\right).\tag{56}$$

From Equations (47) and (48), it is obvious that the derived RCSs can be reduced to some existing results. For example, if only the sway motion of the floating platform is considered, the derived results can be easily reduced to the Ma et al. results [30,31]. If only the sway and surge motions with a dual-frequency model are considered, the derived results agree with those of Ma et al. [34]. If a horizontal oscillation motion model is considered, the derived first-order RCS is consistent with that derived by Yao et al. [36]. In particular, for the case of a stationary ocean platform, the derived results can be readily reduced to the onshore bistatic case [15]. In addition, it should be noted that the first- and second-order ocean surface scattering cross sections for monostatic HF radar incorporating a multi-frequency six DOF oscillation motion model can be easily derived if the bistatic angle is set to zero in Equations (47) and (48).

3. Simulation Results

In this study, using the product of a Pierson–Moskowiz ocean spectral model [38] and a cardioid directional factor [39] as directional ocean wave spectrum, simulations are conducted to analyze the effect of the antenna motion on radar Doppler spectra. The radar operating frequency, range resolution, bistatic angle, and the angle of ellipse normal are set to 5 MHz, 3 km, 45°, and 90°, respectively. The platform parameters are obtained from a deep-water floating platform [40], where the length and width of the platform are, respectively, 240 and 46 m. A single-frequency six DOF oscillation motion for wave frequency (WF) is first considered and the oscillation motion parameters are listed in Table 1. The wind speed and direction are, respectively, 41.12 m/s and 180° with respect to the direction of ellipse normal. A Hamming window is added to smooth the derived RCS curves.

Table 1. Single-frequency six DOF oscillation motion parameters for wave frequency (WF).

Parameters	Sway	Surge	Yaw	Pitch	Roll
Frequency (rad/s)	0.35	0.3	0.4	0.35	0.3
Amplitude	1.192 m	1.785 m	1.336°	1.501°	0.631°

3.1. Comparison with Onshore Case

Comparisons of the RCSs for floating-based monostatic and bistatic radars with those for onshore monostatic and bistatic cases are shown in Figures 4–6. It should be noted that only sway is considered in this part. From Figure 4a, the locations of the Bragg peaks of the onshore monostatic and bistatic HF radars are $\omega_{mB} = \pm\sqrt{2gk_0}$ and $\omega_{bB} = \pm\sqrt{2gk_0}\cos\phi_0$, respectively. It is apparent that the locations of the Bragg peaks of the bistatic HF radar are closer to zero frequency and the amplitudes are lower than those of the monostatic case. This is because a $\cos\phi_0$ term exists in Equation (47). From Figure 4b, it is seen that sway can result in some additional peaks symmetrically appearing in both monostatic and bistatic RCS curves. The amplitudes of these sway-induced peaks are generally lower than those of the Bragg peaks and the locations are respectively at $\omega_{mB} + n_1\omega_1$ and $\omega_{bB} + n_1\omega_1$ for the floating-based monostatic and bistatic cases. However, the locations of the Bragg peaks remain unchanged and the amplitudes are slightly lower than those of onshore cases.

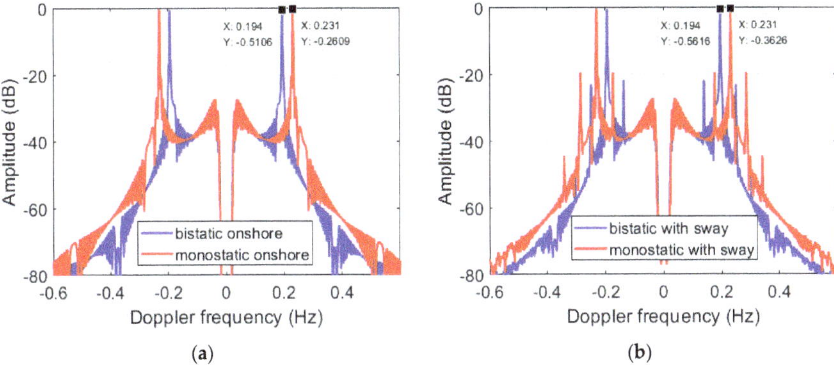

Figure 4. Simulated first-order radar cross sections (RCSs), (**a**) for onshore monostatic and bistatic cases; (**b**) for floating-based monostatic and bistatic cases with sway.

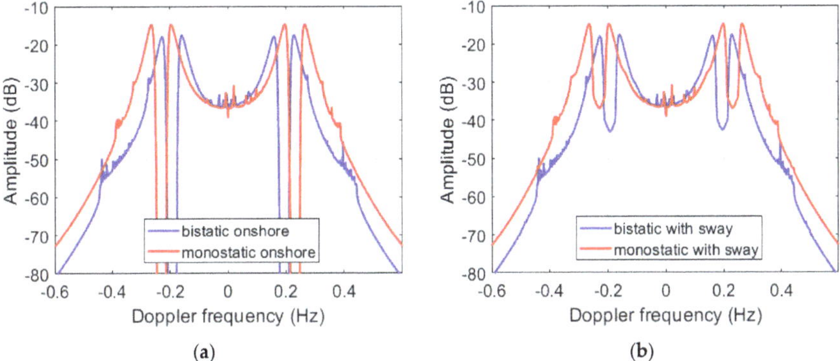

Figure 5. Simulated second-order RCSs, (**a**) for onshore monostatic and bistatic cases; (**b**) for floating-based monostatic and bistatic cases with sway.

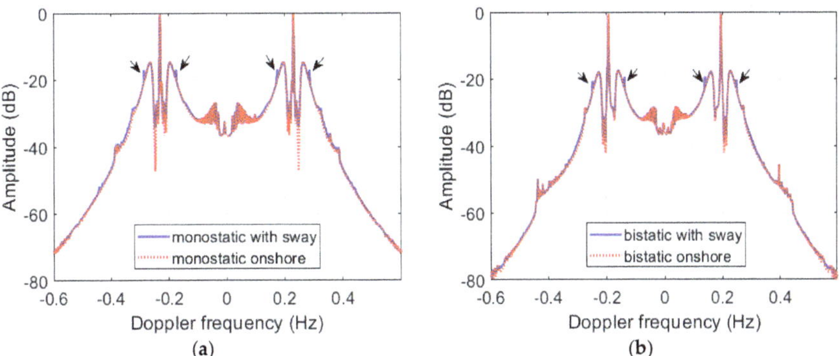

Figure 6. Simulated total RCSs containing first- and second-order RCSs, (**a**) for onshore and floating-based monostatic cases; (**b**) for onshore and floating-based bistatic cases. The black arrows represent additional sway-induced peaks.

As mentioned before, the second-order RCS mainly contains the hydrodynamic and electromagnetic contributions. For onshore bistatic radar, the locations of the hydrodynamic and electromagnetic peaks are respectively given as [15]

$$\omega_{hd} = \pm \sqrt{2}\omega_{bB},$$
(57)

and

$$\omega_{ed} = \pm 2^{3/4}\sqrt{\frac{(1 \pm \sin\phi_0)^{1/2}}{\cos\phi_0}}\omega_{bB}.$$
(58)

For onshore monostatic case, the corresponding locations can be obtained by imposing $\phi_0 = 0$ in Equations (57) and (58). The amplitudes of the second-order RCS for the onshore bistatic case are lower than those of the onshore monostatic case, as shown in Figure 5a. This is because there exists a $\cos\phi_0$ term in Equation (48) compared with the monostatic second-order RCS. Theoretically, additional sway-induced peaks will appear in the second-order RCS curves at frequencies $\omega_{hd} + n_1\omega_1$ and $\omega_{ed} + n_1\omega_1$ for the floating bistatic case. However, from Figure 5b, those additional sway-induced peaks are imperceptible. Although the effect of sway on the second-order RCS is not apparent, the additional peaks appearing in the first-order RCS curve may raise the second-order RCS, as shown in Figure 6.

3.2. Effect of Six DOF Motion on RCS

Figure 7 shows the simulated first-order RCSs for different platform motion models. It can be seen that each one-dimensional oscillation motion will induce additional peaks. The locations and amplitudes of these additional motion-induced peaks are different from each other, which are decided by the frequency and amplitude of corresponding oscillation motion, respectively. From Equation (47), the initial phase of each one-dimensional oscillation motion has no effect on RCSs. Therefore, the oscillation motion of the floating platform can be regarded as frequency modulation of RCS of the onshore bistatic HF radar. From Figure 7d,e, pitch and roll have a smaller effect on RCSs compared to other oscillation motions. This is because the oscillation amplitudes of pitch and roll are relatively small. However, yaw results in more additional peaks appearing in the first-order RCS curve with a small oscillation amplitude and the amplitudes of these yaw-induced peaks are higher with respect to other cases. This is because the radar antenna is generally installed at the edge of the platform (especially a ship) to reduce the electromagnetic effect of platform superstructures on radar Doppler spectra. In this study, the antenna is assumed to be deployed far from the center of rotation. Thus, a small oscillation amplitude of yaw may cause a large horizontal antenna displacement.

When a six DOF oscillation motion model is considered, more additional motion-induced peaks will appear in the first-order RCS curve, which are not only caused by each one-dimensional oscillation motion but also by the combined motion. For such a case, the frequency locations of these motion-induced peaks can be expressed as

$$\omega_d = \omega_{bB} + \sum_{i=1}^{2} n_i\omega_i + \sum_{i=3}^{5} (2n_{i1} + n_{i2})\omega_i.$$
(59)

Therefore, the modulation effect on the first-order RCS of six DOF oscillation motion is significantly greater than that of each one-dimensional oscillation motion.

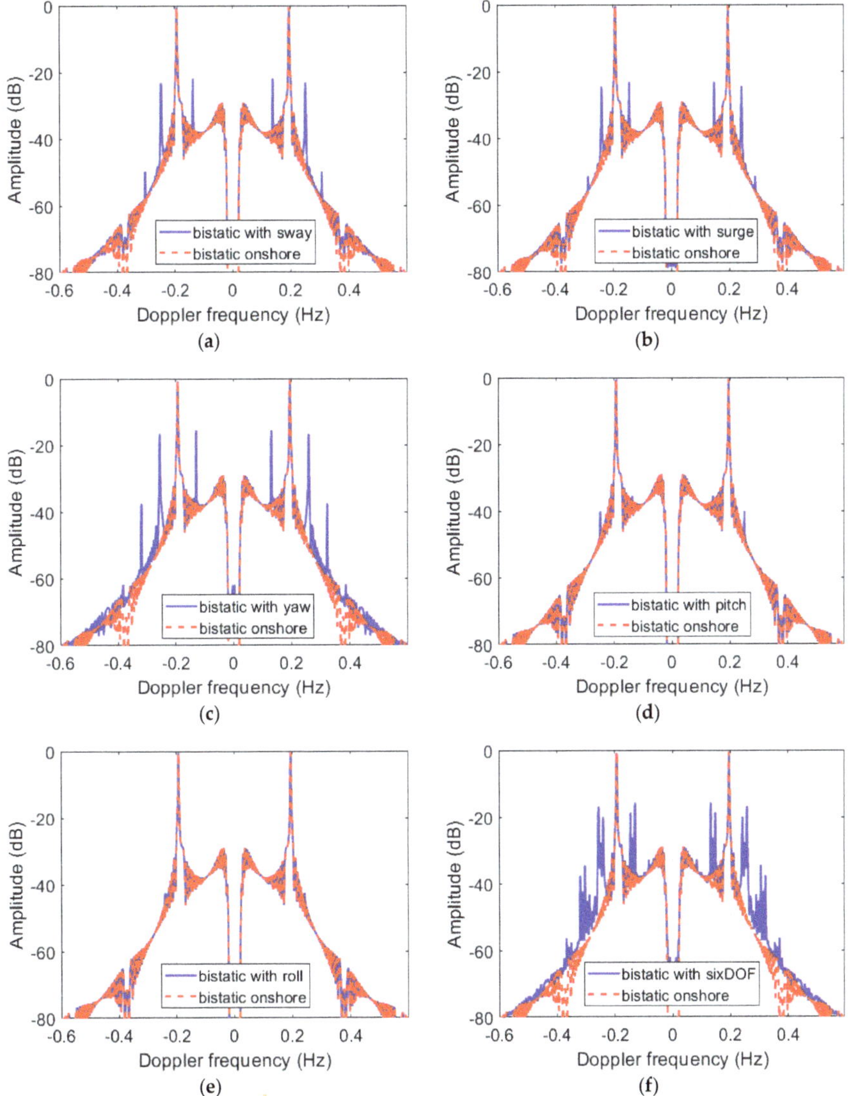

Figure 7. Simulated first-order RCSs for different motions. (**a**) Sway case; (**b**) surge case; (**c**) yaw case; (**d**) pitch case; (**e**) roll case; (**f**) six DOF case.

Figure 8 shows the simulated second-order RCSs for different platform motion models. The additional motion-induced peaks appearing in the second-order RCS curve are not obvious. Similar to the sway case, the effect of surge on the second-order RCS is also small, as shown in Figure 8b. From Figure 8d,e, the effect of pitch and roll on the second-order RCSs may be ignored. However, yaw has an important effect on the second-order RCS due to a larger displacement of the antenna, as shown in Figure 8c. When a six DOF oscillation motion model is considered, more additional peaks caused by each one-dimensional oscillation motion and the combined motion may appear in the second-order RCS curve and the corresponding frequency locations are

$$\omega_d = \omega_{hd} + \sum_{i=1}^{2} n_i \omega_i + \sum_{i=3}^{5} (2n_{i1} + n_{i2}) \omega_i, \tag{60}$$

and

$$\omega_d = \omega_{ed} + \sum_{i=1}^{2} n_i \omega_i + \sum_{i=3}^{5} (2n_{i1} + n_{i2}) \omega_i. \tag{61}$$

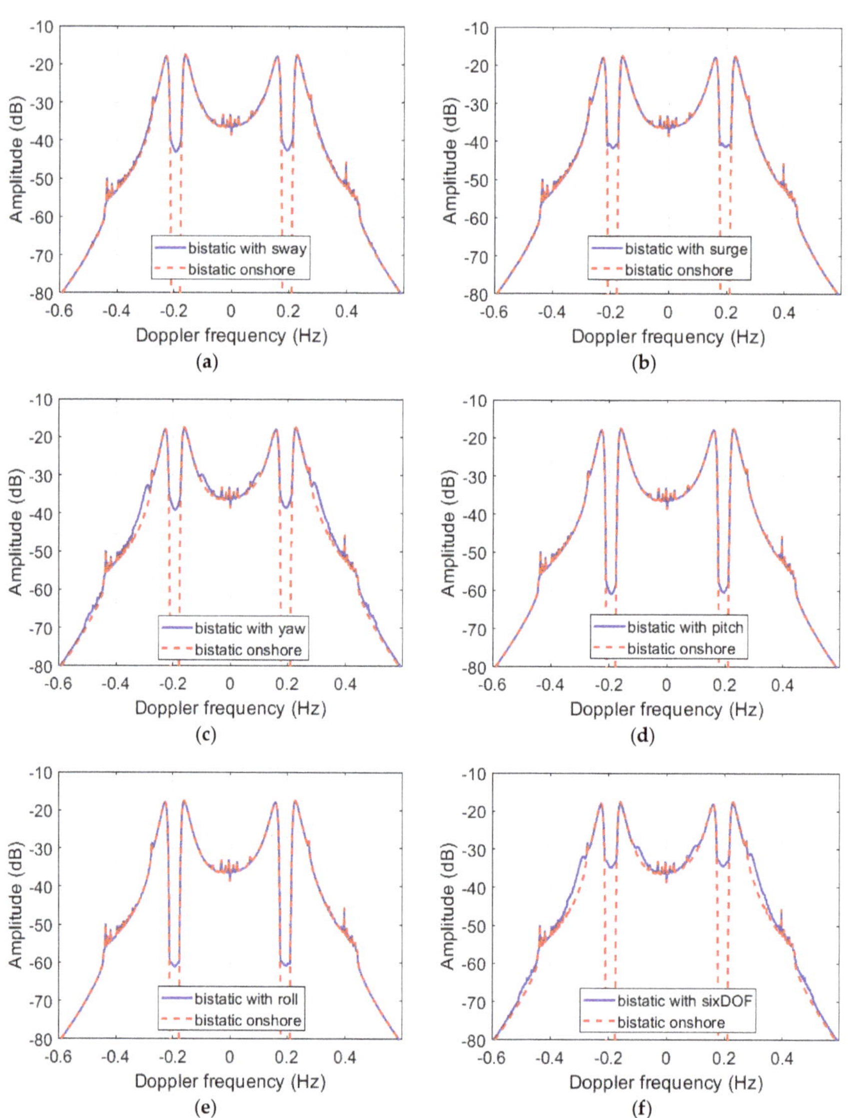

Figure 8. Simulated second-order RCSs for different motions. (**a**) Sway case; (**b**) surge case; (**c**) yaw case; (**d**) pitch case; (**e**) roll case; (**f**) six DOF case.

However, by comparing Figure 8c,f, the RCS curves are basically similar. That is, the modulation effect of yaw on the second-order RCS is dominant.

Figure 9 displays the simulated total RCS containing the first- and second-order RCSs for the bistatic HF radar incorporating a single-frequency six DOF oscillation motion model. From Figure 9, it is seen that the motion-induced peaks appearing in the first-order RCS curve will overlap with the second-order RCS curve and, then, the amplitude of the second-order RCS may be raised. For such a case, the amplitudes of the Bragg peaks are still larger than those of the motion-induced peaks. Compared to the sway case in Figure 6b, more motion-induced peaks with larger amplitude appear in the total RCS curve. Therefore, in practice, just considering one- or two-dimensional oscillation motion is not realistic.

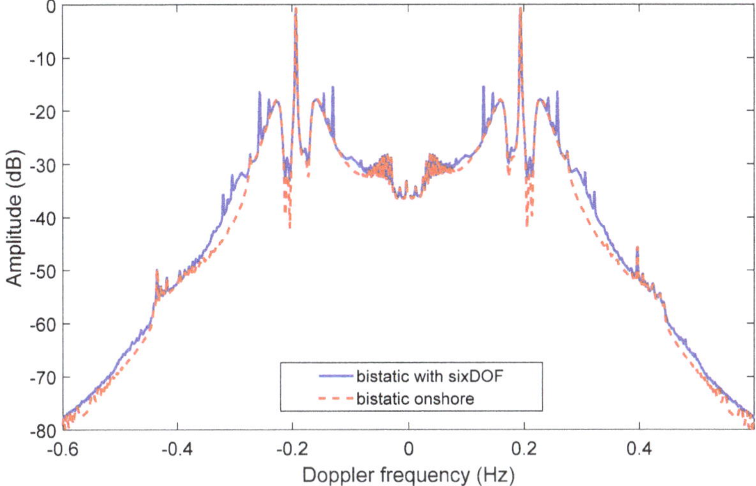

Figure 9. Simulated total RCSs containing first- and second-order RCSs.

3.3. Effect of Dual-Frequency Model on RCS

In an actual marine environment, the frequency of the oscillation motion of a floating platform varies from low frequency (LF) to WF [40]. The WF response at the WF is 0.2–2 rad/s and the LF response at the in-plane resonances is around 0.02 rad/s. Therefore, in order to interpret the characteristics of the sea echo signals more realistically, a multi-frequency six DOF oscillation motion model incorporating LF and WF should be considered. In this study, in order to simply the simulation, a dual-frequency six DOF oscillation motion model incorporating both LF and WF was used. The LF six DOF oscillation motion parameters corresponding to the maximum spectral density of each one-dimensional oscillation motion were selected to simulate RCSs, which are listed in Table 2.

Table 2. Six DOF oscillation motion parameters for low frequency (LF).

Parameters	Sway	Surge	Yaw	Pitch	Roll
Frequency (rad/s)	0.04	0.05	0.03	0.03	0.03
Amplitude	4.32 m	4.27 m	6.87°	0.2°	0.24°

Similar to the WF case, six DOF oscillation motion with LF will also result in some additional peaks. The frequency locations of these motion-induced peaks are extremely close to the Bragg peaks because of a low oscillation motion frequency. When a dual-frequency six DOF oscillation motion model incorporating a LF model and a WF model is considered, the frequency locations of the motion-induced peaks in the first- and second-order RCS curves are, respectively,

$$\omega_d = \omega_{bB} + \sum_{i=1}^{2}\left[\sum_{j=1}^{N_i} n_{i,j}\omega_{i,j}\right] + \sum_{i=3}^{5}\left[\sum_{j=1}^{N_i}\left(2n_{i1,j} + n_{i2,j}\right)\omega_{i,j}\right], \tag{62}$$

$$\omega_d = \omega_{hd} + \sum_{i=1}^{2}\left[\sum_{j=1}^{N_i} n_{i,j}\omega_{i,j}\right] + \sum_{i=3}^{5}\left[\sum_{j=1}^{N_i}\left(2n_{i1,j} + n_{i2,j}\right)\omega_{i,j}\right], \tag{63}$$

and

$$\omega_d = \omega_{ed} + \sum_{i=1}^{2}\left[\sum_{j=1}^{N_i} n_{i,j}\omega_{i,j}\right] + \sum_{i=3}^{5}\left[\sum_{j=1}^{N_i}\left(2n_{i1,j} + n_{i2,j}\right)\omega_{i,j}\right], \tag{64}$$

where $N_i = 2$ represents two frequency components. Figure 10 displays the simulated first- and second-order RCSs for bistatic HF radar incorporating a dual-frequency six DOF oscillation motion model. From Figure 10a, the LF motion-induced peaks appear not only near the Bragg peaks but also near the WF motion-induced peaks, which agrees well with Equation (62). Furthermore, the amplitudes of the Bragg peaks and the WF motion-induced peaks are lower than those of the LF motion-induced peaks due to the modulation effect, which may 'break' Bragg scatter mechanism. A comparison of Figures 8f and 10b shows that the WF motion is the dominant factor affecting the second-order RCS. Figure 10c shows the total RCS containing the first- and second-order RCSs. From Figure 10c, it can be seen that a dual-frequency six DOF oscillation motion may have a more significant effect than a single-frequency case in Figure 9. In addition, the effects of different wind directions, wind speeds, and radar parameters on RCS for bistatic HF radar incorporating a dual-frequency six DOF oscillation motion model are similar to the sway case [30,31] and are not further discussed here.

3.4. Effect of Bistatic Angle on RCS

Figure 11 shows the simulated total RCSs containing the first- and second-order RCSs for different bistatic angles. It is obvious that the Bragg peaks, both the hydrodynamic and electromagnetic peaks for the second-order scatter and additional peaks caused by six DOF oscillation motion move closer to zero Doppler frequency while the bistatic angle is increasing. Furthermore, as the bistatic angle increases, the amplitudes of the second-order RCSs decrease, and the modulation effect caused by the platform motion is weakened. It should be noted that, from Equation (58), the second-order electromagnetic peaks may be far away from the Bragg peaks or even diminished from the total RCS curve for a large bistatic angle, for example $\phi_0 = 85^\circ$ as shown in Figure 11.

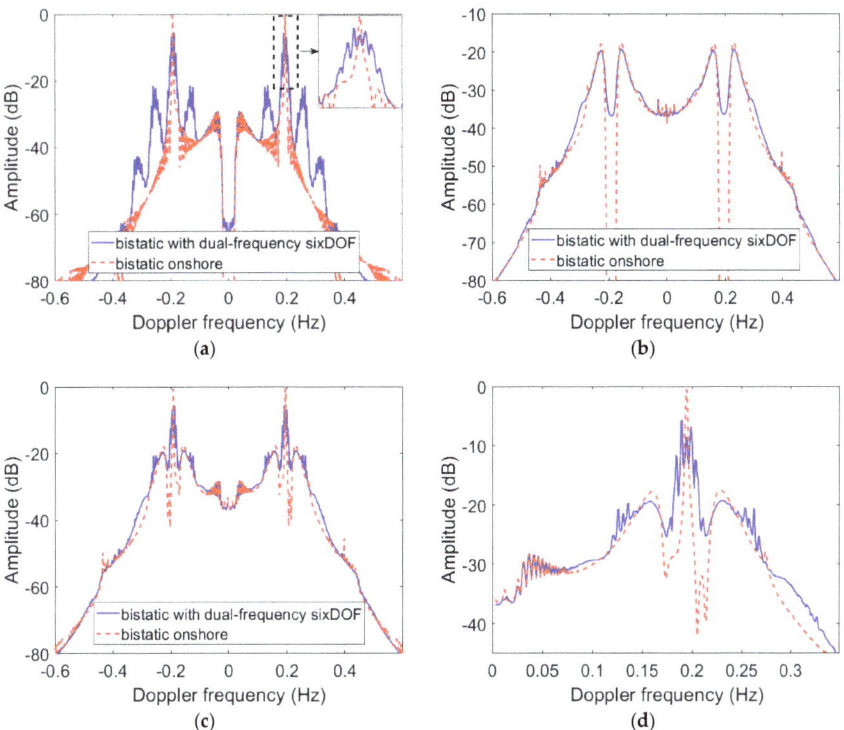

Figure 10. Simulated RCSs for a dual-frequency six DOF oscillation motion model. (**a**) First-order RCS; (**b**) second-order RCS; (**c**) total RCS; (**d**) zoomed in view of the positive Doppler frequency in (c).

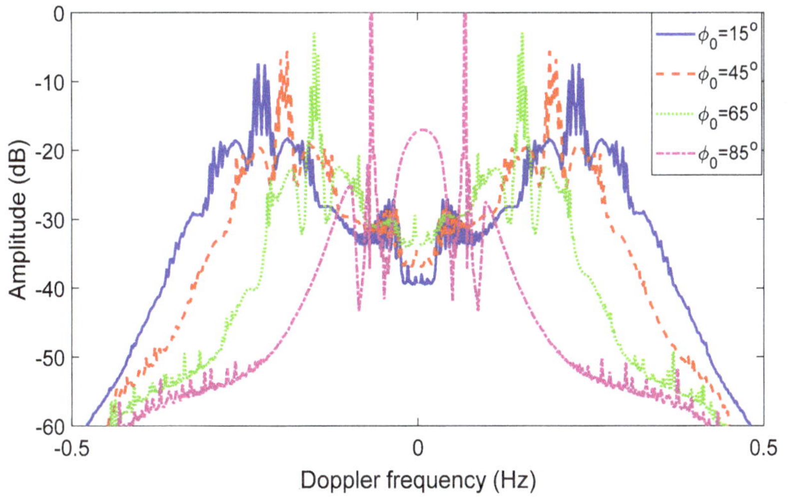

Figure 11. Simulated total RCSs containing first- and second-order RCSs for different bistatic angles.

4. Discussion

On the basis of the preceding simulation results, it may be clearly concluded that a dual-frequency six DOF oscillation motion has a critical effect on ocean surface scattering cross section of bistatic HF radar, which may seriously affect the applications of the bistatic HF radar installed on a floating platform in ocean surface moving target detection and remote sensing of ocean surface dynamics parameters such as wind direction, wind speed, current, and ocean wave parameters. The characteristics (the energy distributions of the Bragg peaks and motion-induced peaks) of radar Doppler spectra depend on the oscillation motion parameters. Wind speed is also one of the factors affecting the oscillation motion parameters. Although the oscillation motion parameters of a large floating platform under a hurricane condition were selected for examples in this study, a small floating platform may yield similar phenomena under a moderate or low sea state. The reliability still needs to be further verified using field data with different floating platforms (shape and size) and different wind speeds in future work.

The motion-induced peaks may mask the moving target echoes and it may be extremely difficult to separate the target echoes from the motion-induced peaks if the target echoes appear near the Bragg peaks. This is because the amplitudes of the motion-induced peaks may be larger than those of the Bragg peaks, which may cause false alarm.

Furthermore, ocean surface wind direction is generally extracted according to the intensity ratio of the positive and negative Bragg peaks. Considering that the energies of the Bragg peaks are modulated by six DOF oscillation motion, the wind direction measurement results may be inaccurate if the modulation effect is ignored. Ocean surface current is generally measured based on the position difference between the theoretical Bragg peaks without ocean surface current and the measured Bragg peaks with ocean surface current. Generally, for the field data, the positions of the Bragg peaks are identified by searching for the strongest peaks in the positive and negative Doppler spectrum. However, due to the modulation effect of six DOF oscillation motion, the energies of the Bragg peaks may be lower than those of the motion-induced peaks, which may severely influence the measurement of ocean surface current. As we all know, the information of ocean surface wind speed and wave parameters is contained in the second-order radar Doppler spectrum. Although the modulation effect of six DOF oscillation motion on the second-order RCS is small, the motion-induced peaks in the first-order RCS may overlap with the second-order RCS, which would severely contaminate the second-order RCS and may have an unfavorable effect on the measurement of wind speed and ocean wave parameters.

Moreover, for six DOF oscillation motion, yaw is a dominant factor affecting RCS. Considering that, the oscillation motion can be regarded as frequency modulation on RCS and the modulation effect depends on the amplitude and frequency of the oscillation motion. Therefore, in order to reduce the effect of yaw on RCS and considering the influence of the floating platform superstructure on electromagnetic scattering, the installation location of the antenna should be at the edge of the floating platform but near the rotation center.

In addition, the modulation effect of six DOF oscillation motion on RCS is reduced with increased bistatic angle. That is, the amplitudes of the motion-induced peaks and the second-order RCS decrease for a large bistatic angle, which may be beneficial for ocean surface moving target detection. However, this case may be adverse for the extraction of ocean surface wind speed and wave parameters from the second-order RCS due to the disappearance of the second-order electromagnetic peaks for a large bistatic angle. Therefore, it is very important to choose a reasonable bistatic angle for different application purposes using a floating platform based bistatic HF radar.

Therefore, in order to improve the performance of moving target detection and the accuracies of ocean surface dynamics parameter measurements, a motion compensation method should be investigated to remove the motion-induced peaks in RCS and to recover the amplitudes of the first- and second-order RCSs in the future.

5. Conclusions

In this paper, the first- and second-order ocean surface cross sections for bistatic HF radar incorporating a multi-frequency six DOF oscillation motion model were theoretically derived. When the bistatic angle is zero, the derived results can be reduced to the monostatic case, and when there is no six DOF oscillation motion, the derived results can be simplified to the onshore bistatic case. Simulations were conducted under different oscillation motion models and different bistatic angles.

Results show that each one-dimensional oscillation motion may induce additional peaks symmetrically appearing the first- and second-order radar Doppler spectra and the combined six DOF oscillation motion may result in more additional peaks. The amplitudes and frequencies of these motion-induced peaks depend on the amplitude and frequency of six DOF oscillation motion. The platform oscillation motion can be viewed as frequency modulation for radar echoes and the modulation effect of six DOF oscillation motion on the first-order radar Doppler spectra is more obvious than that on the second-order radar Doppler spectra. However, the motion-induced peaks appearing in the first-order radar spectra may overlap with the second-order radar spectra, which may raise the second-order radar spectra. It should be noted that yaw is the dominant factor affecting radar Doppler spectra, especially for the second-order spectra. Furthermore, the amplitudes of the Bragg peaks may be lower than those of the motion-induced peaks if a LF six DOF oscillation motion model is considered. This is a very important phenomenon for the application of bistatic HF radar. In addition, the modulation effect of six DOF oscillation motion and amplitudes of the second-order radar Doppler spectra decrease with increasing bistatic angle. For a large bistatic angle, the second-order electromagnetic peaks may be far away from the Bragg peaks or even diminished from radar Doppler spectra. Therefore, if the influences of the platform oscillation motion and bistatic angle on radar Doppler spectra are ignored, it will severely affect the applications of bistatic HF radar in moving target detection and ocean surface dynamics parameter measurements.

Here, the derived results were investigated only with simulated data, the rationality of the derived results should be further validated with field data. Nonetheless, this work provides an important theoretical foundation to determine suitable geometries for the deployment of a platform-based bistatic HF radar.

Author Contributions: Conceptualization, G.Y., J.X.; Methodology, G.Y., J.X., W.H.; Writing—original draft preparation, G.Y.; Writing—review and editing, J.X., W.H.; Software, G.Y.

Funding: This research was funded by National Natural Science Foundation of China key project (61132005), Special Research Foundation for Harbin Science and Technology Innovation Talents (RC2014XK009022), Natural Sciences and Engineering Research Council of Canada Discovery Grants under Grant NSERC RGPIN-2017-04508 and Grant RGPAS-2017-507962.

Acknowledgments: The authors would like to thank anonymous reviewers for their valuable comments in improving the quality of this paper.

Conflicts of Interest: The authors declare no conflict of interest.

References

1. Barrick, D.E. Extraction of wave parameters from measured HF radar sea-echo Doppler spectra. *Radio Sci.* **1986**, *12*, 415–424. [CrossRef]
2. Lipa, B.J.; Barrick, D.E. Extraction of sea state from HF radar sea echo: Mathematical theory and modeling. *Radio Sci.* **1986**, *21*, 81–100. [CrossRef]
3. Khan, R.; Gamberg, B.; Power, D.; Walsh, J.; Dawe, B.; Pearson, W.; Millan, D. Target detection and tracking with a high frequency ground wave radar. *IEEE J. Ocean. Eng.* **1994**, *19*, 540–548. [CrossRef]
4. Hisaki, Y. Nonlinear inversion of the integral equation to estimate ocean wave spectra from HF radar. *Radio Sci.* **1996**, *31*, 25–39. [CrossRef]
5. Heron, M.L.; Prytz, A. Wave height and widn direction from the HF coastal ocean surface radar. *Can. J. Remote Sens.* **2002**, *28*, 385–393. [CrossRef]
6. Huang, W.; Wu, S.; Gill, E.W.; Wen, B.; Hou, J. HF radar wind and wave measurement over the Eastern China Sea. *IEEE Trans. Geosci. Reomote Sens.* **2002**, *40*, 1950–1955. [CrossRef]
7. Emery, B.M.; Washburn, L.; Harlan, J.A. Evaluating radial current measurements from CODAR high-frequency radars with moored current meters. *J. Atmos. Ocean. Technol.* **2004**, *21*, 1259–1271. [CrossRef]
8. Huang, W.; Gill, E.W.; Wu, S.; Wen, B.; Yang, Z.; Hou, J. Measuring surface wind direction by monostatic HF ground-wave radar at the Eastern China Sea. *IEEE J. Ocean. Eng.* **2004**, *29*, 1032–1037. [CrossRef]

9. Wyatt, L.R.; Green, J.J.; Middleditch, A.; Moorhead, M.D.; Howarth, J.; Holt, M.; Keogh, S. Operational wave, current, and wind measurements with the pisces HF radar. *IEEE J. Ocean. Eng.* **2006**, *31*, 819–834. [CrossRef]

10. Jangal, F.; Saillant, S.; Helier, M. Wavelet contribution to remote sensing of the sea and target detection for a high-frequency surface wave radar. *IEEE Geosci. Remote Sens. Lett.* **2008**, *5*, 552–556. [CrossRef]

11. Li, Q.; Zhang, W.; Li, M.; Niu, J.; Wu, Q.M.J. Automatic detection of ship targets based on wavelet transform for HF surface wavelet radar. *IEEE Geosci. Remote Sens. Lett.* **2017**, *14*, 714–718. [CrossRef]

12. Tian, Y.; Wen, B.; Zhou, H.; Wang, C.; Yang, J.; Huang, W. Wave Height Estimation from First-Order Backscatter of a Dual-Frequency High Frequency Radar. *Remote Sens.* **2017**, *9*, 1186. [CrossRef]

13. Shen, W.; Gurgel, K.W. Wind Direction Inversion from Narrow-Beam HF Radar Backscatter Signals in Low and High Wind Conditions at Different Radar Frequencies. *Remote Sens.* **2018**, *10*, 1480. [CrossRef]

14. Walsh, J.; Gill, E.W. An analysis of the scattering of high-frequency electromagnetic radiation form rough surfaces with application to pulse radar operating in backscatter mode. *Radio Sci.* **2000**, *35*, 1337–1359. [CrossRef]

15. Gill, E.W.; Walsh, J. High-frequency bistatic cross sections of the ocean surface. *Radio Sci.* **2001**, *36*, 1459–1475. [CrossRef]

16. Gill, E.W.; Huang, W.; Walsh, J. The effect of bistatic scattering angle on the high frequency radar cross sections of the ocean surface. *IEEE Geosci. Remote Sens. Lett.* **2008**, *5*, 143–146. [CrossRef]

17. Lipa, B.J.; Whelan, C.; Rector, B.; Moorhead, M.D.; Howarth, J.; Holt, M.; Keogh, S. HF radar bistatic measurement of surface current velocities: Drifter comparisons and radar consistency checks. *Remote Sens.* **2009**, *1*, 1190–1211. [CrossRef]

18. Trizna, D.B. A bistatic HF radar for current mapping and robust ship tracking. *Sea Technol.* **2009**, *50*, 29–33.

19. Grosdidier, S.; Forget, P.; Barbin, Y.; Cuerin, C.A. HF bistatic ocean Doppler spectra: Simulation versus experimentation. *IEEE Trans. Geosci. Reomote Sens.* **2014**, *52*, 2138–2148. [CrossRef]

20. Huang, W.; Gill, E.W.; Wu, X.; Li, L. Measurement of sea surface wind direction using bistatic high-frequency radar. *IEEE Trans. Geosci. Reomote Sens.* **2012**, *50*, 4117–4122. [CrossRef]

21. Zhang, J.; Gill, E.W. Extraction of ocean wave spectra from simulated noisy bistatic high-frequency radar data. *IEEE J. Ocean. Eng.* **2006**, *31*, 779–796. [CrossRef]

22. Silva, M.T.; Shahidi, R.; Gill, E.W.; Huang, W. Nonlinear extraction of directional ocean wave spectrum from synthetic bistatic high-frequency surface wave radar data. *IEEE J. Ocean. Eng.* **2019**, in press. [CrossRef]

23. Lipa, B.J.; Barrick, D.E.; Isaacson, J.; Lilleboe, P.M. CODAR wave measurements from a North Sea semisubmersible. *IEEE J. Ocean. Eng.* **1990**, *15*, 119–225. [CrossRef]

24. Howell, R.; Walsh, J. Measurement of ocean wave spectra using a ship-mounted HF radar. *IEEE J. Ocean. Eng.* **1993**, *18*, 306–310. [CrossRef]

25. Gurgel, K.W.; Essen, H. On the performance of a shipborne current mapping HF radar. *IEEE J. Ocean. Eng.* **2000**, *25*, 183–191. [CrossRef]

26. Xie, J.; Yuan, Y.; Liu, Y. Experimental analysis of sea clutter in shipborne HFSWR. *IEE Proc.-Radar Sonar Navig.* **2001**, *148*, 67–71. [CrossRef]

27. Walsh, J.; Huang, W.; Gill, E.G. The first-order high frequency radar ocean surface cross section for an antenna on a floating platform. *IEEE Trans. Antennas Propag.* **2010**, *58*, 2994–3003. [CrossRef]

28. Walsh, J.; Huang, W.; Gill, E.G. The second-order high frequency radar ocean surface cross section for an antenna on a floating platform. *IEEE Trans. Antennas Propag.* **2012**, *60*, 4804–4813. [CrossRef]

29. Sun, M.; Xie, J.; Ji, Z.; Yao, G. Ocean surface cross sections for shipborne HFSWR with sway motion. *Radio Sci.* **2016**, *51*, 1745–1757. [CrossRef]

30. Ma, Y.; Gill, E.W.; Huang, W. First-order bistatic high frequency radar ocean surface cross section for an antenna on a floating platform. *IET Radar Sonar Navig.* **2016**, *10*, 1136–1144. [CrossRef]

31. Ma, Y.; Huang, W.; Gill, E.W. The second-order bistatic high frequency radar ocean surface cross section for an antenna on a floating platform. *Can. J. Remote Sens.* **2016**, *42*, 332–343. [CrossRef]

32. Das, S.N.; Das, S.K. Mathematical modeling of sway, roll and yaw motions: Order-wise analysis to determine coupled characteristics and numerical simulation for restoring moment's sensitivity analysis. *Acta Mech.* **2010**, *213*, 305–322. [CrossRef]

33. Tahar, A.; Kim, M.H. Hull/mooring/riser coupled dynamics analysis and sensitivity study of tanker-based FPSO. *Appl. Ocean Res.* **2003**, *25*, 367–382. [CrossRef]

34. Ma, Y.; Gill, E.W.; Huang, W. Bistatic high-frequency radar ocean surface cross section incorporating a dual-frequency platform motion model. *IEEE J. Ocean. Eng.* **2018**, *43*, 205–210. [CrossRef]
35. Yao, G.; Xie, J.; Huang, W. First-order ocean surface cross-section for shipborne HFSWR incorporating a horizontal oscillation motion model. *IET Radar Sonar Navig.* **2018**, *41*, 970–981. [CrossRef]
36. Yao, G.; Xie, J.; Huang, W. First-order ocean surface cross section for bistatic HFSWR incorporating a horizontal oscillation motion model. In Proceedings of the IEEE Radar Conference, Boston, MA, USA, 22–26 April 2019.
37. Hasselmann, K. On the nonlinear energy transfer in a gravity wave spectrum, part 1, general theory. *J. Fluid Mech.* **1962**, *12*, 481–500. [CrossRef]
38. Pierson, W.J.; Moskowitz, L. A proposed spectral form for fully developed wind sea based on the similarity theory of S. A. Kitaigorodskii. *J. Geophys. Res.* **1964**, *69*, 5181–5190. [CrossRef]
39. Longuet-Higgins, M.S.; Cartwright, D.E.; Smith, N.D. Observation of the directional spectrum of sea waves using the motion of a floating buoy. In *Proceedings of the Conference Ocean Wave Spectra*; Prentice-Hall Inc.: Englewood Cliffs, NJ, USA, 1963; pp. 111–136.
40. Low, Y.M.; Langley, R.S. Time and frequency domain coupled analysis of deepwater floating production systems. *Appl. Ocean Res.* **2006**, *28*, 371–385. [CrossRef]

Communication

Bistatic High-Frequency Radar Cross-Section of the Ocean Surface with Arbitrary Wave Heights

Murilo Teixeira Silva *,†, **Weimin Huang** † and **Eric W. Gill** †

Faculty of Engineering and Applied Science, Memorial University of Newfoundland, St. John's, NL A1C 5S7, Canada; weimin@mun.ca (W.H.); ewgill@mun.ca (E.W.G.)
* Correspondence: murilots@mun.ca
† These authors contributed equally to this work.

Received: 28 December 2019; Accepted: 15 February 2020; Published: 18 February 2020

Abstract: The scattering theory developed in the past decades for high-frequency radio oceanography has been restricted to surfaces with small heights and small slopes. In the present work, the scattering theory for bistatic high-frequency radars is extended to ocean surfaces with arbitrary wave heights. Based on recent theoretical developments in the scattering theory for ocean surfaces with arbitrary heights for monostatic radars, the electric field equations for bistatic high-frequency radars in high sea states are developed. This results in an additional term related to the first-order electric field, which is only present when the small-height approximation is removed. Then, the radar cross-section for the additional term is derived and simulated, and its impact on the total radar cross-section at different radar configurations, dominant wave directions, and sea states is assessed. The proposed term is shown to impact the total radar cross-section at high sea states, dependent on radar configuration and dominant wave direction. The present work can contribute to the remote sensing of targets on the ocean surface, as well as the determination of the dominant wave direction of the ocean surface at high sea states.

Keywords: bistatic radar; HF radar; electromagnetic scattering; radio oceanography

---

1. Introduction

Since the discovery of the Bragg effects on the transmission of electromagnetic fields over the ocean surface in 1955 [1], high-frequency (HF) radars have been largely used in ocean remote sensing, from oceanographic applications (e.g., [2,3]) to target detection and tracking (e.g., [4,5]). HF radars can be presented in two configurations with respect to the relative positions of the transmitter and receiver: Monostatic, where the distance between the transmitter and receiver is much smaller than the distance between them and the scattering object, to the point that they can be considered co-located; and bistatic (or multistatic), where the distance between the transmitter and receiver (or multiple receivers) is comparable to the distance between them and the scattering object. Due to the ubiquity of HF radar systems installed in monostatic configurations and the relative simplicity of geometric and mathematical considerations when compared to a bistatic radar, most of the research developed in the past decades has been dedicated to monostatic HF radar systems.

Although early works on the bistatic radar cross-section of the ocean surface in C-band were published in 1966 [6], the first efforts to implement bistatic HF radars for radio oceanography would only come later in that decade [7–9]. Barrick started exploring the scattering theory of bistatic HF radars in 1970 [10], proposing an expression for the radar cross-section of the ocean surface for HF radars in 1972 [11]. Bistatic scattering coefficients from the ocean surface were later derived by Johnstone [12]. In 1987, Barrick's theory for the radar cross-section of the ocean surface was expanded by Anderson [13], and was later validated through experiments; e.g., [14,15]. In the past two decades, the generalized functions method introduced in [16] has been applied to the development of a scattering theory for bistatic HF radars [17–23], with its validity being experimentally verified [24].

In the development of the scattering theory for the ocean surface in both monostatic and bistatic radar configurations, small-height and small-slope approximations are commonly applied, respectively restricting the scattering analysis to ocean surfaces where wave heights are small compared to the radar wavelength and the surface slopes are sufficiently small [16,25]. In mathematical terms, the small-height approximation limits the scattering analysis of the ocean surface roughness scales $k_0 H_s$ to be much smaller than one, where k_0 is the wavenumber that represents the central radar transmitting frequency ω_0, defined as

$$k_0 = \frac{\omega_0}{c},$$

where c is the speed of light and H_s is the significant wave height of the ocean surface, while the small-slope approximation restricts the surface slope $|\nabla f(x, y; t)|$ to values smaller than unity. Therefore, when the ocean surface violates these restrictions, the validity of the currently-used theory cannot be guaranteed [26], and the development of a scattering theory that would be valid in such circumstances is desirable.

The present work aims to expand the narrow-beam bistatic HF radar scattering theory to ocean surfaces with large roughness scales, allowing arbitrary wave heights. The expression for the electric field scattered from an ocean surface with arbitrary heights and received in a bistatic radar configuration is presented in Section 2, while the radar cross-section expression is derived in Section 3. The simulation results and discussion are presented respectively in Sections 4 and 5, while the concluding remarks are presented in Section 6.

2. First-Order Bistatic Electric Field Scattered from an Ocean Surface with Arbitrary Heights

For the purposes of this work, a conductive rough surface defined as $z = f(x, y; t)$ is considered. $f(x, y; t)$ is a zero-mean, time-varying, two-dimensional random variable representing the ocean surface displacement from the sea level $z = 0$ [27]. In general, the equation for the electromagnetic propagation over a rough surface is defined as [16]

$$\mathcal{NL}^{-1}\left[\frac{\mathcal{LNL}^{-1}\left[2u\mathcal{F}_{xy}(\mathbf{E}_s^{z^-})\right]}{u + jk\Delta}\right] = \mathbf{E}_n^+ + \mathcal{NL}^{-1}\left[\frac{\mathcal{LN}\left[\frac{\nabla_{xy}\left(|\mathbf{n}|E_n^+\right)}{|\mathbf{n}|^2}\right]}{u + jk\Delta}\right], \tag{1}$$

where $\mathbf{n}$ is a vector normal to the surface, understood here to be the ocean surface, and is defined as

$$\mathbf{n} = \hat{\mathbf{z}} - \nabla f(x, y; t). \tag{2}$$

$\mathcal{F}_{xy}(\cdot)$ is the spatial Fourier transform in the xy-plane, $\mathbf{E}_s^{z^-}$ is the source electric field at the point $z = z^- < f(x, y; t)$, $\forall (x, y; t)$, $\mathbf{E}_n^+$ is the normal electric field immediately above the ocean surface, k is the radar wavenumber, u is defined as

$$u = \sqrt{K^2 - k^2}, \tag{3}$$

where K is the surface wavenumber, Δ is the surface impedance, $\mathcal{N}$ is the normalizing operator defined as

$$\mathcal{N}\{\mathbf{A}\} = \hat{\mathbf{n}}\hat{\mathbf{n}} \cdot \mathbf{A}, \ \forall \mathbf{A},$$

and $\mathcal{L}$ is an invertible operator, defined as [16]

$$\mathcal{L}\{\mathbf{A}\} = \mathcal{F}_{xy}\left\{|\mathbf{n}|^2 \mathbf{A} e^{(z^- - f(x,y;t))u}\right\}, \ \forall \mathbf{A}. \tag{4}$$

Now, defining the inverse operator $\mathcal{L}^{-1}$ such that it supports an ocean surface with arbitrary heights [28] and proceeding with the derivations for a vertical dipole source in the far-field region defined as

$$\mathbf{E}_s = \frac{I(\omega)\Delta\ell k^2}{j\omega\epsilon_0}G_0\hat{\mathbf{z}} \equiv C_0 G_0\hat{\mathbf{z}}, \tag{5}$$

where G_0 is the Green's function solution for the Helmholtz equation in free space [16], $\Delta\ell$ is the dipole length, and $I(\omega)$ is the current flowing on the antenna, the first-order electric field for a rough surface with arbitrary heights can be written as [28]

$$\left(E_n^+\right)_1 \sim -jkC_0\frac{e^{\zeta(x_1,y_1;t)}}{(2\pi)^2}\int_{x_1}\int_{y_1}\hat{\boldsymbol{\rho}}_1 \cdot [\nabla_{x_1y_1}f(x_1,y_1;t)]F(\rho_1)F(\rho_2)\frac{e^{-jk(\rho_1+\rho_2)}}{\rho_1\rho_2}dx_1dy_1, \tag{6}$$

where $F(\rho)$ is the Sommerfeld attenuation function as presented in [29], and $\zeta(x_1,y_1;t)$ is the arbitrary height factor, defined as [28]

$$\zeta(x,y;t) \equiv \mathcal{F}_{xy}^{-1}\left\{\mathcal{F}_{xy}\left\{f(x,y;t)\right\}u\right\} = f(x,y;t) \underset{xy}{*} \mathcal{F}_{xy}^{-1}\left\{u\right\}, \tag{7}$$

where $\underset{xy}{*}$ indicates a two-dimensional spatial convolution in the xy-plane. The scattering geometry for the first-order electric field is depicted in Figure 1.

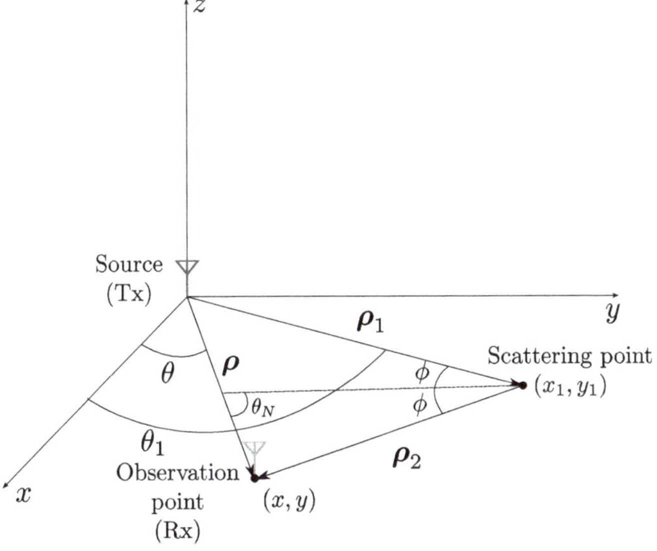

Figure 1. Scattering geometry for the bistatic first-order electric field.

If a Fourier series representation of the ocean surface is considered, the surface displacement $f(x, y; t)$ can be written as [30]

$$f(\boldsymbol{\rho}; t) = \sum_{\mathbf{K}, \omega_\mathbf{K}} f(\mathbf{K}, \omega_\mathbf{K}) e^{j(\mathbf{K} \cdot \boldsymbol{\rho} - \omega_\mathbf{K} t)} = \sum_{\mathbf{K}, \omega_\mathbf{K}} f(\mathbf{K}, \omega_\mathbf{K}) e^{-j\omega_\mathbf{K} t} e^{jK\rho \cos(\theta_\mathbf{K} - \theta)}, \tag{8}$$

with $\mathbf{K} = (K_x, K_y) = (K, \theta_\mathbf{K})$ being the wave vector for the ocean surface and $\omega_\mathbf{K}$ being the angular frequency obtained from the dispersion relation of ocean surface gravity waves [27]. By substituting (8) into (6), expanding the exponential that contains the arbitrary height factor into a power series and applying an asymptotic perturbation expansion to both the Fourier components of the ocean surface and the electric fields using the surface slope as the perturbation parameter [31], it can be shown that the received first-order electromagnetic and second-order hydrodynamic electric fields contain terms equivalent to those for the small-height case, but an extra term appears in the second-order derivation which relates the first-order cross-section and the arbitrary height factor. The new term can be written as [28]

$$(E_n^+)_{12c} \sim \frac{kC_0}{(2\pi)^2} \zeta_1(\boldsymbol{\rho}_1; t) \sum_{\mathbf{K}, \omega_K} f_1(\mathbf{K}, \omega_K) e^{-j\omega_K t} K \iint_{x_1, y_1} \cos(\theta_\mathbf{K} - \theta_1) \frac{F(\rho_1) F(\rho_2)}{\rho_1 \rho_2} e^{-jk\rho_2} e^{j\rho_1 [K \cos(\theta_\mathbf{K} - \theta_1) - k]} \cdot dx_1 dy_1. \tag{9}$$

Comparing the double integrals in (9) with those in the first-order electric field expression in [16,17], it is evident that they are identical. Therefore, following the same procedure detailed in [17] for the first-order bistatic electric field, the following expression is obtained as

$$(E_n^+)_{12c} \sim \frac{kC_0}{(2\pi)^{3/2}} \zeta_1(\boldsymbol{\rho}_1; t) \sum_{\mathbf{K}, \omega_K} f_1(\mathbf{K}, \omega_K) e^{-j\omega_K t} \sqrt{K} e^{j\frac{\rho}{2} \cdot \mathbf{K}} \int_{\rho_s} \frac{F(\rho_1) F(\rho_2)}{\sqrt{\rho_s \left[\rho_s^2 - \left(\frac{\rho}{2}\right)^2\right]}} e^{\mp j\pi/4} \left(\pm\sqrt{\cos\phi}\right) \cdot e^{j\rho_s [\pm K \cos\phi - 2k]} d\rho_s, \tag{10}$$

where ρ_s is defined as

$$\rho_s = \frac{\rho_1 + \rho_2}{2},$$

ϕ is the bistatic angle, defined as the bisection of the angle between $\boldsymbol{\rho}_1$ and $\boldsymbol{\rho}_2$, and $\boldsymbol{\rho}$ is the vector between the transmitter and receiver, shown in Figure 1.

If, similarly to [16,18], an inverse Fourier transform with respect to the radar frequency is applied to (10) while using the associative property of the convolution, the following expression can be obtained:

$$(E_n^+)_{12c}(t) \sim \frac{1}{(2\pi)^{3/2}} \left\{ \mathcal{F}_t^{-1} \{kC_0\} *_t \mathcal{F}_t^{-1} \left\{ \sum_{\mathbf{K}, \omega_K} f_1(\mathbf{K}, \omega_K) e^{-j\omega_K t} \sqrt{K} e^{j\frac{\rho}{2} \cdot \mathbf{K}} \int_{\rho_s} \frac{F(\rho_1) F(\rho_2)}{\sqrt{\rho_s \left[\rho_s^2 - \left(\frac{\rho}{2}\right)^2\right]}} e^{\mp j\pi/4} \right. \right.$$
$$\left. \left. \cdot \left(\pm\sqrt{\cos\phi}\right) e^{j\rho_s [\pm K \cos\phi - 2k]} d\rho_s \right\} \right\} *_t \mathcal{F}_t^{-1} \{\zeta_1(\boldsymbol{\rho}_1; t)\} . \tag{11}$$

Again, it can be easily observed that the time convolution ($*_t$) inside the braces is similar to the one presented in Equation (5) of [18], with the exception of the time-dependent exponential term for the Fourier series expansion of the ocean surface; although this term will not affect the inverse Fourier transform, it might have an effect on the final convolution. It is easy to show that the additional time-dependent exponential term does not affect the resulting expression, since the added terms in the final expression are significantly smaller than the rest of the terms in the expression. Therefore, substituting the resulting expression for the first-order time-varying electric field in [18] for a pulsed radar source, the expression in (11) becomes

$$(E_n^+)_{12c}(t) \sim \left\{ \frac{-j\eta_0 \Delta\ell I_0 k_0^2}{(2\pi)^{3/2}} \sum_{\mathbf{K},\omega_{\mathbf{K}}} f_1(\mathbf{K},\omega_{\mathbf{K}}) \sqrt{K\cos\phi_0} e^{j\omega_{\mathbf{K}}t} e^{jk_0\Delta\rho_s} e^{j\frac{\rho}{2}\cdot\mathbf{K}} e^{j\rho_{0s}(K\cos\phi_0)} \frac{F(\rho_{01},\omega_0)F(\rho_{02},\omega_0)}{\sqrt{\rho_{0s}\left[\rho_{0s}^2-\left(\frac{\rho}{2}\right)^2\right]}} \right.$$
$$\left. \cdot e^{j\pi/4}\Delta\rho_s\, \mathrm{Sa}\left[\frac{\Delta\rho_s}{2}\left(\frac{K}{\cos\phi_0}-2k_0\right)\right] \right\} *_t \mathcal{F}_t^{-1}\{\zeta_{01}(\boldsymbol{\rho}_1;t)\}. \tag{12}$$

where $\mathrm{Sa}(\cdot)$ is the sampling function, defined as

$$\mathrm{Sa}(x) = \frac{\sin x}{x}, \ \forall x,$$

and $\Delta\rho_s$ is the patch width on the ocean surface, defined for a pulse radar as

$$\Delta\rho_s = \frac{c\tau_0}{2},$$

with c being the speed of light and τ_0 being the radar pulse width. Here, it should be noted that the zero-subscripts in ϕ_0, ρ_{01}, and ρ_{02} indicate that the scattering patch is considered to be sufficiently small, allowing variable values at the center of the scattering patch to be taken as representative of their values on the whole patch [32]. Consequently, ρ_{0s} is defined as

$$\rho_{0s} = \frac{c\left(t-\frac{\tau_0}{2}\right)}{2} = \frac{\rho_{01}+\rho_{02}}{2}.$$

On the other hand, the zero-subscripts in k_0, ω_0, and $\zeta_{01}(\boldsymbol{\rho}_1;t)$ indicate that the radar transmitting frequency is considered constant during the radar operation. As explained in [16,18], the Sommerfeld attenuation function does not present significant variations with respect to radar frequency for typical bandwidths in a pulsed HF radar operation, allowing the frequency in $F(\rho)$ to be considered constant and equal to the center-transmitting angular frequency ω_0. Similarly, it is easy to verify through dimensional analysis that variations in the arbitrary height factor $\zeta_1(\boldsymbol{\rho}_1;t)$ with respect to radar frequency are very small, allowing it to be redefined as $\zeta_{01}(\boldsymbol{\rho}_1;t)$, where $k = k_0 = \frac{\omega_0}{c}$.

In order to obtain the expression for the time-varying bistatic electric field over the ocean surface, the arbitrary height factor must be further addressed. From the definition in (7), and taking the first-order expression for the ocean surface expansion appearing in (8), it can be shown that

$$\zeta_{01}(\boldsymbol{\rho}_1;t) = \sum_{\mathbf{K}',\omega_{\mathbf{K}'}} f_1(\mathbf{K}',\omega_{\mathbf{K}'})\sqrt{K'^2-k_0^2}\, e^{-j\omega_{\mathbf{K}'}t} e^{j\boldsymbol{\rho}_1\cdot\mathbf{K}'}. \tag{13}$$

Substituting (13) into (12) and performing the convolution, the time-varying bistatic electric field for arbitrary heights is obtained:

$$(E_n^+)_{12c}(t,t_0) \sim \frac{-j\eta_0 \Delta\ell I_0 k_0^2}{(2\pi)^{1/2}} \sum_{\mathbf{K}',\omega_{\mathbf{K}'}} \sum_{\mathbf{K},\omega_{\mathbf{K}}} f_1(\mathbf{K},\omega_{\mathbf{K}}) f_1(\mathbf{K}',\omega_{\mathbf{K}'})\sqrt{K'^2-k_0^2}\sqrt{K\cos\phi_0}\, e^{-j\omega_{\mathbf{K}'}t}\delta(\omega_{\mathbf{K}}-\omega_{\mathbf{K}'})$$
$$\cdot e^{j\boldsymbol{\rho}_1\cdot\mathbf{K}'} e^{j\omega_{\mathbf{K}}t} e^{jk_0\Delta\rho_s} e^{j\frac{\rho}{2}\cdot\mathbf{K}} e^{j\rho_{0s}(K\cos\phi_0)} \frac{F(\rho_{01},\omega_0)F(\rho_{02},\omega_0)}{\sqrt{\rho_{0s}\left[\rho_{0s}^2-\left(\frac{\rho}{2}\right)^2\right]}} e^{j\pi/4}\Delta\rho_s\, \mathrm{Sa}\left[\frac{\Delta\rho_s}{2}\left(\frac{K}{\cos\phi_0}-2k_0\right)\right], \tag{14}$$

where ρ_{0s}, ρ_{01}, and ρ_{02} are functions of t_0.

Here, a differentiation must be made between the two different time-related variables t and t_0. In [16], Walsh and Gill differentiated between the "time of observation" t_0 and the "experiment time" t for successive pulses, while here, t_0 will be treated as the "radar time" and t as the "ocean surface time". Since radar-dependent events occur on a different time scale from that of the events of the ocean surface, they can be treated as two independent variables, even though both variables refer to time.

Now that (14) has been derived, it is possible to obtain the radar cross-section from the Fourier transform of the autocorrelation of the electric field, as shown in [18].

3. First-Order Radar Cross-Section of the Ocean Surface with Arbitrary Heights

From [18], the autocorrelation for a general electric field, denoted as E_n^+, can be defined as

$$\mathcal{R}(\tau) = \frac{A_r}{2\eta_0}\mathbb{E}\left\{E_n^+(t)\overline{E_n^+(t-\tau)}\right\},\tag{15}$$

such that $\mathcal{R}(0)$ coincides with the average power at the receiver. In (15), A_r is the effective free-space aperture of the receiver, η_0 is the intrinsic impedance of the free space, $\mathbb{E}\{\cdot\}$ is the expected value operator, and the bar over the electric field indicates its conjugate. Expanding (15) into its different perturbation orders and knowing that, for the ocean surface [27]

$$\mathbb{E}\left\{f_m(\mathbf{K},\omega_K)\overline{f_n(\mathbf{K}',\omega_{K'})}\right\} = \begin{cases} S_m(\mathbf{K},\omega_K)d\mathbf{K}d\omega_K, & \text{if } m=n,\ \mathbf{K}=\mathbf{K}',\ \omega_K=\omega_{K'} \\ 0, & \text{otherwise,} \end{cases}$$

where $f_{m,n}(\cdot)$ are the m,n-th-order terms of the asymptotic expansion of the ocean surface and $S_{m,n}(\cdot)$ are their corresponding ocean wave spectra, it is easy to show that the only surviving terms of the autocorrelation are the ones multiplying fields of the same order:

$$\begin{aligned} R(\tau) &= \frac{A_r}{2\eta_0}\mathbb{E}\left\{(E_n^+)_{11}(t)\overline{(E_n^+)_{11}(t-\tau)}\right\} + \frac{A_r}{2\eta_0}\mathbb{E}\left\{(E_n^+)_{12c}(t)\overline{(E_n^+)_{12c}(t-\tau)}\right\} \\ &\quad + \frac{A_r}{2\eta_0}\mathbb{E}\left\{(E_n^+)_{21}(t)\overline{(E_n^+)_{21}(t-\tau)}\right\} + \cdots \\ &\equiv R_{11}(\tau) + R_{12c}(\tau) + R_{21}(\tau) + \cdots, \end{aligned}\tag{16}$$

where

$$R_{11}(\tau) = \frac{A_r}{2\eta_0}\mathbb{E}\left\{(E_n^+)_{11}(t)\overline{(E_n^+)_{11}(t-\tau)}\right\},$$

$$R_{12c}(\tau) = \frac{A_r}{2\eta_0}\mathbb{E}\left\{(E_n^+)_{12c}(t)\overline{(E_n^+)_{12c}(t-\tau)}\right\},$$

and

$$R_{21}(\tau) = \frac{A_r}{2\eta_0}\mathbb{E}\left\{(E_n^+)_{21}(t)\overline{(E_n^+)_{21}(t-\tau)}\right\}$$

are respectively the autocorrelations of the first-order electric field, second-order correction to the first-order electric field, and second-order hydrodynamic electric field, with the first-order and second-order hydrodynamic electric fields defined as in [17]. Therefore, the correction term for the bistatic electric field for the ocean surface with arbitrary heights presented here does not affect the form of the radar cross-section expressions previously derived in [18]. Thus, taking the autocorrelation of (14) according to the expression presented in (16) and proceeding with the derivations, the following expression is obtained:

$$\begin{aligned} R_{12c}(\tau) = \frac{A_r\pi^2\eta_0|\Delta\ell I_0|^2 k_0^4|F(\rho_{01},\omega_0)F(\rho_{02},\omega_0)|^2}{2(2\pi)^3\rho_{0s}\left[\rho_{0s}^2-\left(\frac{\rho}{2}\right)^2\right]}\sum_{m=\pm1}\iint S(m\mathbf{K})S(mK)|K^2-k_0^2|e^{-j\omega_K\tau}\frac{K^{\frac{7}{2}}}{\sqrt{8}} \\ \cdot\Delta\rho_s^2\,\mathrm{Sa}\left[\frac{\Delta\rho_s}{2}\left(\frac{K}{\cos\phi_0}-2k_0\right)\right]dKd\theta_\mathbf{K}. \end{aligned}\tag{17}$$

From [18], it is known that

$$\frac{dP(\omega_d)}{dA} = \frac{A_r\eta_0|\Delta\ell I_0|^2 k_0^2|F(\rho_{01},\omega_0)F(\rho_{02},\omega_0)|^2}{16(2\pi)^3(\rho_{01}\rho_{02})^2}\sigma(\omega_d),\tag{18}$$

where $\mathcal{P}(\omega_d)$ is the power spectral density of the electric field, defined as the Fourier transform of the autocorrelation with respect to τ, and $\sigma(\omega_d)$ is the radar cross-section of the scattering object. After obtaining the power spectral density from the autocorrelation in (17), knowing from the bistatic scattering geometry that $\theta_\mathbf{K} = \theta_N$, where θ_N is the direction normal to the scattering ellipse at the scattering patch, and that [18]

$$\frac{\Delta\rho_s d\theta_N}{\rho_{0s}\left[\rho_{0s}^2 - \left(\frac{\rho}{2}\right)^2\right]} = \frac{dA}{(\rho_{01}\rho_{02})^2},$$

the second-order correction to the first-order bistatic radar cross-section for an ocean surface with arbitrary heights can be obtained by comparison with (18) as

$$\sigma_{12c}(\omega_d) = 2^5\pi^3 k_0^2\Delta\rho \sum_{m=\pm 1} S(m\mathbf{K})S(m\mathbf{K})\left|K^2 - k_0^2\right|\frac{K^4}{g}\cos\phi_0\,\mathrm{Sa}\left[\frac{\Delta\rho_s}{2}\left(\frac{K}{\cos\phi_0} - 2k_0\right)\right]. \quad (19)$$

4. Simulation Results

In order to assess the impact of the newly-derived term in the total radar cross-section of the ocean surface, simulations were conducted in both high and low sea states. In both cases, two different bistatic configurations were chosen, with two different dominant wave directions, such that both the effects of wave directions and bistatic configurations on the results could be analyzed.

Before proceeding with the radar cross-section simulations, the validity of the second-order correction to the first-order term must be investigated, since the small-slope approximation still applies to (19). For this purpose, a number of total mean-square slope models proposed in the literature were used to compute the root-mean-square slope for the different ocean conditions used in the simulations. These models were developed empirically, using ocean surface measurements obtained with different instruments such as aerial photographs [33] and GPS-R [34]. The resulting total root-mean-square slopes for each of the simulated meteorological conditions are presented in Table 1, where $U_{19.5}$ is the wind speed measured at 19.5 m above the ocean surface.

Table 1. Total root-mean-square slopes for simulated meteorological conditions using different slope models.

MSS Slope Model	Total Root-Mean-Square Slope	
	$U_{19.5} = 10$ m/s	$U_{19.5} = 20.7$ m/s
Cox and Munk (1954) [33]	0.22206	0.31477
Wu (1990) [35]	0.22319	0.30573
Hwang (2005) [36]	0.22252	0.32197
Katzberg, Torres and Ganoe (2006) [37]	0.15098	0.18104
Gleason et al. (2018) [34]	0.16119	0.18041

For the radar cross-section simulations, the Pierson–Moskowitz (PM) spectrum was chosen as the nondirectional spectral model of the ocean surface [38], with a cosine-power model for the directional factor [39] using the frequency-dependent wave-spreading factor proposed in [40]. Figures 2 and 3 present the radar cross-section simulation results for low and high sea states, respectively. In the presented results, σ_{11} and σ_{2P} respectively indicate the first- and second-order radar cross-sections of the ocean surface.

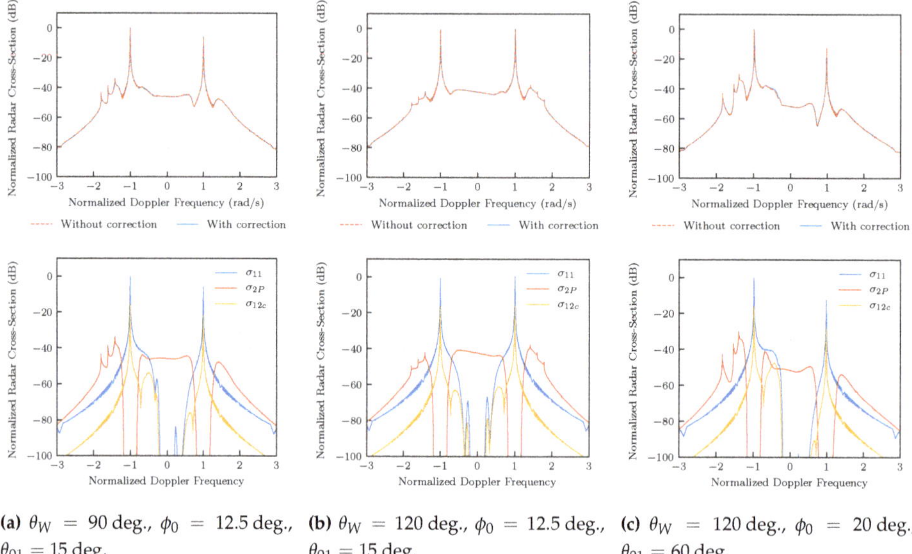

(a) $\theta_W = 90$ deg., $\phi_0 = 12.5$ deg., $\theta_{01} = 15$ deg.

(b) $\theta_W = 120$ deg., $\phi_0 = 12.5$ deg., $\theta_{01} = 15$ deg.

(c) $\theta_W = 120$ deg., $\phi_0 = 20$ deg., $\theta_{01} = 60$ deg.

Figure 2. Total bistatic radar cross-section and its components for three different bistatic geometries and wave directions at low sea states. Wind speed $U_{19.5} = 10$ m/s, radar frequency $f_0 = 13.385$ MHz, and roughness scale $k_0 H_s = 0.60$.

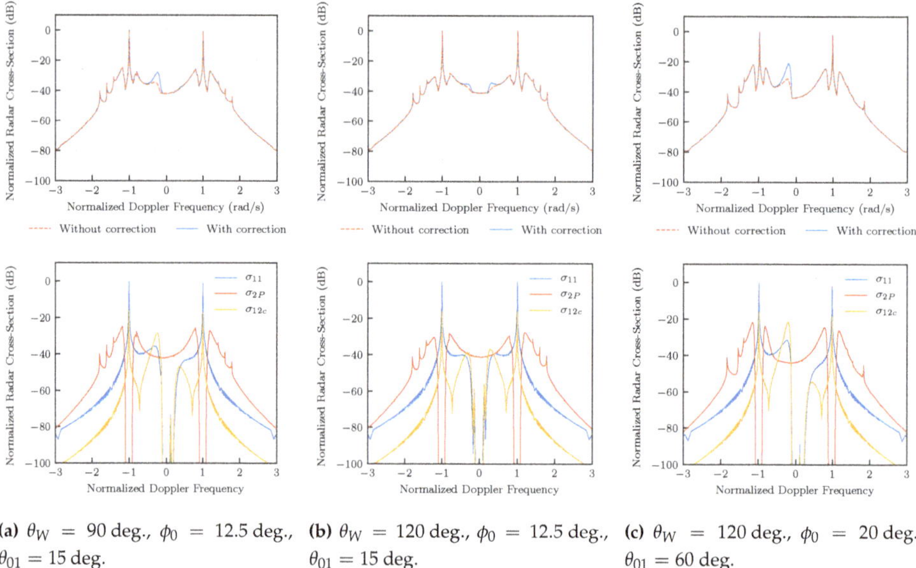

(a) $\theta_W = 90$ deg., $\phi_0 = 12.5$ deg., $\theta_{01} = 15$ deg.

(b) $\theta_W = 120$ deg., $\phi_0 = 12.5$ deg., $\theta_{01} = 15$ deg.

(c) $\theta_W = 120$ deg., $\phi_0 = 20$ deg., $\theta_{01} = 60$ deg.

Figure 3. Total bistatic radar cross-section and its components for three different bistatic geometries and wave directions at high sea states. Wind speed $U_{19.5} = 20.7$ m/s, radar frequency $f_0 = 13.385$ MHz, and roughness scale $k_0 H_s = 2.56$.

5. Discussion

In [26], the transition zone for the validity of the asymptotic perturbation method for roughness scales was defined for perturbation parameters between 0.4 and 0.7. Using the same intervals for the perturbation parameter chosen in the present work, it can be observed that all of the empirical models shown in Table 1 yielded total root-mean-square slopes below the lower bound of the transition zone, meaning that the validity condition for the perturbation theory has not been violated under either of the meteorological conditions used in the simulations.

In observing the results in Figures 2 and 3, it is clear that the proposed cross-section has little to no impact on the total cross-section at low sea states, with a maximum difference of less than 0.1 dB, as shown in Figure 2b. This is an expected result, since the traditional second-order scattering theory is still valid for $k_0 H_s < 0.7$, even though the roughness scale is within the transition zone [26]. However, when observing the total radar cross-section at high sea states in Figure 3, where the roughness scale is above the upper limit of the transition zone, the correction term has an evident impact on the total radar cross-section, with a maximum difference of 8 dB in Figure 3c.

The enhanced part in the total cross-section depends on the dominant wave direction of the ocean surface θ_W, as observed when comparing Figure 3a with Figure 3b, and on the bistatic radar configuration, as evidenced in analyzing Figure 3b,c. The maximum difference between the total cross-sections with and without the additional term on the presented results is 8 dB.

In addition, the effects of the additional term are mostly evident in the central part of the total cross-section, at Doppler frequencies close to 0 rad/s; this is due to natural limitations on the steepness of large ocean waves, as well as to restrictions on the wave slope that are still imposed in the present analysis.

6. Conclusions and Future Work

In the present work, the scattered electric field and radar cross-section for an ocean surface with arbitrary heights using a narrow-beam bistatic HF radar have been derived. Previously derived electric field expressions presented in [17] for the small-height condition still appear in the final result, but due to the removal of the small-height approximation, a new term appears. The new term is interpreted as a correction to the first-order cross-section for arbitrary heights. The radar cross-section due to the correction term is also derived following the procedure presented in [18], and is then simulated for low and high sea states, showing an impact on the total radar cross-section for high sea states. A similar analysis is currently under review for the monostatic case.

The present work shows that changes in the dominant wave direction of the ocean surface and the bistatic configuration impact the contribution of the correction term to the radar cross-section. Since the other terms of the radar cross-section are saturated at sufficiently high sea states, the correction term may provide input for determining the dominant wave direction at high sea states. More extensive work needs to be dedicated to the understanding of interactions between the bistatic configuration and dominant wave direction, as well as the correction term, determining thresholds above which the correction term should be considered, and how different bistatic configurations can be used in observing the ocean surface at large roughness scales.

It can also be noted that the simulations in the current work were carried out using a Pierson–Moskowitz spectral model, which implies the assumption of a fully-developed sea and does not depend on fetch [38]. Since changes in fetch are known to affect the mean squared slope of the ocean surface [27], the work of determining how changes in fetch affect the correction term, allowing its application to developing sea states, is ongoing.

Since the present work represents the first attempt to overcome the small-height constraint imposed on previously derived bistatic high-frequency radar cross-sections of the ocean surface, no practical applications of the proposed theory have been suggested in the present work, as more work must be done in the future for these applications to be devised. In addition, due to the lack of narrow-beam bistatic HF radar data available to the authors, which were measured under conditions that violate the limiting roughness scales of the theories proposed in the literature, or the lack of any other derived model for radar cross-section of the ocean surface at large roughness scales, a validation of the presented theory is not available in the current work. To validate the current theory, collaboration from the radio oceanography community is necessary, as multiple observations need to be conducted by users of narrow-beam bistatic HF radars at ocean conditions that violate the small-height assumption. In addition, efforts to establish bistatic operation of existing monostatic HF radars are ongoing on the coastline of Placentia Bay, Newfoundland, Canada in order to collect data for the validation of the presented theory.

Author Contributions: Conceptualization: M.T.S., W.H., and E.W.G.; methodology: M.T.S.; formal analysis: M.T.S.; investigation: M.T.S.; resources: W.H. and E.W.G.; data curation: M.T.S.; writing—original draft preparation: M.T.S.; writing—review and technical editing: M.T.S., W.H., and E.W.G.; visualization: M.T.S.; project administration: W.H.; supervision: W.H. and E.W.G.; funding acquisition: W.H. and E.W.G. All authors have read and agreed to the published version of the manuscript.

Funding: This work was supported by the Natural Sciences and Engineering Research Council of Canada (NSERC) under Discovery Grants to W. Huang (NSERC RGPIN-2017-04508 and RGPAS-2017-507962) and E. W. Gill (NSERC RGPIN-2015-05289).

Conflicts of Interest: The authors declare no conflict of interest.

References

1. Crombie, D.D. Doppler spectrum of sea echo at 13.56 Mc./s. *Nature* **1955**, *175*, 681–682. [CrossRef]
2. Barrick, D.E. HF radio oceanography—A review. *Bound-Lay. Meteorol.* **1978**, *13*, 23–43. [CrossRef]
3. Hardman, R.L.; Wyatt, L.R. Inversion of HF radar Doppler spectra using a neural network. *J. Mar. Sci. Eng.* **2019**, *7*, 255. [CrossRef]
4. Sun, W.; Huang, W.; Ji, Y.; Dai, Y.; Ren, P.; Zhou, P.; Hao, X. A vessel azimuth and course joint re-estimation method for compact HFSWR. *IEEE Trans. Geosci. Remote Sens.* **2019**, 1–11. [CrossRef]
5. Sun, W.; Huang, W.; Ji, Y.; Dai, Y.; Ren, P.; Zhou, P. Vessel tracking with small-aperture compact high-frequency surface wave radar. In Proceedings of the OCEANS 2019—Marseille, Marseille, France, 17–20 June 2019; pp. 1–4. [CrossRef]
6. Pidgeon, V.W. Bistatic cross section of the sea. *IEEE Trans. Antennas Propag.* **1966**, *14*, 405–406. [CrossRef]
7. Peterson, A.M.; Teague, C.C.; Tyler, G.L. Bistatic-radar observation of long-period, directional ocean-wave spectra with Loran A. *Science* **1970**, *170*, 158–61. [CrossRef]
8. Teague, C.C. Bistatic-radar techniques for observing long-wavelength directional ocean-wave spectra. *IEEE Trans. Geosci. Electron.* **1971**, *9*, 211–215. [CrossRef]
9. Teague, C.C.; Vesecky, J.F.; Fernandez, D.M. HF radar instruments, past to present. *Oceanography* **1997**, *10*, 40–44.[CrossRef]
10. Barrick, D.E. The interaction of HF/VHF radio waves with the sea surface and its implications. In *AGARD Conference Proceedings No. 77 on Electromagnetics of the Sea*; Aerospace Research & Development (AGARD), Ed.; National Atlantic Threaty Organization (NATO): Washington, DC, USA, 1970; number AGARD-CP-77-70, pp. 18-1–18-25.
11. Barrick, D.E. Remote sensing of sea state by radar. In *Remote Sensing of the Troposphere*; Derr, V., Ed.; U.S. Department of Commerce—National Oceanic and Atmospheric Administration: Boulder, CO, USA, 1972; Chapter 12, pp. 12-1–12-46.
12. Johnstone, D.L. Second-Order Electromagnetic and Hydromagnetic Effects in High-Frequency Radio-Wave Scattering from the Sea. Ph.D. Thesis, Stamford, CA, USA, 1975.
13. Anderson, S.J.; Anderson, W.C. Bistatic HF scattering from the ocean surface and its application to remote sensing of seastate. In Proceedings of the 1987 IEEE APS International Symposium, Blacksburg, VA, USA, 15–19 June 1987.
14. Anderson, S.J.; Fuks, I.M.; Praschifka, J. Multiple scattering of HF radio waves propagating across the sea surface. *Wave. Random Media* **1998**, *8*, 283–302. [CrossRef]

15. Anderson, S.J. Directional wave spectrum measurement with multistatic HF surface wave radar. In *Taking the Pulse of the Planet: The Role of Remote Sensing in Managing the Environment. Proceedings (Cat. No.00CH37120)*; IEEE: Honolulu, HI, USA, 2000. [CrossRef]

16. Walsh, J.; Gill, E.W. An analysis of the scattering of high-frequency electromagnetic radiation from rough surfaces with application to pulse radar operating in backscatter mode. *Radio Sci.* **2000**, *35*, 1337–1359. [CrossRef]

17. Gill, E.W.; Walsh, J. Bistatic form of the electric field equations for the scattering of vertically polarized high-frequency ground wave radiation from slightly rough, good conducting surfaces. *Radio Sci.* **2000**, *35*, 1323–1335. [CrossRef]

18. Gill, E.W.; Walsh, J. High-frequency bistatic cross sections of the ocean surface. *Radio Sci.* **2001**, *36*, 1459–1475. [CrossRef]

19. Gill, E.W.; Huang, W.; Walsh, J. On the development of a second-order bistatic radar cross section of the ocean surface: A high-frequency result for a finite scattering patch. *IEEE J. Ocean. Eng.* **2006**, *31*, 740–750. [CrossRef]

20. Chen, S.; Huang, W.; Gill, E.W. First-order bistatic high-frequency radar power for mixed-path ionosphere-ocean propagation. *IEEE Geosci. Remote Sens. Lett.* **2016**, *13*, 1940–1944.[CrossRef]

21. Ma, Y.; Gill, E.W.; Huang, W. First-order bistatic high-frequency radar ocean surface cross-section for an antenna on a floating platform. *IET Radar Sonar Navig.* **2016**, *10*, 1136–1144. [CrossRef]

22. Ma, Y.; Huang, W.; Gill, E.W. The second-order bistatic high-frequency radar cross section of ocean surface for an antenna on a floating platform. *Can. J. Remote Sens.* **2016**, *42*, 332–343. [CrossRef]

23. Ma, Y.; Gill, E.W.; Huang, W. Bistatic high-frequency radar ocean surface cross section incorporating a dual-frequency platform motion model. *IEEE J. Ocean. Eng.* **2018**, *43*, 205–210. [CrossRef]

24. Grosdidier, S.; Forget, P.; Barbin, Y.; Gúerin, C.A. HF bistatic ocean Doppler spectra: simulation versus experimentation. *IEEE Trans. Geosci. Remote Sens.* **2014**, *52*, 2138–2148. [CrossRef]

25. Barrick, D.E. Theory of HF and VHF propagation across the rough sea, 1, the effective surface impedance for a slightly rough highly conducting medium at grazing incidence. *Radio Sci.* **1971**, *6*, 517–526. [CrossRef]

26. Wyatt, L.R. High order nonlinearities in HF radar backscatter from the ocean surface. *IEE Proc. Radar Sonar Navig.* **1995**, *142*, 293. [CrossRef]

27. Massel, S.R. *Ocean Surface Waves: Their Physics and Prediction*, 3rd ed.; World Scientific Publishing: Singapore, 2017; p. 776.

28. Silva, M.T.; Gill, E.W.; Huang, W. First-order high-frequency scattering for ocean surfaces with large roughness scales. Presented at the 28th Annual Newfoundland Electrical and Computer Engineering Conference (NECEC 2019), St. John's, NL, Canada, 19 November 2019.

29. Walsh, J. Asymptotic expansion of a Sommerfeld integral. *Electron. Lett.* **1984**, *20*, 746. [CrossRef]

30. Weber, B.L.; Barrick, D.E. On the nonlinear theory for gravity waves on the ocean's surface. Part I: derivations. *J. Phys. Oceanogr.* **1977**, *7*, 3–10. [CrossRef]

31. Hasselmann, K. On the non-linear energy transfer in a gravity-wave spectrum Part 1. General theory. *J. Fluid Mech.* **1962**, *12*, 481. [CrossRef]

32. Gill, E.W. The Scattering of High Frequency Electromagnetic Radiation from the Ocean Surface: An Analysis Based on a Bistatic Ground Wave Radar Configuration. Ph.D. Thesis, Memorial University of Newfoundland, St. John's, NL, Canada, 1999.

33. Cox, C.; Munk, W. Measurement of the roughness of the sea surface from photographs of the sun's glitter. *J. Opt. Soc. Am.* **1954**, *44*, 838. [CrossRef]

34. Gleason, S.; Zavorotny, V.U.; Akos, D.M.; Hrbek, S.; PopStefanija, I.; Walsh, E.J.; Masters, D.; Grant, M.S. Study of surface wind and mean square slope correlation in hurricane Ike with multiple sensors. *IEEE J. Sel. Topics Appl. Earth Observ. Remote Sens.* **2018**, *11*, 1975–1988. [CrossRef]

35. Wu, J. Mean square slopes of the wind-disturbed water surface, their magnitude, directionality, and composition. *Radio Sci.* **1990**, *25*, 37–48. [CrossRef]

36. Hwang, P.A. Wave number spectrum and mean square slope of intermediate-scale ocean surface waves. *J. Geophys. Res.* **2005**, *110*. [CrossRef]

37. Katzberg, S.J.; Torres, O.; Ganoe, G. Calibration of reflected GPS for tropical storm wind speed retrievals. *Geophys. Res. Lett.* **2006**, *33*. [CrossRef]

38. Pierson, W.J.; Moskowitz, L. *A Proposed Spectral form for Fully Developed Wind Seas Based on the Similarity Theory of S. A. Kitaigorodskii*; Technical Report, U.S. Naval Oceanographic Office: New York, NY, USA, 1963.

39. Longuet-Higgins, M.S. The effect of non-linearities on statistical distributions in the theory of sea waves. *J. Fluid Mech.* **1963**, *17*, 459. [CrossRef]

40. Mitsuyasu, H.; Tasai, F.; Suhara, T.; Mizuno, S.; Ohkusu, M.; Honda, T.; Rikiishi, K. Observations of the directional spectrum of ocean waves using a cloverleaf buoy. *J. Phys. Oceanogr.* **1975**, *5*, 750–760. [CrossRef]

 *remote sensing*

Article

Measuring the Directional Ocean Spectrum from Simulated Bistatic HF Radar Data

Rachael L. Hardman [1,†], **Lucy R. Wyatt** [1,2,*,†] **and Charles C. Engleback** [1]

[1]   School of Mathematics and Statistics, University of Sheffield, Sheffield S10 2TN, UK;
      rlhardman1@sheffield.ac.uk (R.L.H.); charlie@engleback.com (C.C.E.)
[2]   Seaview Sensing Ltd., Sheffield S10 3GR, UK
*   Correspondence: l.wyatt@sheffield.ac.uk
†   These authors contributed equally to this work.

Received: 19 December 2019; Accepted: 14 January 2020; Published: 18 January 2020

Abstract: HF radars are becoming important components of coastal operational monitoring systems particularly for currents and mostly using monostatic radar systems where the transmit and receive antennas are colocated. A bistatic configuration, where the transmit antenna is separated from the receive antennas, offers some advantages and has been used for current measurement. Currents are measured using the Doppler shift from ocean waves which are Bragg-matched to the radio signal. Obtaining a wave measurement is more complicated. In this paper, the theoretical basis for bistatic wave measurement with a phased-array HF radar is reviewed and clarified. Simulations of monostatic and bistatic radar data have been made using wave models and wave spectral data. The Seaview monostatic inversion method for waves, currents and winds has been modified to allow for a bistatic configuration and has been applied to the simulated data for two receive sites. Comparisons of current and wave parameters and of wave spectra are presented. The results are encouraging, although the monostatic results are more accurate. Large bistatic angles seem to reduce the accuracy of the derived oceanographic measurements, although directional spectra match well over most of the frequency range.

Keywords: HF radar; remote sensing; inversion; radar cross section; bistatic radar; directional wave spectrum

---

1. Introduction

Coastal high frequency or HF radar (3–30 MHz) is a tool that has enabled users to remotely measure ocean currents, winds and waves, in real-time, since the initial observation of Crombie [1] in 1955, who realised the relationship between ocean wave and HF radar Doppler spectra (such as that shown in Figure 1). The measurements are important in a number of coastal engineering topics, including testing of and assimilation in operational wave models, sea vessel navigation, land/beach erosion, designing offshore structures, and in supporting marine activities. Additionally, collecting ocean data over a long period of time can be useful in climate changes studies; the same data can also be used to assess the potential of a coastal region to become a wave/wind farm as shown by Wyatt [2]. With such important applications, the accuracy of the measurements is imperative.

The majority of the existing theory is for monostatic radar, where the transmitter and receiver are co-located. Ocean surface current measurements from such a radar are robust, and hundreds of radars provide real-time current measurements all over the world; see the work of Paduan and Washburn [3] for an introduction to the subject. Methods for measuring wind direction and variability are also reliable, such as the method of Wyatt et al. [4], where the maximum likelihood method is used fit the wind distribution to the available radar data. Ocean wave measurements are also possible, however, they can be less robust as they are more vulnerable to noise (for limitations see

Wyatt et al. [5]). Notable methods of Lipa [6,7], Howell and Walsh [8], and Hisaki [9] are all significant in the history of measuring the wave spectrum. Another key method is the Seaview method, presented by Wyatt [10], and Green and Wyatt [11], which will be explained in Section 3.

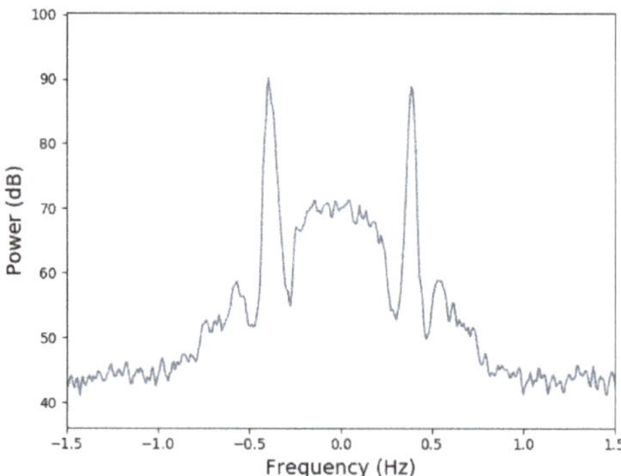

Figure 1. Example of a radar Doppler spectrum measured by a bistatic HF radar on the south coast of France on 09/07/2014 00:01. Radar data provided by Celine Quentin, University of Toulon.

To obtain the ocean measurements using radar data, many researchers use the *radar cross section* of the ocean surface, which models what the radar output will be for a given ocean state and radar frequency. For monostatic radars, the most commonly used radar cross section is that of Barrick [12,13], derived in 1972, based on the perturbation method of Rice [14]. The expression is split into its first and second order components; the first order radar cross section, $\sigma^{(1)}(\omega)$, is due to resonance between the emitted radio waves and ocean waves of a particular length and direction, and the second order radar cross section, $\sigma^{(2)}(\omega)$, is due to double scattering of the emitted radio waves from two ocean waves and the non linear combination of the same two ocean waves. In full, for radar wavenumber k_0 and ocean spectrum $S(\vec{k})$,

$$\sigma^{(1)}(\omega) = 2^6 \pi k_0^4 \sum_{m=\pm 1} S(m\vec{k}_B)\delta(\omega - m\omega_B),$$ (1)

where $k_B = 2k_0$, is known as the *Bragg wavenumber* and

$$\omega_B = \sqrt{2gk_0 \tanh(2k_0 d)}$$ (2)

is known as the *Bragg frequency*, for ocean depth d and gravity g. The second order term is given by

$$\sigma^{(2)}(\omega) = 2^6 \pi k_0^4 \sum_{m,m'=\pm 1} \iint_{-\infty}^{\infty} |\Gamma_T|^2 S(m\vec{k}_1)S(m'\vec{k}_2)\delta(\omega - m\omega_1 - m'\omega_2)\,dp\,dq,$$ (3)

where $\vec{k}_1$ and $\vec{k}_2$, with respective angular frequencies ω_1 and ω_2, are the two contributing wave vectors, defined by the relationship $\vec{k}_1 + \vec{k}_2 = \vec{k}_B$. The $|\Gamma_T|^2$ term is known as the *coupling coefficient* and contains the mathematics of the nonlinear combinations of the waves and, as such, is a function of $\vec{k}_1$ and $\vec{k}_2$. More detail on the monostatic coupling coefficient is given by Lipa and Barrick [15].

Numerical methods like that of Holden and Wyatt [16] can be used to simulate monostatic Doppler spectra for given ocean conditions and radar settings, using Equations (24) and (28); a comparison of a radar measured Doppler spectrum and a simulated Doppler spectrum, using co-spatial and co-temporal wave buoy data as input to the simulation, is shown in Figure 2 where good agreement is shown.

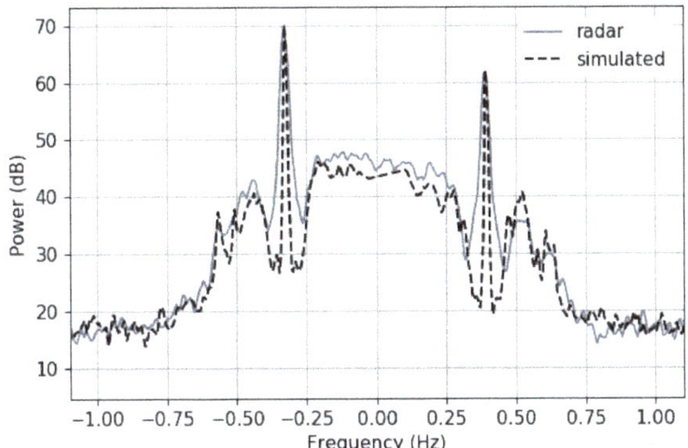

Figure 2. Comparison of measured and simulated monostatic Doppler spectra. The simulated Doppler spectrum has been generated using wave buoy data, measured at the same time and place as the radar Doppler spectrum. Radar and buoy data provided by Daniel Conley, Univeresity of Plymouth

Recently, bistatic radar—where the transmitter and receiver are separated by a notable distance—is on the ascendancy. Therefore, conversely to monostatic radar which receives backscatter, the detected radio waves in a bistatic radar have been scattered at a non-zero angle. A traditional coastal HF radar site consists of two monostatic radars, each providing data from ocean backscatter. However, a third dataset can be obtained at no additional cost if one of the receivers also receives bistatic scatter from the other transmitter. In this case, the radar site is called multistatic. Each different radar setup is shown in Figure 3. The advantages of employing a bistatic/multistatic radar setup are that (1) it can reduce the cost of setting up/maintaining a HF radar and (2) it can increase spatial coverage and data quality as shown by Whelan & Hubbard [17].

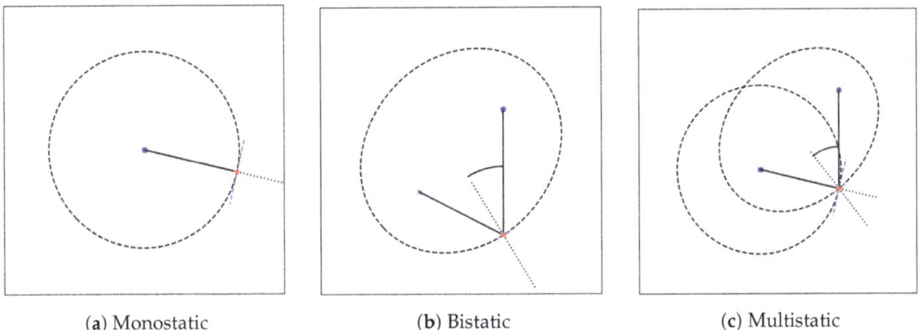

(a) Monostatic   (b) Bistatic   (c) Multistatic

Figure 3. Comparison of (**a**) monostatic (receiver and transmitter colocated), (**b**) bistatic (receiver and transmitter separated) and (**c**) multistatic (**b** with extra receiver) radar geometries. In each case, the transmitter is shown by the blue cross and the receiver is shown by the blue circle (in the monostatic case, is also at the same location as the transmitter). An example scatter point is shown by the red star and the path the signal takes is shown by the solid black line. The line of constant range for each particular range is shown by the dashed black line and the angle marker shown represents the bistatic angle.

In this paper, the aim is to obtain directional wave measurements from bistatic radar data. Previously, for bistatic radar, Zhang and Gill [18] developed an inversion algorithm to obtain the nondirectional wave spectrum from bistatic radar data and, when tested on simulated data, they obtained good results. Silva [19] has also presented results of simulated bistatic HF radar data inversion where the directional wave spectrum was estimated using Tikhonov regularization. They achieved good results for simulated data, however, the method is limited as a model is assumed for the direction of the spectrum, and this assumption may not always be appropriate.

In the existing numerical methods for extracting wave measurements from monostatic radar data, the aim is to invert Equation (3), to obtain the ocean spectrum $S(\vec{k})$ in terms of the measured $\sigma(\omega)$. Anderson [20] stated that the existing algorithms should work for a bistatic system if the inverted radar cross section is changed to the bistatic expression. In this work, we test this hypothesis and modify the Seaview method to measure the directional wave spectrum from bistatic HF radar data. Therefore, to do this, the bistatic radar cross section must be known.

In 1975, Johnstone [21] presented a bistatic radar cross section of the ocean surface and then, in 2001, Gill and Walsh [22] presented an alternative expression. However, under monostatic conditions, neither of the expressions reduce exactly to the monostatic term of Barrick [12,13] (which the Seaview method depends on). The derivation of Johnstone appears to have an error which causes the difference in the resulting expressions; Gill and Walsh followed a more complicated method, however, it has been shown that the monostatic form of their radar cross section is similar to Barrick's and it is, therefore, unnecessary to change the existing operational inversion programs to use theirs instead. Another recent derivation is given in Chen et al. [23].

A bistatic radar cross section that reduces exactly to the monostatic term of Barrick [12,13] would be beneficial to systems based on Barrick's expression (such as the Seaview inversion) as, in the radar coverage area, the bistatic angle can vary between 0° and 90°, so the discontinuity between the monostatic and bistatic radar cross section expressions would cause a discontinuity in the inversion program used and perhaps, then, the results. Therefore, in this work, we follow the method of Barrick, whilst retaining the bistatic angle, to derive the bistatic radar cross section of the ocean surface. A reviewer of this paper has drawn our attention to similar work by Hisaki and Tokuda [24] who allowed for a finite scattering area and showed that their equations reduced to those of Barrick for an infinite scattering area and a monostatic geometry.

We begin with an overview of the derivation of the bistatic radar cross section in Section 2.1, before presenting the numerical solution of the resulting expression in Section 2.2. Details of the Seaview inversion method (for which details of the cross-section equations and numerical simulations are a pre-requisite) are then given in Section 3. The results of the modified Seaview inversion, when tested on simulated bistatic data, are given in Section 4 and these are discussed in Section 5 which also includes some concluding remarks.

2. Materials and Methods

2.1. Bistatic Radar Cross Section of the Ocean Surface

To derive the bistatic radar cross section of the ocean surface, we follow the method of Barrick [12], where the equivalent monostatic radar cross section was derived. In his work, Barrick used the perturbation analysis of Rice [14] where, by assuming small waveheights and slopes, the electric field scattered from the ocean surface, $\vec{E}_s$, was calculated. They key points of the derivation follow, however, more details can be found in the work of Hardman [25].

The value of $\vec{E}_s$ depends on both the incident radio waves and the properties of the scattering surface. Firstly, the incident waves will propagate as vertically polarised ground waves. Secondly, as the ocean varies in both time and space, by assuming that these variations are periodic and that the surface is of infinite extent, we can define the surface, $f(x, y, t)$, as a Fourier series expansion, such that

$$z = f(x, y, t) = \sum_{mnl=-\infty}^{\infty} P(m, n, l)e^{ia(mx+ny)-iwlt}, \tag{4}$$

for wavenumber $a = 2\pi/L$, angular frequency $w = 2\pi/T$ and Fourier coefficients $P(m, n, l)$ which are dependent on the integers m, n and l.

The electric field scattered from this surface, will also be periodic with the same fundamental spatial and temporal periods and hence Rice defined $\vec{E}_s$ as a Fourier series. For a perfectly conducting flat surface, the exact solution can be found. Therefore, we perturb the solution around the flat surface, with ordering parameter $k_0 f$ (for radar wavenumber k_0), using Maxwell's equations and the tangential boundary condition to obtain the first and second order Fourier coefficients. Details of the calculations for vertically polarised waves can be found in the work of Hardman [25]; for horizontally polarised waves, the details are provided by Rice [14]. The resulting scattered electric field has components

$$
\begin{aligned}
E_x = 2 \sum_{mnl} E(m,n,z,l) e^{-i\omega_0 t} & \left[i(k_0 - am)P(m-v,n,l) \right. \\
& \left. + \sum_{qrs} \left\{ a^2(m-q)(v-q)k_0 + (k_0 - am)b^2(q,r) \right\} Q(m,n,l,q,r,s) \right],
\end{aligned}
\tag{5}
$$

$$
\begin{aligned}
E_y = 2a \sum_{mnl} E(m,n,z,l) e^{-i\omega_0 t} & \left[-inP(m-v,n,l) \right. \\
& \left. + \sum_{qrs} \left\{ a(n-r)(v-q)k_0 - nb^2(q,r) \right\} Q(m,n,l,q,r,s) \right]
\end{aligned}
\tag{6}
$$

$$
\begin{aligned}
E_z = 2e^{ik_0 x} e^{-i\omega_0 t} + 2 \sum_{mnl} \frac{E(m,n,z,l)}{b(m,n)} e^{-i\omega_0 t} & \left[\left(-i(a(m-v)k_0 + b^2(m,n)) \right) P(m-v,n,l) \right. \\
& + \sum_{qrs} \left\{ (a^3(q-v)(m^2+n^2-qm-rn)k_0) \right. \\
& \left. \left. + a(a(m^2+n^2)-mk_0)b^2(q,r) \right\} Q(m,n,l,q,r,s) \right],
\end{aligned}
\tag{7}
$$

where ω_0 is the angular frequency of the emitted radio waves,

$$
E(m,n,z,l) = e^{i(a(mx+ny)+b(m,n)z)} e^{-iwlt},
$$

in which

$$
b(m,n) = \begin{cases} (k_0^2 - a^2 m^2 - a^2 n^2)^{1/2} & \text{if } m^2 + n^2 < k_0^2/a^2 \\ i(a^2 m^2 + a^2 n^2 - k_0^2)^{1/2} & \text{if } m^2 + n^2 > k_0^2/a^2 \end{cases},
$$

and

$$
Q(m,n,l,q,r,s) = \frac{P(q-v,r,s)P(m-q,n-r,l-s)}{b(q,r)}.
$$

By definition, the electric field in Equations (5)–(7) corresponds to the scattering of infinite plane waves from a surface of infinite extent. To transform the fields to finitely scattered fields, which is necessary as only a portion of the whole ocean will be illuminated by the radar, we can use the equation presented by Johnstone [21] who followed the work of Stratton [26]. He showed that if the electric field is known on a finite section of an infinitely large volume such as the surface S_1 on the hemisphere shown in Figure 4, then the scattered electric field at a point (x', y', z') inside of the volume can be calculated.

The equation is given by

$$\vec{E}(x',y',z') = \frac{e^{ik_0R}}{4\pi R}\int_{S_1}\left\{\left[\left(\frac{\partial E_x}{\partial z}-\frac{\partial E_z}{\partial x}\right)\vec{a}_x+\left(\frac{\partial E_y}{\partial z}-\frac{\partial E_z}{\partial y}\right)\vec{a}_y\right]_{z=0}\right.$$
$$+ik_0\left[E_x\cos\theta\vec{a}_x+E_y\cos\theta\vec{a}_y-(E_x\sin\theta\cos\varphi+E_y\sin\theta\sin\varphi)\vec{a}_z\right]_{z=0}$$
$$+\left[\left(\frac{\partial E_z}{\partial x}-\frac{\partial E_x}{\partial z}\right)\sin\theta\cos\varphi+\left(\frac{\partial E_z}{\partial y}-\frac{\partial E_y}{\partial z}\right)\sin\theta\sin\varphi\right]_{z=0}$$
$$\left.\cdot\left[\sin\theta\cos\varphi\vec{a}_x+\sin\theta\sin\varphi\vec{a}_y+\cos\theta\vec{a}_z\right]\right\}e^{-i\vec{k}_0\cdot\rho}dS_1, \tag{8}$$

where ρ is the vector (x,y,z) on S_1, (R,θ,φ) are the radius, polar angle and azimuthal angle measured from the origin to (x',y',z') and $\vec{k}_0$ is the radar wavevector given by

$$\vec{k}_0 = k_0(\sin\theta\cos\varphi, \sin\theta\sin\varphi, \cos\theta).$$

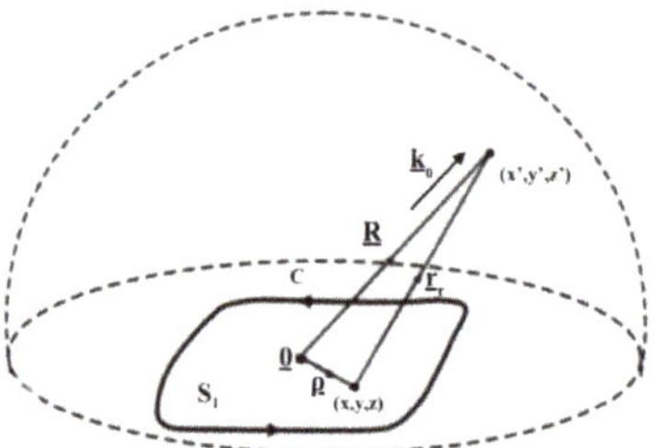

Figure 4. The finite scattering surface, S_1, with boundary C, as part of a hemispherical surface. The vector ρ denotes the position (x,y,z) on S_1; the vector $\vec{r}_r$ is the vector from (x,y,z) to some distant point (x',y',z') where the scattered electric field is desired. The vector $\vec{k}_0$ is the radar wavevector in the direction of the scattered radio wave, $\vec{R}$.

Note that the integral can be evaluated on any plane and $z=0$ is used for convenience. However, as pointed out by Hisaki and Tokuda [24], this choice is in fact the infinite scattering surface limit. Evaluating the integral on $z=f$ as they did changes the resulting power spectrum of the scatter but differences are very small at the Doppler frequencies used for inversion so $z=0$ is sufficient for this work. To find the scattered field at a point (x',y',z'), the values for E_x, E_y and E_z from Equations (5)–(7) are substituted into Equation (8) and the integral is calculated. The vertically polarised component, $E_\theta(t)$, is then identified in the resulting expression (as these are the radio waves that the receiver will detect) such that

$$E_\theta(t) = \frac{ie^{ik_0R}}{2\pi R}L^2\sum_{mnl}\left\{B(t)[-ix_1P(m-v,n,l)+\sum_{qrs}x_2Q]+C(t)[-iy_1P(m-v,n,l)+\sum_{qrs}y_2Q]\right.$$
$$\left.+D(t)[-iz_1P(m-v,n,l)+\sum_{qrs}z_2Q]\right\}, \tag{9}$$

where,

$$x_1 = am - k_0; \qquad x_2 = a^2(m-q)(v-q)k_0+(k_0-am)b^2(q,r)$$
$$y_1 = an; \qquad y_2 = a^2(n-r)(v-q)k_0-anb^2(q,r)$$
$$z_1 = a(m-v)k_0+b^2(m,n); \qquad z_2 = \left[a^3(q-v)(m^2+n^2-qm-rn)k_0+a\left(a(m^2+n^2)-mk_0\right)b^2(q,r)\right].$$

and,

$$B(t) = \sum_{mnl} \cos\varphi(k_0 + b(m,n)\cos\theta)\operatorname{sinc}(XR)\operatorname{sinc}(YR)e^{-i(wl+\omega_0)t} \tag{10}$$

$$C(t) = \sum_{mnl} \sin\varphi(k_0 + b(m,n)\cos\theta)\operatorname{sinc}(XR)\operatorname{sinc}(YR)e^{-i(wl+\omega_0)t} \tag{11}$$

$$D(t) = \sum_{mnl} -\cos\theta(am\cos\varphi + an\sin\varphi)\frac{\operatorname{sinc}(XR)\operatorname{sinc}(YR)}{b(m,n)}e^{-i(wl+\omega_0)t}, \tag{12}$$

for $XR = \dfrac{L}{2}(am - k_0\sin\theta\cos\varphi)$ and $YR = \dfrac{L}{2}(an - k_0\sin\theta\sin\varphi)$.

In the scattered electric field in Equation (9), the first order components (namely, the terms including a single P term) represent the single scattering of one electromagnetic wave, to the receiver, from one ocean wave. The second order components (which include a factor of Q) represent doubly scattered electromagnetic waves, to the receiver, from two single ocean waves. The order of the ocean wave, currently denoted by $P(m-v,n,l)$, has not yet been considered and is assumed to be first order. However, in making such an assumption, a second order contribution from first order scattering from second order oceans waves is missed, where a second order ocean wave is the result of the nonlinear interaction between two first order ocean waves.

To allow for the second order hydrodynamic effects in shallow water, Barrick and Lipa [27] used a perturbation method to relate the second order coefficients $P^{(2)}(\vec{k},\omega)$, of a surface defined by $z = \sum_{\vec{k},\omega} P(\vec{k},\omega)e^{i\vec{k}\cdot\vec{r}-i\omega t}$, to the first order coefficients $P^{(1)}(\vec{k},\omega)$. Their method involved expanding the surface height Fourier coefficients around the flat surface, i.e.,

$$P(\vec{k},\omega) = P^{(1)}(\vec{k},\omega) + P^{(2)}(\vec{k},\omega) + \ldots,$$

alongside boundary conditions from the equations of motion, also expanded to second order. They showed that for ocean waves with wavevectors $\vec{k}_1$ and $\vec{k}_2$, with corresponding angular frequencies ω_1 and ω_2 (related by the dispersion relation of ocean waves given by $\omega = \sqrt{gk\tanh(kd)}$),

$$P^{(2)}(\vec{k}'',\omega'') = \sum_{\vec{k}_1\vec{k}_2}\sum_{\omega_1\omega_2}\Gamma_H(\vec{k}_1,\omega_1,\vec{k}_2,\omega_2)P^{(1)}(\vec{k}_1,\omega_1)P^{(1)}(\vec{k}_2,\omega_2), \tag{13}$$

where $\vec{k}'' = \vec{k}_1 + \vec{k}_2$, $\omega'' = \omega_1 + \omega_2$, and

$$\Gamma_H = \frac{1}{2}\Bigg\{ k_1\tanh(k_1 d) + k_2\tanh(k_2 d) + \frac{\omega''}{g}\frac{(\omega_1^3\operatorname{csch}^2(k_1 d) + \omega_2^3\operatorname{csch}^2(k_2 d))}{(\omega''^2 - gk''\tanh(k''d))}$$
$$+ \frac{(k_1 k_2\tanh(k_1 d)\tanh(k_2 d) - \vec{k}_1\cdot\vec{k}_2)}{\sqrt{k_1 k_2\tanh(k_1 d)\tanh(k_2 d)}}\left(\frac{gk''\tanh(k''d) + \omega''^2}{gk''\tanh(k''d) - \omega''^2}\right) \Bigg\}, \tag{14}$$

is called the *hydrodynamic coupling coefficient*.

To include the second order hydrodynamic effects in the scattered electric field, we expand $P(m-v,n,l)$ into $P^{(1)}(m-v,n,l) + P^{(2)}(m-v,n,l)$ and then substitute in the value of $P^{(2)}(m-v,n,l)$ using Equation (13) (by letting $\vec{k}'' = (m-v,n)$ and $\omega = l$), and so Equation (9) becomes

$$E_\theta(t) = \frac{ie^{ik_0 R}}{2\pi R}L^2\sum_{mnl}\Bigg\{ -i\zeta(t)P^{(1)}(m-v,n,l) + \sum_{qrs}[-i\zeta(t)\Gamma_H b(p,q) + \xi(t)]\,Q(m,n,l,q,r,s) \Bigg\}, \tag{15}$$

where for brevity

$$\zeta(t) = B(t)x_1 + C(t)y_1 + D(t)z_1$$

and

$$\zeta(t) = B(t)x_2 + C(t)y_2 + D(t)z_2.$$

Following Johnstone [21], we finally calculate the radar cross section by substituting Equation (15) into

$$\sigma(\omega) = \lim_{R\to\infty} 4\pi R^2 \frac{\mathcal{F}\left[\langle E_\theta(t_1)E_\theta^*(t_2)\rangle\right]}{L^2}, \tag{16}$$

where R is the distance from the scatter patch to the receiver and $\mathcal{F}$ denotes a Fourier transform, with definition

$$\mathcal{F}[f(t)] = \frac{1}{2\pi}\int_\infty^\infty f(t)e^{-i\omega t}\,dt. \tag{17}$$

To calculate Equation (16), the following properties of the surface height Fourier coefficients are used:

- The Fourier coefficients are normally distributed about zero; hence

$$\langle P(m,n,l)\rangle = 0. \tag{18}$$

- As the surface is real, $f(x,y,t)$ is equal to $f^*(x,y,t)$, which is true when

$$P(-m,-n,-l) = P^*(m,n,l). \tag{19}$$

- From Thomas [28],

$$\langle P_1 P_2 P_3 \rangle = 0 \tag{20}$$

and

$$\langle P_1 P_2 P_3 P_4 \rangle = \langle P_1 P_2 \rangle \langle P_3 P_4 \rangle + \langle P_1 P_3 \rangle \langle P_2 P_4 \rangle + \langle P_1 P_4 \rangle \langle P_2 P_3 \rangle. \tag{21}$$

- The surface roughness spectrum $S(p,q,wl)$, found by utilising the Wiener–Khinchin theorem, is related to the surface height Fourier coefficients by

$$\langle P(m,n,l)P(q,r,s)\rangle = \begin{cases} \dfrac{(2\pi)^3 S(p,q,wl)}{L^2 T} & \text{if } q,r,s = -m,-n,-l \\ 0 & \text{if else,} \end{cases} \tag{22}$$

where $p = am$ and $q = an$.

Then, substituting $E_\theta(t)$ from Equation (15) into Equation (16) and using Equation (20) leads to

$$\sigma(\omega) = \frac{1}{\pi}\mathcal{F}\left[L^2 \sum_{\substack{mnl \\ m'n'l'}} \left\{ \zeta(t_1)\zeta'^*(t_2) \left\langle P^{(1)}(m-v,n,l)P^{(1)'*}(m'-v,n',l') \right\rangle \right. \right.$$
$$\left. \left. + \sum_{\substack{qrs \\ q'r's'}} \left\{ \left[-i\zeta(t_1)b(q,r)\Gamma_H + \xi(t_1)\right]\left[i\zeta'^*(t_2)b^*(q',r')\Gamma'_H + \xi'^*(t_2)\right]\langle QQ'^*\rangle \right\} \right\} \right], \tag{23}$$

where the arguments of Q are implied. The calculation of Equation (23) can be separated into its first and second order terms, such that

$$\sigma(\omega) = \sigma^{(1)}(\omega) + \sigma^{(2)}(\omega),$$

where the first order radar cross section, $\sigma^{(1)}(\omega)$, is defined by the term including the average $\left\langle P^{(1)}(m-v,n,l)P^{(1)'*}(m'-v,n',l')\right\rangle$, and the second order, $\sigma^{(2)}(\omega)$, including $\langle QQ'^* \rangle$. To calculate each of $\sigma^{(1)}(\omega)$ and $\sigma^{(2)}(\omega)$, the properties in Equations (18)–(22) are used to enforce restrictions on the Fourier coefficients and to introduce the roughness spectrum $S(\vec{k})$. The mathematical details are spared here, but can be found in the work of Hardman [25].

2.1.1. First Order

The first order radar cross section is given by

$$\sigma^{(1)}(\omega) = 2^5 \pi k_0^4 \cos^4 \varphi_{bi} \sum_{m=\pm 1} S(m\vec{k}_B)\delta(\omega - m\omega_B), \tag{24}$$

defined at the *Bragg frequencies*, $\pm \omega_B$, where

$$\omega_B = \sqrt{2gk_0 \cos \varphi_{bi} \tanh(2k_0 d \cos \varphi_{bi})}, \tag{25}$$

for the bistatic angle, φ_{bi}, which is shown in Figures 3 and 5, and is related to the azimuthal scatter angle φ by

$$\varphi_{bi} = \frac{1}{2}(\pi - \varphi). \tag{26}$$

The value of $\sigma^{(1)}(\omega)$ depends on the ocean spectrum contribution for the *Bragg wavevector*, $\vec{k}_B$, which travels in the elliptical normal direction from the scatter point (as shown in Figures 3 and 5), and is defined by

$$\vec{k}_B = -2k_0 \cos \varphi_{bi}(\cos \varphi_{bi}, - \sin \varphi_{bi}). \tag{27}$$

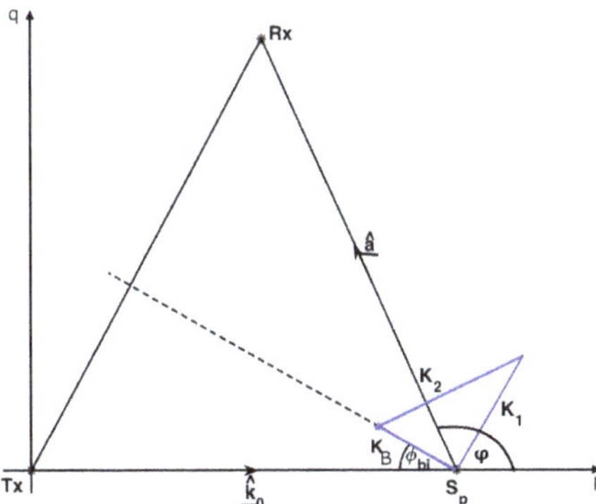

Figure 5. Scattering geometry for a bistatic radar where T_x, S_p and R_x denote the transmitter, scatter patch and receiver respectively, φ_{bi} is the bistatic angle, $\vec{k}_0$ is the radar wavevector and, p and q are spatial wavenumbers, with p in the direction of the emitted radio wave.

2.1.2. Second Order

The second order bistatic radar cross section is

$$\sigma^{(2)}(\omega) = 2^5 \pi k_0^4 \cos^4 \varphi_{bi} \sum_{m,m'=\pm 1} \iint |\Gamma_E - i\Gamma_H|^2 S(m\vec{k}_1)S(m'\vec{k}_2)\delta(\omega - m\omega_1 - m'\omega_2) \, dp \, dq, \tag{28}$$

for wavevector pairs $\vec{k_1}$ and $\vec{k_2}$ (with respective angular frequencies ω_1 and ω_2) such that

$$\vec{k_1} + \vec{k_2} = \vec{k_B}.$$

Explicitly,

$$\vec{k_1} = (p - k_0, q) \quad \text{and} \quad \vec{k_2} = (-k_0 \cos(2\varphi_{bi}) - p, k_0 \sin(2\varphi_{bi}) - q), \tag{29}$$

and Γ_E is the *electromagnetic coupling coefficient* given by

$$\Gamma_E = \frac{1}{2^2 \cos^2 \varphi_{bi}} \left(\frac{a_1}{b_1 - k_0 \triangle} + \frac{a_2}{b_2 - k_0 \triangle} \right), \tag{30}$$

where

$$\triangle = 0.011 - 0.012i, \tag{31}$$

is the normalized surface impedance derived by Barrick [29] and

$$a_1 = -k_{1x}(\vec{k_2} \cdot \vec{a}) - 2 \cos^2 \varphi_{bi} \left(-k_2^2 + 2k_0(\vec{k_2} \cdot \vec{a}) \right), \tag{32}$$

$$a_2 = -k_{2x}(\vec{k_1} \cdot \vec{a}) - 2 \cos^2 \varphi_{bi} \left(-k_1^2 + 2k_0(\vec{k_1} \cdot \vec{a}) \right), \tag{33}$$

$$b_1 = \sqrt{-k_2^2 + 2k_0(\vec{k_2} \cdot \vec{a})} \tag{34}$$

and

$$b_2 = \sqrt{-k_1^2 + 2k_0(\vec{k_1} \cdot \vec{a})} \tag{35}$$

(noting that both b_1 and b_2 can be real or imaginary depending on the argument), where $\vec{a}$ is a unit vector in the direction of the receiver from the scattering patch (see Figure 5), namely,

$$\vec{a} = (-\cos(2\varphi_{bi}), \sin(2\varphi_{bi})).$$

2.1.3. Monostatic Conditions

When $\varphi_{bi} = 0$, i.e., under monostatic conditions, the first and second order radar cross section expressions given in Equations (24) and (28), respectively, are equivalent to the commonly used first order monostatic radar cross section of Lipa and Barrick [15], except for a factor of 2. The difference is due to differing Fourier transform definitions, however, the factor is ultimately not important as when the inverse Fourier transform (which will also be different by a factor of 2) is taken to find the power in the spectrum, the two terms will be equal. Furthermore, when inverting the expression, the whole spectrum is normalised to removed the effects of propagation over the ocean and so the factor is again inconsequential.

2.2. Numerical Solution

The method for finding the numerical solution of the bistatic radar cross section given in Equations (24) and (28), is analogous to the method of Holden and Wyatt [16].

2.2.1. First Order

Due to the delta function in Equation (24), the first order contribution to the Doppler spectrum will appear as two peaks at $\pm \omega_B$, defined in Equation (25). The contribution comes from two wave vectors, $\pm \vec{k}_B$, travelling in the direction of the Bragg bearing, both toward and away from the radar set up. As Crombie [1] hypothesised (for monostatic radar), these particular signals are amplified by resonance. As k_B includes a factor of $\cos \varphi_{bi}$, for one radar using a single carrier frequency, there will be a number of different Bragg waves, dependent on the radar beam range and angle. Scattering from locations where $\varphi_{bi} \to 90°$ are referred to as *forward scatter* and when this occurs, both ω_B and k_B tend to 0. Consequently, the wavelengths of the Bragg waves in this region become infinitely long and in addition, the $\cos^4 \varphi_{bi}$ factors in Equations (24) and (28) will tend to zero, leading to a low SNR.

2.2.2. Second Order

The second order radar cross section contribution, given in Equation (28), is due to double electromagnetic scattering from two first order ocean waves, with wavevectors $\vec{k}_1$ and $\vec{k}_2$, and the nonlinear interaction between the same two ocean waves. The wave vector pair sum to give the Bragg wavevector $\vec{k}_B$ and, theoretically, large numbers of pairs exist in the p, q wavenumber plane; an example pair can be seen in Figure 6. To find the values of $\vec{k}_1$ and $\vec{k}_2$, the solution to the delta function in Equation (28), such that

$$\omega - m\sqrt{gk_1 \tanh(k_1 d)} - m'\sqrt{gk_2 \tanh(k_2 d)} = 0 \tag{36}$$

is sought. For each value of ω, the set of solutions of $\vec{k}_1$ and $\vec{k}_2$ defines a frequency contour in the p, q plane. As m and m' can both take the values of 1 or -1, Equation (36) has four different forms.

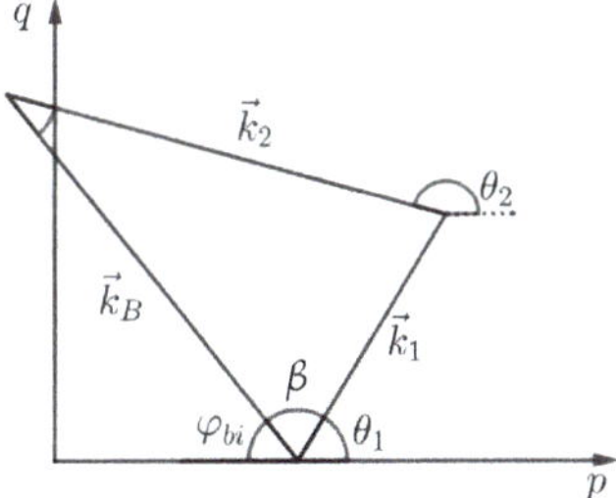

Figure 6. Geometry of the second order scattering wave vectors, $\vec{k}_1$ and $\vec{k}_2$, at angles θ_1 and θ_2, respectively.

Case $m = m'$:

Squaring Equation (36) gives

$$\omega^2 = g \left(k_1 \tanh(k_1 d) + k_2 \tanh(k_2 d) + 2\sqrt{k_1 \tanh(k_1 d)k_2 \tanh(k_2 d)} \right)$$

and it can be shown that

$$\omega^2 \geq (k_1 + k_2) \tanh((k_1 + k_2)d).$$

Now, as the sum of two sides of a triangle is greater than the third, $k_1 + k_2 > 2k_0 \cos \varphi_{bi}$, and therefore

$$\omega^2 > 2gk_0 \cos \varphi_{bi} \tanh(2gk_0 \cos \varphi_{bi} d).$$

Taking the square root gives

$$|\omega| > \sqrt{2gk_0 \cos \varphi_{bi} \tanh(2gk_0 \cos \varphi_{bi} d)},$$

which is equal to the bragg frequency ω_B given in Equation (25). This leads to two conditions;

$$\begin{cases} \omega > \omega_B & m = m' = 1 \\ \omega \leq -\omega_B & m = m' = -1. \end{cases}$$

Figure 7 shows the frequency contours, defined by Equation (36). At frequencies close to the Bragg frequencies the contours are circular in shape, centred around the Bragg frequency. As ω increases, the contours become less circular, until $|\omega| = 2\sqrt{gk_0 \cos \varphi_{bi} \tanh(k_0 d \cos \varphi_{bi})}$, shown in white in Figure 7, where they separate.

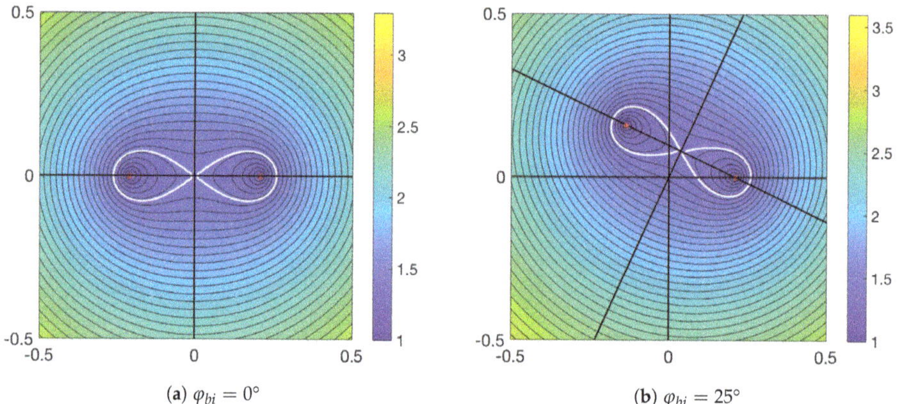

(a) $\varphi_{bi} = 0°$ (b) $\varphi_{bi} = 25°$

Figure 7. The frequency contours of Equation (36) for two values of φ_{bi}, (a) monostatic $\varphi_{bi} = 0$ and (b) bistatic angle $\varphi_{bi} = 25°$ when $m = m' = 1$. The normalised frequency, $\eta = \omega/\omega_B$, is shown by the colour, in the p, q plane.

When $\varphi_{bi} = 0$, the contours are symmetrical about the p and q axes, however, when $\varphi_{bi} > 0$, the contours rotate clockwise in the p, q plane, becoming symmetrical about some other axes, say p' and q', shown by the additional black lines in Figure 7b.

Case $m \neq m'$:

When $m \neq m'$, the square of Equation (36) is

$$\omega^2 = g\left(k_1 \tanh(k_1 d) + k_2 \tanh(k_2 d) - 2\sqrt{k_1 \tanh(k_1 d) k_2 \tanh(k_2 d)}\right),$$

and it can be shown that

$$\omega^2 \leq g\left((k_2 - k_1) \tanh((k_2 - k_1)d)\right).$$

In the right hand plane, when $\vec{k_1}$ and $\vec{k_2}$ lie in opposite directions along the p' axis, meeting at a point past the bragg frequency, $k_2 - k_1$ reaches its maximum value of $2k_0 \cos \varphi_{bi}$. Therefore,

$$\omega^2 \leq g(2k_0 \cos \varphi_{bi} \tanh(2k_0 \cos \varphi_{bi} d))$$

and this leads to the conditions

$$\begin{cases} 0 < \omega \leq \omega_B & m = -1, m' = 1 \\ -\omega_B < \omega \leq 0 & m = 1, m' = -1. \end{cases}$$

In the left hand plane, where $k_2 < k_1$, the result is reversed giving

$$\begin{cases} 0 < \omega \leq \omega_B & m = 1, m' = -1 \\ -\omega_B < \omega \leq 0 & m = -1, m' = 1. \end{cases}$$

Figure 8 shows the normalised frequency contours when $m \neq m'$ for different values of φ_{bi}. Like when $m = m'$, the contours are symmetric about the p' and q' axes, however, in this case, they do not cross the q' axis for any frequency. The contours depict small circles around the Bragg frequencies, growing in size and becoming less circular as the frequency approaches zero.

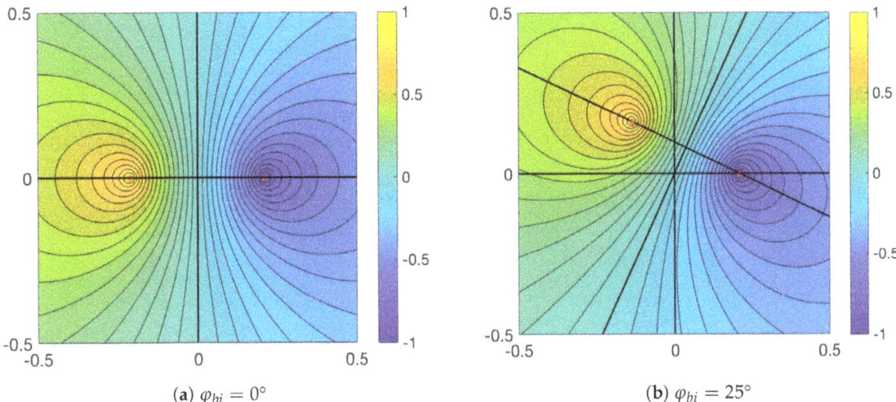

(a) $\varphi_{bi} = 0°$ (b) $\varphi_{bi} = 25°$

Figure 8. The frequency contours of Equation (36) shown for $m \neq m'$ (where $m = 1$) with two different values for φ_{bi}: (a) monostatic $\varphi_{bi} = 0$ and (b) bistatic angle $\varphi_{bi} = 25°$. The colour shows the value of the normalised frequency, $\eta = \omega/\omega_B$, in the p, q plane.

As the frequency contours for all four possible combinations of m and m' are symmetrical about the q' axis, the integration in Equation (28) can be taken over one half of the symmetric plane and doubled. Therefore, we integrate Equation (28) over the right hand p' plane, and double the result and hence

$$\sigma_2(\omega) = 2^6 \pi k_0^4 \cos^4 \varphi_{bi} \sum_{m,m'=\pm 1} \int_{-\infty}^{\infty} \int_{q'}^{\infty} |\Gamma_E - i\Gamma_H|^2 S(m\vec{k_1}) S(m'\vec{k_2}) \delta(\omega - m\omega_1 - m'\omega_2) \, dp \, dq. \quad (37)$$

In the right hand p' plane, where $k_1 \leq k_2$, we can calculate the integral in polar coordinates k_1 and θ_1, where θ_1 is the angle between $\vec{k_1}$ and the p axis, as shown Figure 6. Explicitly,

$$\sigma_2(\omega) = 2^6 \pi k_0^4 \cos^4 \varphi_{bi} \sum_{m,m'=\pm 1} \int_{-\theta_L^-}^{\theta_L^+} \int_0^{\infty} |\Gamma_E - i\Gamma_H|^2 S(m\vec{k_1}) S(m'\vec{k_2}) \delta(\omega - m\omega_1 - m'\omega_2) k_1 \, dk_1 \, d\theta_1, \quad (38)$$

where θ_L^- and θ_L^+ are the integration limits of θ_1 and vary with frequency as well as φ_{bi}. In general, we find that the integration limits are

$$\theta_L^- = -(\pi + \varphi_{bi}) \quad \text{and} \quad \theta_L^+ = \pi - \varphi_{bi}, \quad (39)$$

however a particular set of ω values require different limits. This happens when $m = m'$ and $|\omega| > 2\sqrt{k_0 g \cos \varphi_{bi} \tanh(k_0 d \cos \varphi_{bi})}$, as the frequency contours cross the q' axis and no longer complete full rotations in the right hand p' plane. By symmetry, when the contours cross the q' axis, k_1 and k_2 are the same length. Therefore, when $k_2 = k_1$ and $m = m'$, the delta constraint of Equation (36) becomes

$$\omega = \pm 2\sqrt{g k_1 \tanh(k_1 d)} \tag{40}$$

and then squaring Equation (40) gives

$$\frac{\omega^2}{4g} = k_1 \tanh(k_1 d), \tag{41}$$

which can be solved numerically for k_1.

Introducing a term, β, as the angle between $\vec{k_1}$ and $\vec{k_B}$ (see Figure 6), the integration limits are given by

$$\theta_L^+ = \pi - \varphi_{bi} - \beta \quad \text{and} \quad \theta_L^- = -\left(2\pi - \theta_L^+ - 2\beta\right). \tag{42}$$

Therefore, as $\beta = \cos^{-1}\left(\dfrac{k_B}{2k_1}\right)$, where the solution for k_1 from Equation (41) is used,

$$\theta_L^+ = \pi - \varphi_{bi} - \cos^{-1}\left(\frac{k_B}{2k_1}\right) \quad \text{and} \quad \theta_L^- = -\left(\pi + \varphi_{bi} - \cos^{-1}\left(\frac{k_B}{2k_1}\right)\right). \tag{43}$$

In terms of k_1 and θ_1,

$$k_2 = \sqrt{k_1^2 + k_B^2 + 2k_B k_1 \cos(\theta_1 + \varphi_{bi})} \quad \text{and} \quad \theta_2 = \pi + \theta_1 - \cos^{-1}\left(\frac{k_1 + k_B \cos(\theta_1 + \varphi_{bi})}{k_2}\right).$$

To calculate $\sigma^{(2)}(\omega)$, we now reduce the double integral in Equation (38) to a single integral using the delta function. By defining

$$y_s = \sqrt{k_1}$$
$$h(y_s, \theta_1) = m y_s \sqrt{g \tanh(y_s^2 d)} + m' \sqrt{g k_2 \tanh(k_2 d)}$$
$$I(y_s, \theta_1) = 2^7 \pi |\Gamma_T|^2 k_0^4 \cos^4 \varphi_{bi} S(m\vec{k_1}) S(m'\vec{k_2}) y_s^3,$$

Equation (38) can be written as

$$\sigma_2(\omega) = \int_{-\theta_L^-}^{\theta_L^+} \int_0^\infty I(y_s, \theta_1) \delta\left(\omega - h(y_s, \theta_1)\right) dy_s \, d\theta_1. \tag{44}$$

Now, in order to integrate over the delta function, Equation (44) should have an integration variable of h. Therefore, we calculate

$$\sigma_2(\omega) = \int_{-\theta_L^-}^{\theta_L^+} \int_0^\infty I(y_s, \theta_1) \delta\left(\eta - h(y_s, \theta_1)\right) \left|\frac{\partial y_s}{\partial h}\right|_\theta dh \, d\theta_1, \tag{45}$$

where

$$\left|\frac{\partial h}{\partial y_s}\right|_{\theta_1} = \sqrt{g}\left\{ m\left(\sqrt{\tanh(y_s^2 d)} + \frac{y_s^2 d(\operatorname{sech}^2(y_s^2 d))}{\sqrt{\tanh(y_s^2 d)}}\right) \right.$$
$$\left. + \frac{m'(y_s^3 + y_s k_B \cos(\theta_1 + \varphi_{bi}))}{k_2^{3/2}}\left\{\sqrt{\tanh(k_2 d)} + k_2 d\frac{\operatorname{sech}^2(k_2 d)}{\sqrt{\tanh(k_2 d)}}\right\}\right\},$$

whose reciprocal is $\left|\dfrac{\partial y_s}{\partial h}\right|_{\theta_1}$. To integrate over the delta function, the solution, y^*, to

$$\omega - h(y^*, \theta_1) = 0 \tag{46}$$

is required, which can be found using a numerical method. For timely convergence in the numerical method, a good initial guess for y^* is important. As the solution for shallow water should not be too different to that for deep water, we find an initial solution for the deep water case and use that, as a starting point, for the shallow water case. The deep water equation can be solved exactly in two cases:

- When $mm' = 1$ and $\theta_1 = -\varphi_{bi}$, the solution of $f(y)$ is

$$y_0^* = \frac{\omega^2 - gk_B}{2m\sqrt{g\omega}}. \tag{47}$$

- When $mm' = -1$ and $\theta_1 = \pi - \varphi_{bi}$, the solution is

$$y_0^* = \frac{m\omega + \sqrt{2gk_B - \omega^2}}{2\sqrt{g}}. \tag{48}$$

Upon finding y^*, the second order cross section calculation in Equation (45) reduces to

$$\sigma_2(\omega) = \int_{-\theta_L^-}^{\theta_L^+} I(y_s, \theta_1)\left|\frac{\partial y_s}{\partial h}\right|_{\theta_1}\bigg|_{y_s = y^*} d\theta_1, \tag{49}$$

which can be calculated using a numerical integration method. For speed and convergence we update the value of y_0^* to the previously found solution for y^*, as θ_1 incrementally increases.

2.2.3. Electromagnetic Singularities

The electromagnetic coupling coefficient given in Equation (30), contains two singularities; either when b_1 or b_2 is equal to zero. The singularities lie on two circles in the p, q plane, shown in Figure 9; explicitly,

$$p^2 + q^2 = k_0^2 \tag{50}$$

and

$$(k_0 \cos\varphi - p + k_0)^2 + (k_0 \sin\varphi - q)^2 = k_0^2. \tag{51}$$

Each singularity will be most prominent when a frequency contour is tangential to the singular circle. In order to find the frequencies that this is true for, the solutions for p and q such that the gradient of the frequency contour expression of Equation (36) is equal to the gradient of the circle functions of Equations (50) and (51) are sought. Knowledge of the geometry of the contours and the radii of the circles is exploited to find the solutions for p and q and then the solutions are substituted into Equation (36) to give the tangential frequencies. The four solutions for p and q are

$$p = k_0 \sin \varphi_{bi} \text{ and } q = k_0 \cos \varphi_{bi}, \tag{52}$$

$$p = -k_0 \sin \varphi_{bi} \text{ and } q = -k_0 \cos \varphi_{bi}, \tag{53}$$

$$p = k_0 \sin \varphi_{bi} + 2k_0 \sin^2 \varphi_{bi} \text{ and } q = k_0 \cos \varphi_{bi} + 2k_0 \sin \varphi_{bi} \cos \varphi_{bi} \tag{54}$$

and

$$p = -k_0 \sin \varphi_{bi} + 2k_0 \sin^2 \varphi_{bi} \text{ and } q = -k_0 \cos \varphi_{bi} + 2k_0 \sin \varphi_{bi} \cos \varphi_{bi}. \tag{55}$$

Substituting the solutions for p and q, from Equations (52)–(55) into Equation (36) gives two distinct tangential frequencies:

$$\omega = 2^{3/4} \omega_B \sqrt{\frac{\sqrt{1 \pm \sin \varphi_{bi}}}{\cos \varphi_{bi}}} \frac{\sqrt{\tanh(d\sqrt{2k_0^2(1 \pm \sin \varphi_{bi})})}}{\sqrt{\tanh(2k_0 \cos \varphi_{bi} d)}}, \tag{56}$$

where the solution with the $+$ signs is for the p and q in Equations (52) and (53), and the $-$ signs for the solutions of p and q in Equations (54) and (55).

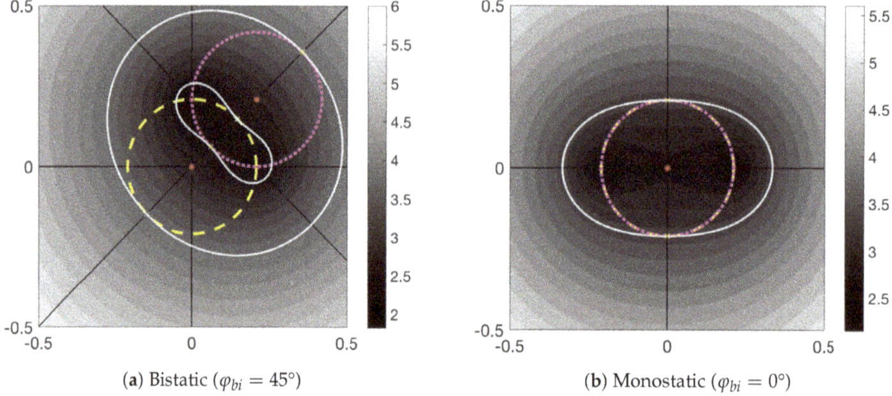

(a) Bistatic ($\varphi_{bi} = 45°$)   (b) Monostatic ($\varphi_{bi} = 0°$)

Figure 9. Contours in the p, q plane defined by Equation (36) when $m = m' = 1$. (**a**) Bistatic case with bistatic angle $\varphi_{bi} = 45°$, (**b**) monostatic case. The electromagnetic singularities are shown for both monostatic and bistatic radars. The yellow dashed circle shows the singularities defined by Equation (50) and the magenta dotted circle shows those defined by Equation (51). In the monostatic case, Equations (50) and (51) are equal and hence both circles are in the same location. The white contours highlight the frequencies tangential to the circles.

These values for ω are highlighted in Figure 7 by the white contours and are shown to be tangential to the circles expressed in Equations (50) and (51). Both singularities are highlighted in a simulated Doppler spectrum in Figure 10 by dashed vertical lines. A low amplitude Gaussian noise spectrum has been added as can be seen at the extremities of the plot. For deep water, or when $d \to \infty$, the values for ω become

$$\omega = 2^{3/4}\omega_B \sqrt{\frac{\sqrt{1 \pm \sin \varphi_{bi}}}{\cos \varphi_{bi}}},$$ (57)

which agree with the results of Gill & Walsh [22].

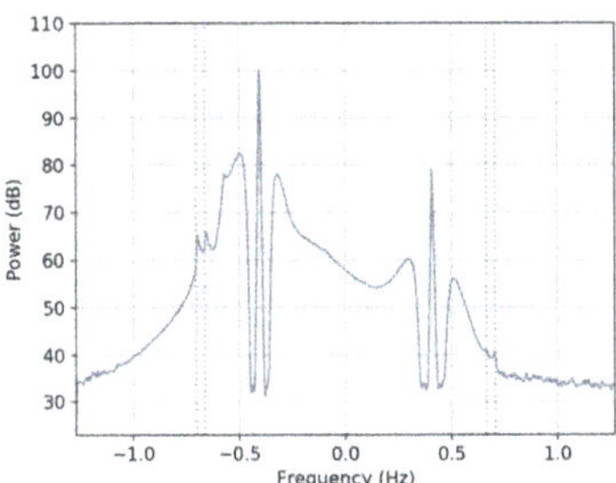

Figure 10. A simulated bistatic Doppler spectrum showing the electromagnetic singularities.

2.2.4. Currents

An ocean current affects how an ocean wave propagates, both in direction and speed. The change in speed means that, because of the Doppler effect, the entire spectrum is subject to an additional shift, $\triangle\omega$. The additional shift is

$$\triangle\omega = 2k_0 v_E(\varphi) \cos \varphi_{bi},$$ (58)

where $v_E(\varphi)$ is the component of the current velocity in the elliptical normal direction for beam angle φ.

3. Inversion of $\sigma(\omega)$ to Measure the Directional Wave Spectrum

The aim of inversion is to obtain the ocean wave directional spectrum, $S(\vec{k})$, from the power spectrum of the measured radar cross section, $\sigma(\omega)$ using Equations (24) and (28). The power spectrum of the backscattered radar signal is proportional to $\sigma(\omega)$ and therefore in principle needs to be calibrated to account for antenna gains, propagation losses and other factors. To avoid this the problem is usually framed in terms of the ratio of the second and first order backscatter power spectra. A number of inversion methods have been published e.g., [6,8,9,15,16,30]. Here we use the method of Wyatt [10,11] which is referred to in this paper as the Seaview inversion method since Seaview Sensing Ltd has an exclusive license from the University of Sheffield to commercialise the software package and continue its development.

The Seaview Inversion Method

The method used makes the assumption that the first order Bragg wave, $\vec{k_B}$, is generated by the local wind and that, by limiting the Doppler frequency range used in the inversion, the waves contributing to the second order scatter can be separated into long waves $\vec{k_1}$ and short waves $\vec{k_2}$ the latter also being locally wind-driven. These short wind waves are then modelled with a Pierson–Moskowitz spectrum [31], using an initial waveheight estimate obtained using an empirical method [32], and a sech^2 directional distribution [33] the parameters of which are determined using a wind direction estimation model [4].

The method is iterative and is initialised assuming the wind–wave model applies at all wavenumbers. The ratio of the Equations (1) and (3), $\frac{\sigma_2(\omega)}{\sigma_1(\omega)}$, is then integrated to provide a simulated Doppler spectrum ratio which is compared with the measured Doppler spectrum ratio to obtain the difference between them at each Doppler frequency. The wave spectrum is then updated taking into account the calculated Doppler spectrum ratio difference and the value of the coupling coefficient at the long wave vector wavenumber, $\vec{k_1}$ relative to the maximum along the Doppler frequency contour. The iteration continues until convergence is achieved or a specified number of iterations, usually 100, has been reached. The quality of the convergence is measured by a quantity that reflects the difference between the measured and simulated ratio. The solution can be very different from the initialising spectrum and often shows bimodality in frequency or direction or both due to the presence of swell or changing wind conditions. More details of the method can be found in Wyatt [10], and Green and Wyatt [11]. The maximum frequency (or equivalently wavenumber) of the long waves that can be measured is dependant on the radio frequency [5,34]. The minimum frequency depends on the quality of the radar data which impacts on the ability to clearly separate first and second order parts of the measured Doppler spectrum. An independent validation of the method applied to monostatic data is presented in [35].

The inversion software, providing surface current, wind direction and wave information, was written for monostatic radar configurations only. This has been extended to bistatic configurations using the analysis in Sections 2.1 and 2.2 above with some modifications to account for a difference in the coordinate system used in the Seaview software. The bistatic extension is not yet part of the commercial package. It has been tested using simulated data, using the methods given in Section 2.2, of two types: (a) two monostatic radars (in order to check that the bistatic extension to the Seaview package provides the same results for zero bistatic angle as the original monostatic package); (b) one monostatic and one bistatic radar sharing a transmitter at the monostatic site. Tests using two bistatic radars with a transmitter between the two sites are in progress but the results are not ready to be reported on at this time. The two receiver sites are those of the University of Plymouth wavehub WERA radar [36,37] and a limited number of cells from the coverage grid for that radar have been used to provide different bistatic angles for the simulations. Using an existing configuration made it easier to provide data for the inversions in standard Seaview formats. The wave parameters used in each simulation are the same at all cells and propagations losses and any antenna effects are not included so the only differences at each cell are the bistatic angle and the Bragg wave bearing relative to wave,wind and current directions.

The wave parameters for the different simulations are described in Table 1. They include modelled and buoy-measured wave spectra.

Table 1. Wave, wind and current parameters used for the Doppler spectra simulations. The buoy data are not separated into wind–waves and swell but their peak period and direction are included in the swell columns.

Case	Type	Wind–Wave Hs m	θ_w	Spread	Wind Speed m/s	Current Speed cm/s	θ_c	Swell Hs m	θ_s	Tp s	Spread
1	Model	3.07	0.0	3.0	12.0	1.4	70.0				
2	Model	3.07	30.0	2.0	12.0	1.8	90.0	3.0	140.0	13.2	10.0
3	Buoy	1.72				1.8	90.0		68.24	12.8	
4	Buoy	1.72				1.8	90.0		148.16	6.74	
5	Buoy	2.70				1.8	90.0		109.01	14.22	
6	Buoy	5.87				1.0	245.0		162.36	9.85	

4. Results

Figures 11 and 12 show inverted surface current speed and direction, wind direction and significant waveheight and spectral peak direction maps for the model cases in Table 1 for the monostatic and bistatic configurations. For the bimodal case 2, Figure 13 shows the long (swell, here defined as waves in 0.05-0.1Hz band) and short (wind–wave, 0.1–0.2Hz) contributions separately to confirm that the latter are aligned with the wind. Figures 14–17 show current, wind and wave maps for the buoy cases. Figures 18–21 show sample directional spectra compared with those measured by the buoy and used in the simulations to provide a qualitative validation of the radar measured spectra. They are from 4 selected locations to cover the key parameter ranges expected to be important in the accuracy of the inversion. The key parameters are the bistatic angles and the difference in angle between the two Bragg directions (a minimum of 30° is required for monostatic processing [38]) and are presented in Table 2. Note that for some of the locations the Bragg angle difference is below the suggested monostatic threshold. Cell 1664 is the one on the left of the top row in the maps, cells 3116, 3128 and 3140 are the lower three going south along the column to the east of $-5°36'$.

There is generally good agreement both for both the standard monostatic and the bistatic case. Differences will be discussed further in the next section.

Table 2. Configuration parameters at selected cells.

Cell Number	Configuration	Bistatic Angle	Angle between Braggs
1664	monostatic	0	40.6
	1 mono, 1 bistatic	20.4	20.3
3116	monostatic	0	60.0
	1 mono, 1 bistatic	30.1	29.9
3128	monostatic	0	77.8
	1 mono, 1 bistatic	39.1	38.8
3140	monostatic	0	99.6
	1 mono, 1 bistatic	50.1	49.5

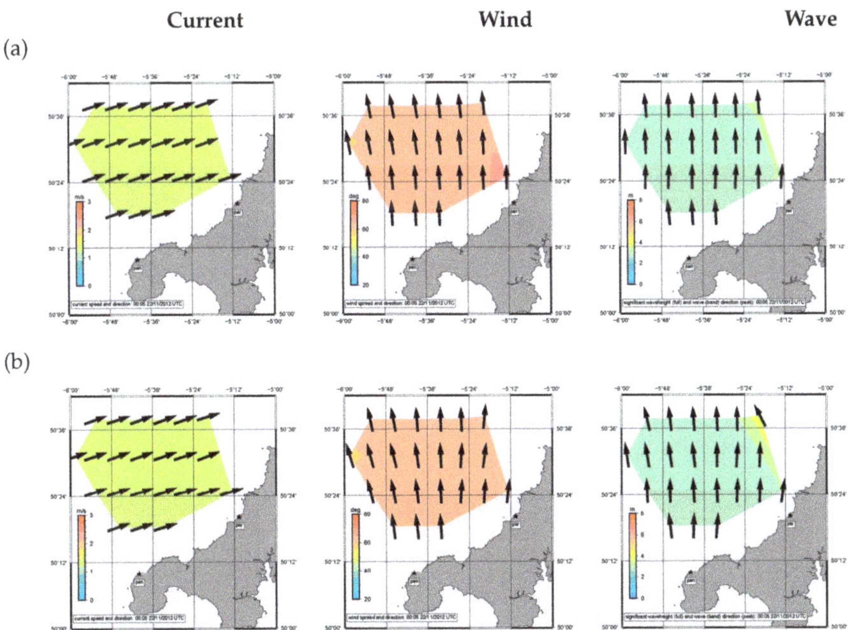

Figure 11. Inverted data for case 1. (a) monostatic (b) 1 bistatic. Current speed and direction on left, shortwave directional spreading and wind direction, centre, significant waveheight and peak direction, right.

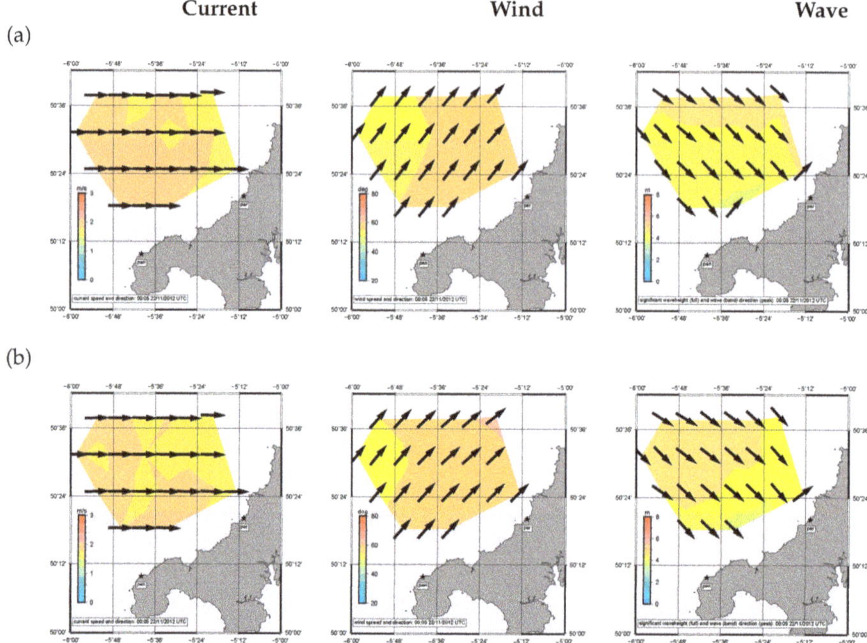

Figure 12. Inverted data for case 2. (**a**) monostatic (**b**) 1 bistatic.

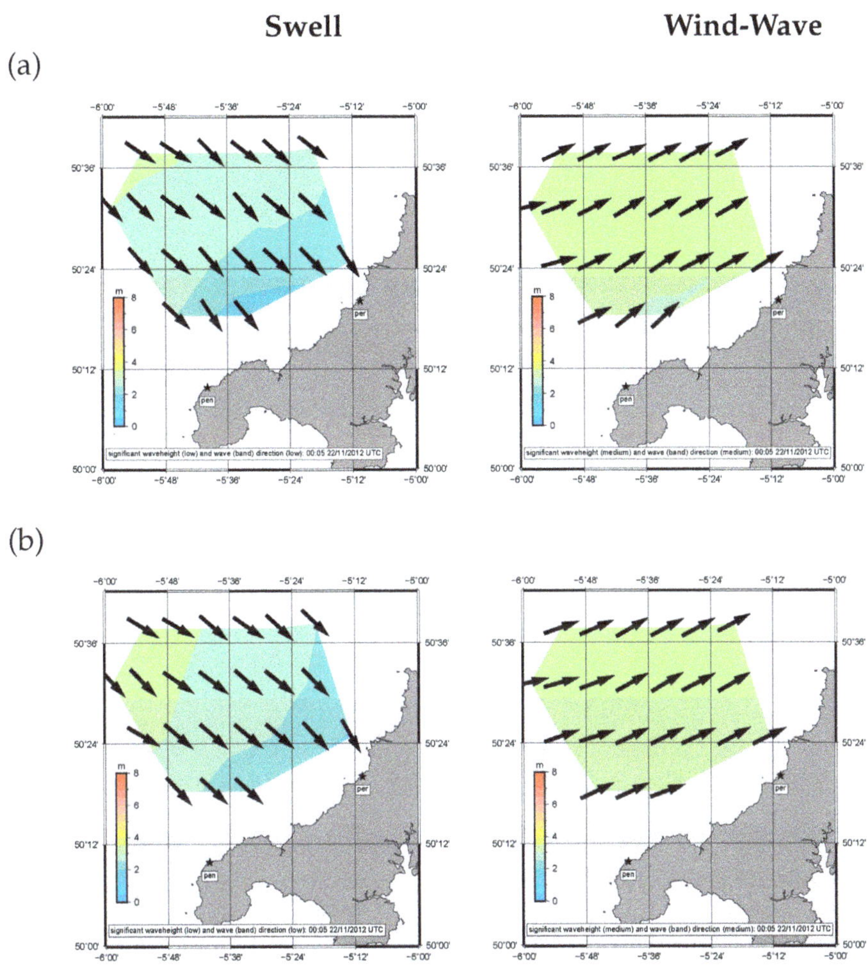

Figure 13. Inverted swell and wind wave components for case 2 (**a**) monostatic (**b**) 1 bistatic.

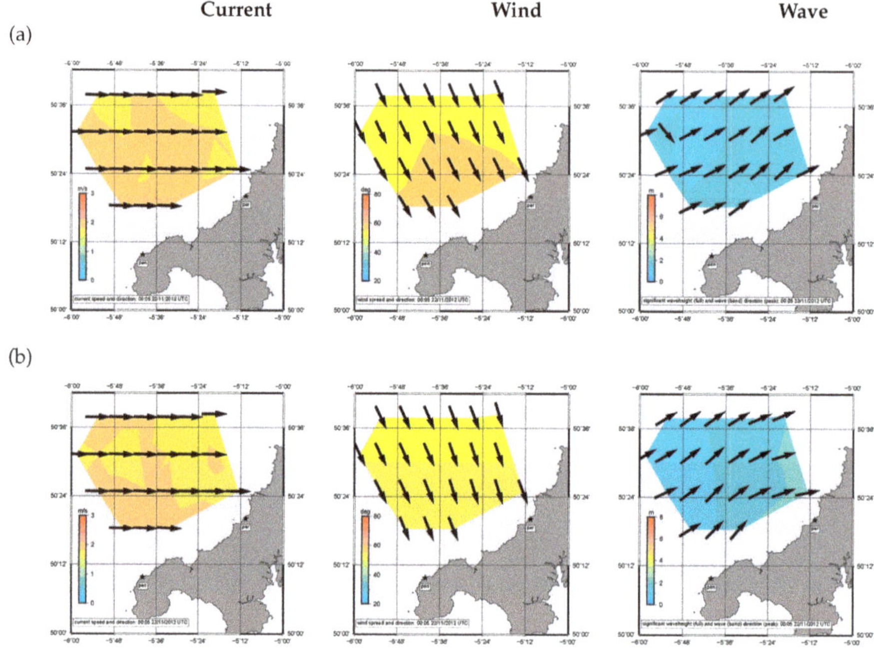

Figure 14. Inverted data for case 3. (**a**) monostatic (**b**) 1 bistatic.

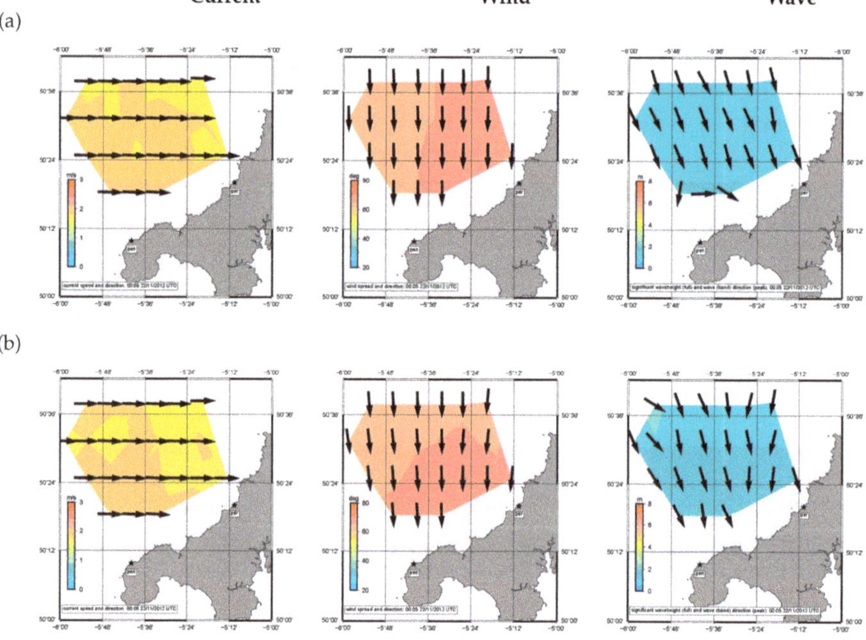

Figure 15. Inverted data for case 4. (**a**) monostatic (**b**) 1 bistatic.

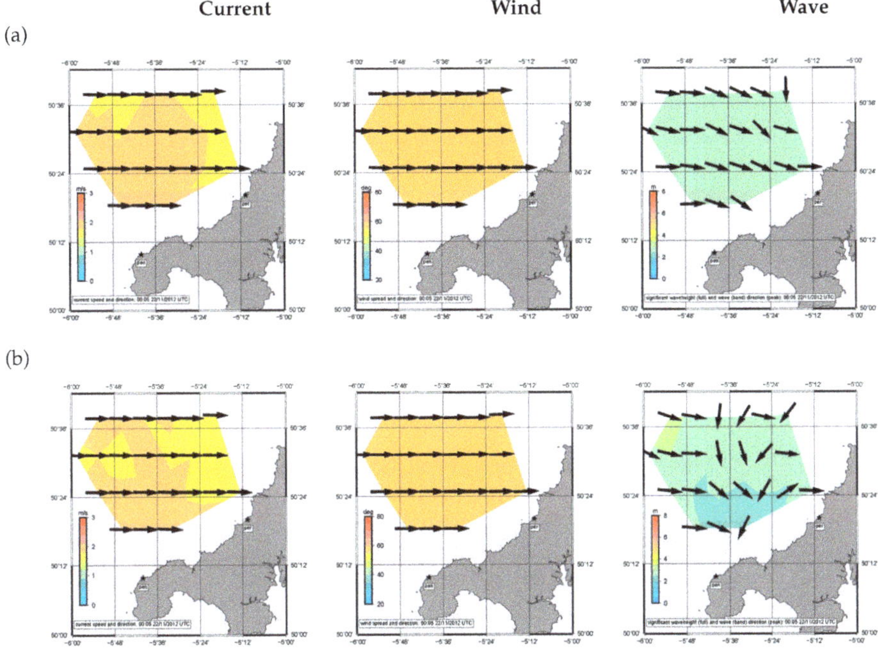

Figure 16. Inverted data for case 5. (**a**) monostatic (**b**) 1 bistatic.

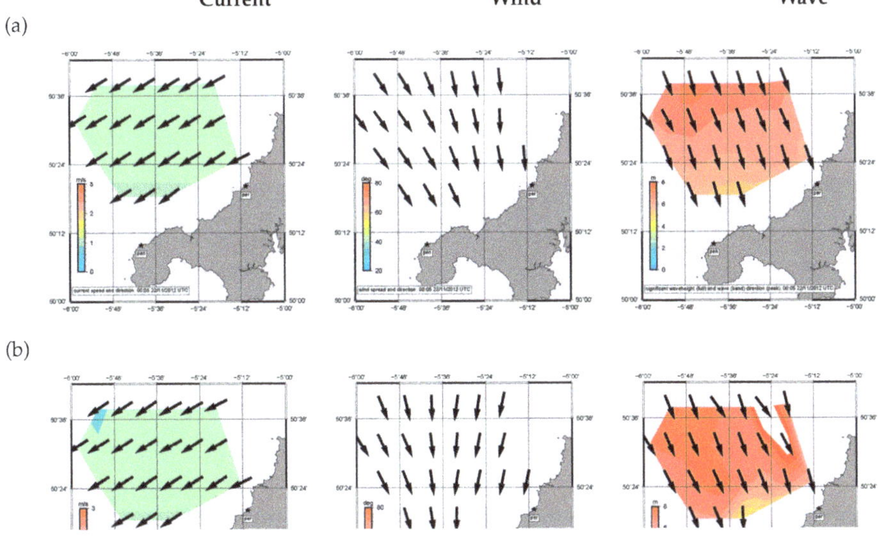

Figure 17. Inverted data for case 6. (**a**) monostatic (**b**) 1 bistatic.

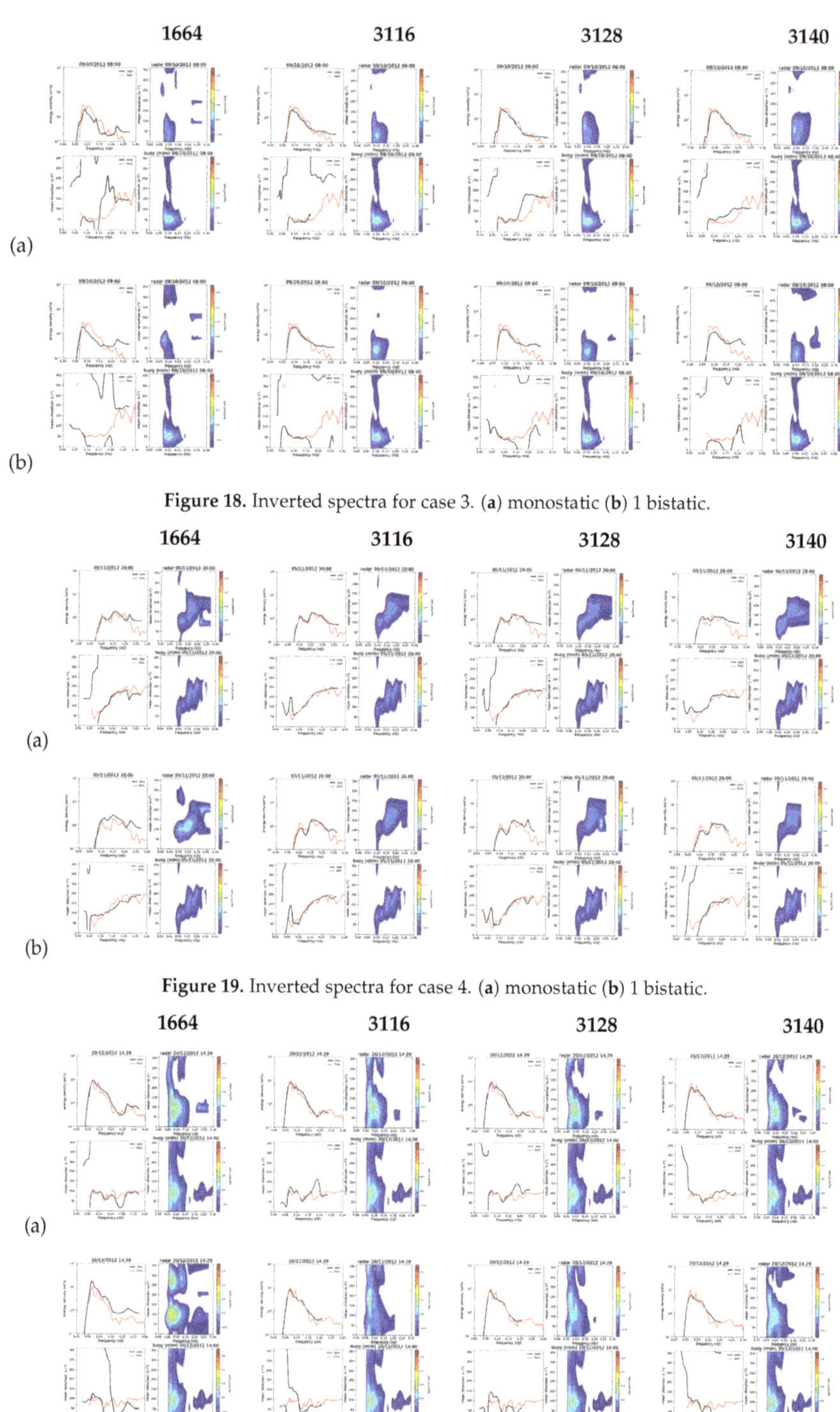

Figure 18. Inverted spectra for case 3. (**a**) monostatic (**b**) 1 bistatic.

Figure 19. Inverted spectra for case 4. (**a**) monostatic (**b**) 1 bistatic.

Figure 20. Inverted spectra for case 5. (**a**) monostatic (**b**) 1 bistatic.

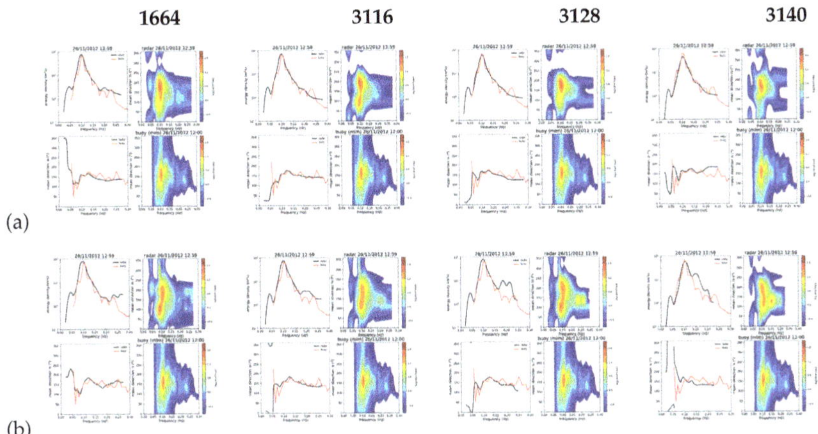

Figure 21. Inverted spectra for case 6. (**a**) monostatic (**b**) 1 bistatic.

Scatter plots and statistics of the comparisons for currents are presented in Figure 22 and for waves in Figure 23. The data are colour-coded according to the bistatic angle with red being the largest bistatic angle. There is some dependence of the accuracy of the wave measurements on this parameter as can be seen in the right hand column of Figure 23. Most of the larger differences in peak period and peak direction are associated with the bimodal model case where the swell and wind–waves peaks were of similar magnitude and small differences in these magnitudes can lead to differences in peak identification. This is also evident when comparing Figure 12 with Figure 13.

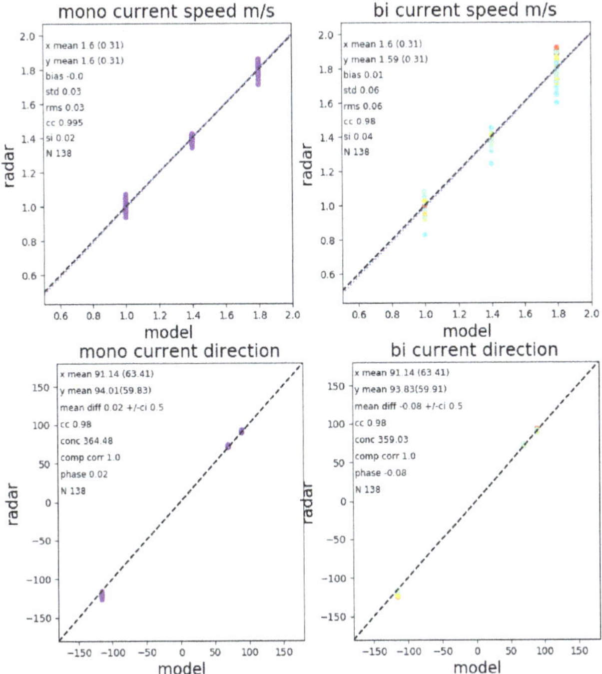

Figure 22. Scatter plots and statistics of the current measurements. These are colour-coded with the bistatic angle.

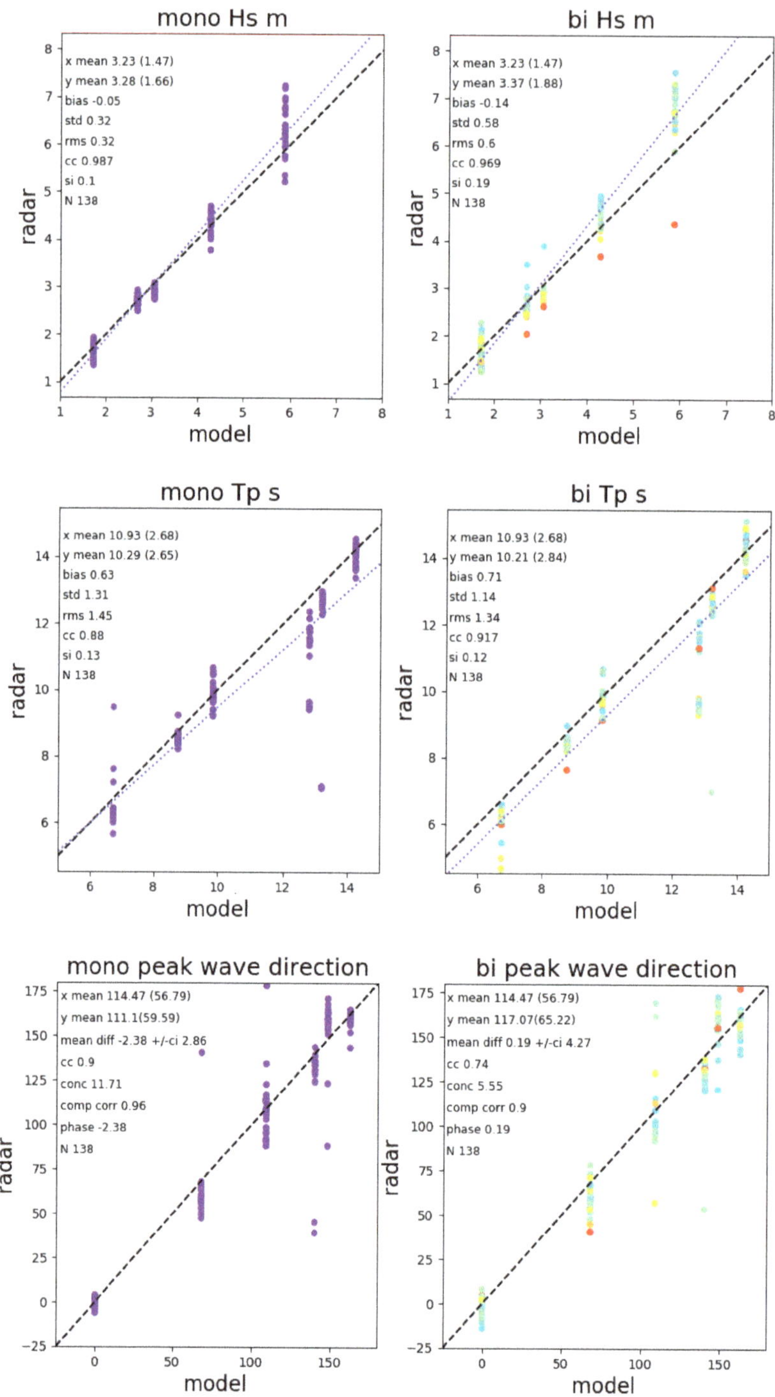

Figure 23. Scatter plots and statistics of the wave parameter measurements, colour-coded with the bistatic angle.

5. Discussion

Monostatic and bistatic radar data have been simulated using the backscatter cross-section formulations developed in Section 2. The Seaview software package has been modified to include bistatic configurations and inverted to provide current, wind direction and wave measurements.

Currents have been obtained with good accuracy and consistency over many different bistatic and Bragg angles as evidenced in the scatter plots, Figure 22, and maps.

Wind directions are consistent with modelling except for the case shown in Figure 17. Note that the colour-coding in the wind plots is the derived directional spreading of the short waves which we haven't attempted to validate at this point. In Figure 17 this goes beyond the expected maximum (80°) for this parameter. The reason is that the simulation used a wind direction that was roughly aligned with the Bragg direction and the smaller Bragg peak was mostly lost in the simulated noise level. The Seaview algorithm has difficulty estimating a wind direction accurately in these circumstances which, in our experience, rarely occur in measured data. It is interesting to note that waves are still measurable in these conditions confirming that the inversion result is independent of the initial guess which uses the wind direction.

The wave inversions are not as uniform as those for currents and winds. The significant waveheights are in reasonable agreement, Figure 23, although there is some evidence of overestimation at the highest simulated waveheight. That figure also shows that waveheight is underestimated for the largest bistatic angles of 64°. More work is needed to determine a bistatic angle threshold for accurate wave measurement. Case 5 shows particularly noisy peak wave directions, Figure 16, including at the selected cells for which directional spectra are shown in Figure 20. While the frequency spectra (top left in each case) show good agreement, the low frequency part of the mean directions as a function of frequency (bottom left in each case) are not good at the low frequency peaks for three of these cases particuarly for the bistatic case. The inversion seems to be oversensitive to noise at these frequencies, which correspond with Doppler frequencies near the first order peak, and this needs further work as has been noted in many other applications of this method ([34,39]). Although some locations, including cell 1664, have Bragg angle differences in the bistatic case that are below the suggested monostatic angle threshold, there is no evidence that the results are worse there. This aspect also needs further work.

The directional spectra in Figures 18–21 use log scales for both the frequency spectra (top left) and the directional spectra (right column) so that differences at both high and low amplitudes can be identified. Apart from the low frequency issue referred to above, the shape of the spectra are in good agreement in all cases. As mentioned above, the maximum frequency for the radar measurements is variable and depends on geometrical factors such as Bragg direction relative to wind direction. At the frequency (12.355 MHz) of the examples presented her, the monostatic cases have maximum frequencies in the range of 0.22–0.283 Hz and the bistatic in the range of 0.187–0.277 Hz.

6. Conclusions

In this paper, the theory for the interpretation of bistatic radar Doppler spectra in terms of currents, winds and waves has been reviewed and methods to simulate and then invert such data have been developed. As far as the authors are aware this is the first time that ocean wave directional spectra, to a maximum frequency that depends on the geometrical parameters and without any prior assumptions about the shape of those spectra, have been obtained from bistatic, albeit only simulated, data. The next step will be to apply the method to measured radar data.

Current, wind and wave measurements from bistatic radar data have been obtained with reasonable accuracy. The statistics for the bistatic cases are not quite as good as the monostatic cases, although they are biased by the large bistatic angle cases. The exact limits on bistatic angle and angle between Braggs still need to be determined but are expected to be about 60° and $< \approx 30°$ respectively.

The results in this paper compare a monostatic configuration with a combined monostatic and bistatic configuration with one transmitter at one of the receive sites. Work on a configuration involving two bistatic radars with one transmitter located between the two sites is in progress. This will help to determine suitable configurations for bistatic radar installations for oceanographic measurements.

Author Contributions: Conceptualization, R.L.H., L.R.W.; methodology, R.L.H., L.R.W.; software, C.C.E., L.R.W., R.L.H.; validation, R.L.H., L.R.W.; formal analysis, R.L.H., L.R.W.; investigation, R.L.H., L.R.W.; resources, R.L.H., L.R.W.; data curation, R.L.H., L.R.W.; writing–original draft preparation, R.L.H., L.R.W.; writing–review and editing, L.R.W.; visualization, R.L.H., L.R.W.; supervision, L.R.W.; project administration, L.R.W.; funding acquisition, R.L.H., L.R.W. All authors have read and agreed to the published version of the manuscript.

Funding: This research was partly funded by the UK Engineering and Physical Sciences Research Council (grant number: X/008627-14). The contributions of L.R.W. and C.C.E. were funded by The University of Sheffield and the APC by the UKRI block grant administered by UoS.

Conflicts of Interest: R.L.H. and C.C.E. declare no conflict of interest. L.R.W. is the Technical Director of Seaview Sensing Ltd but her role in this paper is as a University of Sheffield Professor and, as such, makes only scientific inputs and judgements on the work. The funders had no role in the design of the study; in the collection, analyses, or interpretation of data; in the writing of the manuscript, or in the decision to publish the results.

References

1. Crombie, D.D. Doppler spectrum of sea echo at 13.56 Mc./s. *Nature* **1955**, *175*, 681–682. [CrossRef]
2. Wyatt, L.R. Wave and Tidal Power measurement using HF radar. In Proceedings of the Oceans 2007-Europe, Aberdeen, UK, 18–21 June 2007; pp. 1–5.
3. Paduan, J.D.; Washburn, L. High-frequency radar observations of ocean surface currents. *Annu. Rev. Mar. Sci.* **2013**, *5*, 115–136. [CrossRef] [PubMed]
4. Wyatt, L.R.; Ledgard, L.; Anderson, C. Maximum-likelihood estimation of the directional distribution of 0.53-Hz ocean waves. *J. Atmos. Ocean. Technol.* **1997**, *14*, 591–603. [CrossRef]
5. Wyatt, L.R. Limits to the inversion of HF radar backscatter for ocean wave measurement. *J. Atmos. Ocean. Technol.* **2000**, *17*, 1651–1666. [CrossRef]
6. Lipa, B.J. Derivation of directional ocean-wave spectra by integral inversion of second-order radar echoes. *Radio Sci.* **1977**, *12*, 425–434. [CrossRef]
7. Lipa, B.J. Inversion of second-order radar echoes from the sea. *J. Geophys. Res. Ocean.* **1978**, *83*, 959–962. [CrossRef]
8. Howell, R.; Walsh, J. Measurement of ocean wave spectra using narrow-beam HE radar. *IEEE J. Ocean. Eng.* **1993**, *18*, 296–305. [CrossRef]
9. Hisaki, Y. Nonlinear inversion of the integral equation to estimate ocean wave spectra from HF radar. *Radio Sci.* **1996**, *31*, 25–39. [CrossRef]
10. Wyatt, L.R. A relaxation method for integral inversion applied to HF radar measurement of the ocean wave directional spectrum. *Int. J. Remote Sens.* **1990**, *11*, 1481–1494. [CrossRef]
11. Green, J.J.; Wyatt, L.R. Row-action inversion of the Barrick–Weber equations. *J. Atmos. Ocean. Technol.* **2006**, *23*, 501–510. [CrossRef]
12. Barrick, D.E. First-order theory and analysis of MF/HF/VHF scatter from the sea. *IEEE Trans. Antennas Propag.* **1972**, *20*, 2–10. [CrossRef]
13. Barrick, D.E. Remote Sensing of Sea State by Radar. In *Remote Sensing of the Troposphere*; Derr, V.E., Ed.; GPO: Washington, DC, USA, 1972; Chapter 12.
14. Rice, S.O. Reflection of electromagnetic waves from slightly rough surfaces. *Commun. Pure Appl. Math.* **1951**, *4*, 351–378. [CrossRef]
15. Lipa, B.J.; Barrick, D.E. Extraction of sea state from HF radar sea echo: Mathematical theory and modeling. *Radio Sci.* **1986**, *21*, 81–100. [CrossRef]
16. Holden, G.J.; Wyatt, L.R. Extraction of sea state in shallow water using HF radar. In *IEE Proceedings F (Radar and Signal Processing)*; IET: London, UK, 1992; Volume 139, pp. 175–181.
17. Whelan, C.; Hubbard, M. Benefits of multi-static on HF Radar networks. In Proceedings of the OCEANS'15 MTS/IEEE Washington, Washington, DC, USA, 19–22 October 2015; pp. 1–5.
18. Zhang, J.; Gill, E.W. Extraction of ocean wave spectra from simulated noisy bistatic high-frequency radar data. *IEEE J. Ocean. Eng.* **2006**, *31*, 779–796. [CrossRef]

19. Silva, M.T. A Nonlinear Approach to Ocean Wave Spectrum Extraction From Bistatic HF-Radar Data. Master's Thesis, Faculty of Engineering and Applied Science, Memorial University of Newfoundland, St. John's, NL, Canada, 2017.

20. Anderson, S. Directional wave spectrum measurement with multistatic HF surface wave radar. In Proceedings of the IEEE 2000 International Geoscience and Remote Sensing Symposium, Honolulu, HI, USA, 24–28 July 2000; Volume 7, pp. 2946–2948.

21. Johnstone, D.L. *Second-Order Electromagnetic and Hydrodynamic Effects in High-Frequency Radio-Wave Scattering from the Sea*; Technical Report, DTIC Document; Stanford University: Stanford, CA, USA, 1975.

22. Gill, E.; Walsh, J. High-frequency bistatic cross sections of the ocean surface. *Radio Sci.* **2001**, *36*, 1459–1475. [CrossRef]

23. Chen, Z.; Li, J.; Zhao, C.; Ding, F.; Chen, X. The Scattering Coefficient for Shore-to-Air Bistatic High Frequency (HF) Radar Configurations as Applied to Ocean Observations. *Remote Sens.* **2019**, *11*, 2978. [CrossRef]

24. Hisaki, Y.; Tokuda, M. VHF and HF sea echo Doppler spectrum for a finite illuminated area. *Radio Sci.* **2001**, *36*, 425–440.

25. Hardman, R.L. Remote Sensing of Ocean Winds and Waves With Bistatic HF Radar. Ph.D. Thesis, School of Mathematics and Statistics, University of Sheffield, Sheffield, UK, 2019.

26. Stratton, J.A. *Electromagnetic Theory*; John Wiley & Sons: Hoboken, NJ, USA, 2007.

27. Barrick, D.E.; Lipa, B.J. The second-order shallow-water hydrodynamic coupling coefficient in interpretation of HF radar sea echo. *IEEE J. Ocean. Eng.* **1986**, *11*, 310–315. [CrossRef]

28. Thomas, J.B. *An Introduction to Statistical Communication Theory*; Wiley: Hoboken, NJ, USA, 1969.

29. Barrick, D.E. Theory of HF and VHF propagation across the rough sea, 1, The effective surface impedance for a slightly rough highly conducting medium at grazing incidence. *Radio Sci.* **1971**, *6*, 517–526. [CrossRef]

30. Hashimoto, N.; Tokuda, M. A Bayesian approach for estimating directional spectra with HF radar. *Coast. Eng. J.* **1999**, *41*, 137–149. [CrossRef]

31. Pierson, W.J.; Moskowitz, L. A proposed spectral form for fully developed wind seas based on the similarity theory of SA Kitaigorodskii. *J. Geophys. Res.* **1964**, *69*, 5181–5190. [CrossRef]

32. Wyatt, L.R. An evaluation of wave parameters measured using a single HF radar system. *Can. J. Remote Sens.* **2002**, *28*, 205–218. [CrossRef]

33. Donelan, M.A.; Hamilton, J.; Hui, W. Directional spectra of wind-generated ocean waves. *Phil. Trans. R. Soc. Lond. A* **1985**, *315*, 509–562. [CrossRef]

34. Wyatt, L.R.; Green, J.J.; Middleditch, A. HF radar data quality requirements for wave measurement. *Coast. Eng.* **2011**, *58*, 327–336. [CrossRef]

35. Lopez, G.; Conley, D.C. Comparison of HF Radar Fields of Directional Wave Spectra against in Situ Measurements at Multiple Locations. *J. Mar. Sci. Eng.* **2019**, *7*, 217. [CrossRef]

36. Lopez, G.; Conley, D.C.; Greaves, D. Calibration, validation and analysis of an empirical algorithm for the retrieval of wave spectru from HF radar sea echo. *J. Atmos. Ocean. Technol.* **2016**, *33*, 245–261. [CrossRef]

37. Hardman, R.; Wyatt, L. Inversion of HF radar Doppler spectrum using a neural network. *J. Mar. Sci. Eng.* **2019**, *7*, 255. [CrossRef]

38. Wyatt, L.R.; Holden, G.J. Limits in direction and frequency resolution for HF radar ocean wave directional spectra measurement. *Glob. Atmos. Ocean. Syst.* **1994**, *2*, 265–290.

39. Wyatt, L.R. Measuring the ocean wave directional spectrum 'First Five' with HF radar. *Ocean. Dyn.* **2019**, *69*, 123–144. [CrossRef]

Article

Coast–Ship Bistatic HF Surface Wave Radar: Simulation Analysis and Experimental Verification

Yonggang Ji [1,*], Jie Zhang [1], Yiming Wang [1], Chao Yue [1], Weichun Gong [1], Junwei Liu [1], Hao Sun [1], Changjun Yu [2] and Ming Li [3]

[1] Laboratory of Marine Physics and Remote Sensing, First Institute of Oceanography, Ministry of Natural Resources, No. 6 Xianxialing Road, Qingdao 266061, China; zhangjie@fio.org.cn (J.Z.); wangyiming@fio.org.cn (Y.W.); yuechao@fio.org.cn (C.Y.); gongweichun@fio.org.cn (W.G.); liujunwei@fio.org.cn (J.L.); sunhao@fio.org.cn (H.S.)

[2] School of Information and Electrical Engineering, Harbin Institute of Technology at Weihai, Weihai 264209, China; yuchangjun@hit.edu.cn

[3] College of Engineering, Ocean University of China, Qingdao, Shandong 266100, China; limingneu@ouc.edu.cn

* Correspondence: jiyonggang@fio.org.cn; Tel.: +86-532-88967114

Received: 29 November 2019; Accepted: 31 January 2020; Published: 2 February 2020

Abstract: The coast–ship bistatic high-frequency surface wave radar (HFSWR) not only has the anti-interference advantages of the coast-based bistatic HFSWR, but also has the advantages of maneuverability and an extended detection area of the shipborne HFSWR. In this paper, theoretical formulas were derived for the coast–ship bistatic radar, including the first-order sea clutter scattering cross-section and the Doppler frequency shift of moving targets. Then, simulation results of the first-order sea clutter spectrum under different operating conditions were given, and the range of broadening of the first-order sea clutter spectrum and its influence on target detection were investigated. The simulation results show the broadening ranges of the right sea clutter spectrum and left sea clutter spectrum were symmetric when the shipborne platform was anchored, whereas they were asymmetric when the shipborne platform was underway. This asymmetry is primarily a function of platform velocity and radar frequency. Based on experimental data of the coast–ship bistatic HFSWR conducted in 2019, the broadening range of the sea clutter and the target frequency shift were analyzed and compared with simulation results based on the same parameter configuration. The agreement of the measured results with the simulation results verifies the theoretical formulas.

Keywords: bistatic HFSWR; shipborne HFSWR; first-order sea clutter; experiment verification

---

1. Introduction

The high-frequency surface wave radar (HFSWR), also known as the HF surface over-the-horizon radar, operates in the 3–30 MHz frequency band at wavelengths between 100 m and 10 m, respectively. The HFSWR can provide additional information on maritime traffic because it can detect targets over the horizon, has continuous temporal coverage, and can estimate vessel velocity based on Doppler data [1,2]. Currently, most HFSWR systems operate in a monostatic mode that requires the collocation of the transmitter and the receiver. This raises practical issues in terms of the coastal space required for installation of both the transmitting and the receiving antenna arrays, as well as the problem of mutual interference between antennas. These issues can be overcome by resorting to a bistatic radar system in which the transmitter and the receiver are located some distance apart [3–5]. In a bistatic HFSWR system, the receiver has robust anti-active directional jamming and anti-destruction characteristics because of the physical separation of the transmitter and the receiver, giving it unique advantages and potential regarding anti-electronic interference. A shipborne HFSWR system has the advantage of flexibility and it can increase the radar detection range beyond that of an onshore HFSWR. A system in which one of the transmitting or receiving stations is installed on the coast and the other is placed on a ship forms a coast–ship bistatic HFSWR system. Such systems can be classified either as a coast-transmit ship-receive (CTSR) bistatic HFSWR or as a ship-transmit coast-receive (STCR) bistatic HFSWR depending on whether the receiving station is placed on the ship or the coast, respectively.

A CTSR bistatic HFSWR has the advantages of anti-stealth, anti–interference, and no onboard electromagnetic radiation because the ship carrying the receiver can move to areas far from the coast. Compared with STCR systems, CTSR systems can exploit fully the flexibility of a shipborne platform and further expand the radar detection range by adjusting the attitude of the shipborne platform and changing the radar system configuration, e.g., adjusting the radar spindle angle. However, the azimuth resolution of such systems is reduced because shipboard platforms are limited by the size of the platform and thus the radar receiving station aperture is typically limited to ≤100 m. In an STCR bistatic HFSWR system, the shipborne equipment comprises only the transmitter, and there is no need to consider the deployment of a receiver, signal processor, or other equipment. In addition, the antenna aperture of the radar system is not limited by the size of the ship, which means that a large aperture-receiving antenna array could be installed onshore to improve azimuthal resolution. However, the fixed nature of the coast-based receiving station means that the receiving array spindle angle cannot be changed, which limits the detection range of the radar system to a certain extent. In addition, a transmitter/receiver–receiver radar system could be formed by adding a second coast-based/shipborne transmitter–receiver monostatic radar on the same basis as the coast–ship bistatic radar. Then, the detection performance and positioning accuracy of the marine target could be improved through fusion of the results of the two systems.

In their research into onshore and shipborne bistatic HFSWR systems, Gill and Walsh derived analysis relevant to onshore bistatic HFSWRs [6,7] based on the equation for the first-order radar echo spectrum derived by Barrick and by Lipa and Barrick for an onshore multistatic HFSWR [8,9]. Based on the space–time distribution of the first-order radar echo spectrum of an onshore bistatic HFSWR system, Xie et al. proposed and verified a spreading model through simulation and experiment [10]. In research of shipborne HFSWRs, both Walsh et al. and Khoury and Guinvarch derived the first-order ocean surface cross section of a monostatic HFSWR system on a moving platform [11–13]. In addition, Xie et al. studied the first-order ocean surface cross-section for a shipborne HFSWR both theoretically and through experiment [14,15]. In research of coast–ship bistatic HFSWRs, Li et al., Chen et al., and Liu et al. all presented the first-order sea clutter based on system configuration and analyzed the mechanism of broadening [16–19]. It is worth noting that the HFSWR system used in their research employed multichannel transmitting arrays mounted on the coast and only one receiving antenna mounted on a ship, whereas the HFSWR system of consideration in the present study employed one transmitting antenna and multichannel receiving arrays. Zhu et al. analyzed the influence of bistatic angle and shipborne platform motion on the characteristics of the broadened sea clutter spectrum [20]. Most previous research on coast–ship bistatic HFSWR systems focused on simulation analysis, with few reports on experimental verification.

The main characteristic of the coast–ship bistatic HFSWR is that it combines the advantages (and disadvantages) of monostatic coast-based and shipborne HFSWR systems. The radar spectrum of the coast–ship bistatic radar, including the first-order sea clutter and moving target echoes, will change with variation of the motion of the shipborne platform and the existence of the bistatic angle. A frequency shift of the first-order sea clutter spectrum will lead to expansion of the first-order sea clutter, and the blind area, due to the broadening of the first-order sea clutter spectrum, could cause difficulty in detecting targets within it, significantly reducing the overall detection performance. Therefore, it is necessary to investigate the mechanism of broadening of the first-order sea clutter spectrum and its influence on target detection. A comprehensive understanding of the bistatic spreading mechanism of the first-order sea clutter spectrum is essential for study of bistatic clutter suppression in target monitoring.

The remainder of this paper is organized as follows. In Section 2, the theoretical formulas of the first-order sea clutter scattering cross-section and the Doppler frequency shift of moving ship targets for the two types of coast–ship bistatic radar system are derived. In Section 3, simulation results of the first-order sea clutter spectrum under different operating conditions are presented and the influence on target detection is analyzed. In Section 4, the coast–ship bistatic HFSWR experiment is introduced and the derived results interpreted. Finally, brief conclusions are outlined in Section 5.

2. Spreading Mechanism of the First-Order Sea Clutter Spectrum and Moving Targets

The geometry of the propagation path of a coast–ship bistatic HFSWR system is shown in Figure 1.

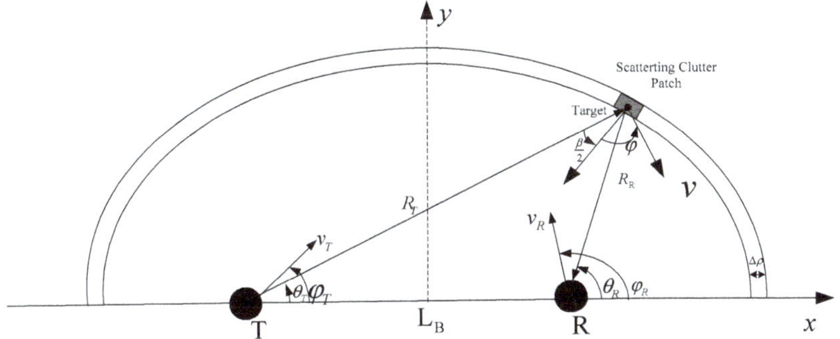

Figure 1. The two-dimensional coordinate system confined to a bistatic plane formed by transmitter site **T** and receiver site **R**. L$_B$ is the distance of the baseline range between **T** and **R** that is coincident with the x-axis. All angles are described anticlockwise relative to the x-axis. To consider the two types of coast–ship bistatic high-frequency surface wave radar (HFSWR) systems simultaneously, both the transmitting station and the receiving station were set as shipborne platforms. The transmitter and receiver have projected velocity vectors of magnitude v_T and v_R and aspect angles φ_T and φ_R with respect to the x-axis, respectively. For a given moving target, its velocity vector projected onto the bistatic plane has magnitude v and aspect angle φ referenced to the bistatic bisector. The distance from the target to **T** and **R** is R_T and R_R, respectively, and θ_T and θ_R describe the angles of R_T and R_R with respect to the x-axis, respectively. The bistatic angle is β, and $\Delta\rho$ is the range resolution.

It is known that the Doppler frequency f_d between a scattered signal and the direct signal of a bistatic radar system is proportional to the temporal rate of change of the total path length of the scattered signal [21]. Thus, the Doppler frequency of a moving target at the moving transmitting station and moving receiving station of a bistatic HFSWR system can be calculated as follows:

$$f_d = \frac{1}{\lambda}\frac{d\Delta R}{d\Delta t} \tag{1}$$

where ΔR is the total range of the target relative to both the transmitting station and the receiving station. For a bistatic HFSWR, $\Delta R = \Delta R_T + \Delta R_R$, where

$$\Delta R_T = v\cos(\varphi + \frac{\beta}{2})\Delta t + v_T\cos(\varphi_T - \theta_T)\Delta t \tag{2}$$

and

$$\Delta R_R = v\cos(\varphi - \frac{\beta}{2})\Delta t + v_R\cos(\varphi_R - \theta_R)\Delta t. \tag{3}$$

Then,

$$
\begin{aligned}
f_d &= \tfrac{1}{\lambda}v\left[\cos\left(\varphi+\tfrac{\beta}{2}\right)+\cos\left(\varphi-\tfrac{\beta}{2}\right)+\tfrac{1}{\lambda}v_T\cos(\varphi_T-\theta_T+v_R\cos(\varphi_R-\theta_R)\right] \\
&= \tfrac{2v}{\lambda}\cos(\varphi)\cos\left(\tfrac{\beta}{2}\right)+\left[\tfrac{v_R}{\lambda}\cos(\varphi_R-\theta_R)+\tfrac{v_T}{\lambda}\cos(\varphi_T-\theta_T)\right]
\end{aligned}
\tag{4}
$$

It can be seen from Equation (4) that the Doppler frequency of a moving target is the sum of the target motion, transmitter motion, and receiver motion [3], i.e., $f_d = f_{d1} + f_{d2}$, where $f_{d1} = \frac{2v}{\lambda}\cos(\varphi)\cos\left(\frac{\beta}{2}\right)$ is the Doppler shift caused by the motion of the target itself, and $f_{d2} = \frac{v_R}{\lambda}\cos(\varphi_R-\theta_R)+\frac{v_T}{\lambda}\cos(\varphi_T-\theta_T)$ is the Doppler frequency caused by the motion of the shipborne mobile platform, which includes that of the transmitting station and the receiving station.

When only the receiving station is placed on a moving shipborne platform, the Doppler frequency of a moving target in a CTSR bistatic HFSWR can be calculated as follows:

$$
f_{dCTSR} = \frac{2v}{\lambda}\cos(\varphi)\cos\left(\frac{\beta}{2}\right)+\frac{v_R}{\lambda}\cos(\varphi_R-\theta_R)
\tag{5}
$$

Similarly, the Doppler frequency of a moving target in an STCR bistatic HFSWR can be calculated as follows:

$$
f_{dSTCR} = \frac{2v}{\lambda}\cos(\varphi)\cos\left(\frac{\beta}{2}\right)+\frac{v_T}{\lambda}\cos(\varphi_T-\theta_T)
\tag{6}
$$

For the first-order sea clutter spectrum of an HFSWR, Gill [6] derived the first-order sea surface scattering cross section of a coast-based bistatic HFSWR:

$$
\sigma_1(\omega_d) = 2^4\pi k_0^2 \sum_{m=\pm1} S\left(m\vec{K}\right)\frac{K^{5/2}\cos\left(\frac{\beta}{2}\right)}{\sqrt{g}}\Delta\rho Sa^2\left[\frac{\Delta\rho}{2}\left(\frac{K}{\cos\left(\frac{\beta}{2}\right)}-2k_0\right)\right]
\tag{7}
$$

where $\vec{K}$ is the first-order sea wave vector with magnitude K ($\vec{K} = 2\vec{k_0}$, where $k_0 = \frac{2\pi}{\lambda}$ is a wave number), g is the acceleration due to gravity, ω_d is the Doppler radian frequency, $\omega_d^2 = \left(2\pi\sqrt{\frac{g}{\pi\lambda}}\right)^2 = gK$, $\delta(\cdot)$ is the Dirac delta function, $m = \pm1$ means the Doppler shift of the wave, λ is the radar wavelength, and $S(\cdot)$ is the ocean spectrum given as a function of the wave vector. As the $Sa^2[Mx]$ function satisfy: [6] $\lim\limits_{M\to\infty} Sa^2[Mx] = \pi\delta(x)$, then

$$
\lim_{\frac{\Delta\rho}{2\cos\left(\frac{\beta}{2}\right)}\to\infty} Sa^2\left[\frac{\Delta\rho}{2\cos\left(\frac{\beta}{2}\right)}\left(K-2k_0\cos\left(\frac{\beta}{2}\right)\right)\right]= \frac{2\pi\cos\left(\frac{\beta}{2}\right)}{\Delta\rho}\delta[K-2k_0\cos\left(\frac{\beta}{2}\right)]
\tag{8}
$$

Thus Equation (7) can be simplified further into the following form:

$$
\sigma_1(\omega_d) = 2^5\pi^2 k_0^2 \sum_{m=\pm1} S\left(m\vec{K}\right)\frac{K^{5/2}\cos^2\left(\frac{\beta}{2}\right)}{\sqrt{g}}\delta\left(\omega_d+m\sqrt{2gk_0\cos\left(\frac{\beta}{2}\right)}\right).
\tag{9}
$$

Then, the first-order Bragg frequencies can be derived as follows according to the property of the Dirac delta function $\delta(\cdot)$:

$$
f_B = \pm\sqrt{2gk_0\cos\left(\frac{\beta}{2}\right)}.
\tag{10}
$$

When the motion of the shipborne mobile platform is taken into account, the added Doppler shift caused by platform motion on the first-order Bragg frequency should be the same as that on a moving target. Then, the first-order Bragg frequencies are given as

$$f_B = \pm \sqrt{2gk_0\cos\left(\frac{\beta}{2}\right)} + \left[\frac{v_R}{\lambda}\cos(\varphi_R - \theta_R) + \frac{v_T}{\lambda}\cos(\varphi_T - \theta_T)\right]. \tag{11}$$

Thus, the first-order sea surface scattering cross section of CTSR and STCR bistatic HFSWR systems can be expressed as

$$\sigma_{1CTSR}(\omega_d) = A.\delta\left[\omega_d + m\sqrt{2gk_0\cos(\frac{\beta}{2})} + \frac{v_R}{\lambda}\cos(\varphi_R - \theta_R)\right] \tag{12}$$

and

$$\sigma_{1STCR}(\omega_d) = A.\delta\left[\omega_d + m\sqrt{2gk_0\cos(\frac{\beta}{2})} + \frac{v_T}{\lambda}\cos(\varphi_T - \theta_T)\right] \tag{13}$$

where $A = 2^5\pi^2k_0^2\sum_{m=\pm1}S\left(m\vec{K}\right)\frac{K^{5/2}\cos^2(\frac{\beta}{2})}{\sqrt{g}}$. Accordingly, their first-order Bragg frequencies are given as follows:

$$f_{BCTSR} = \pm \sqrt{2gk_0\cos\left(\frac{\beta}{2}\right)} + \frac{v_R}{\lambda}\cos(\varphi_R - \theta_R) \tag{14}$$

$$f_{BSTCR} = \pm \sqrt{2gk_0\cos\left(\frac{\beta}{2}\right)} + \frac{v_T}{\lambda}\cos(\varphi_T - \theta_T) \tag{15}$$

It should be noted that the motion of the shipborne platform mentioned here refers primarily to forward navigation movement. When the sway of the ship platform caused by sea waves and wind is considered [7,12], the periodic motion of the shipborne platform can affect the first-order sea clutter spectrum and the target echoes. In addition, surface currents can lead to Doppler shift of the sea clutter spectrum. However, it is inappropriate to use only a single velocity of surface current for different sea areas with different directions of arrival. Therefore, the effects of surface currents were not considered in the model proposed.

3. Simulation Analysis of the First-Order Sea Clutter Spectrum for a Coast–Ship Bistatic HFSWR

As can be seen from the above, the Doppler shift of both the first-order sea clutter spectrum and the moving target echo were affected by the radar frequency, bistatic angle, and velocity of the shipborne platform. To quantitatively analyze the characteristics of the broadening of the sea clutter spectrum and the related blind area, simulation analysis of the first-order sea clutter spectrum for a coast–ship bistatic HFSWR was studied under conditions of different frequencies, different platform velocities, and different elliptical rates. In the following simulation, it was assumed that heading of the shipborne platform was consistent with the navigation direction.

It must be highlighted that the focus of this paper was on the application of bistatic radar target monitoring. For this reason, the Doppler shift of the first-order sea clutter spectrum and the related influence on the blind area are expressed with base units of velocity magnitude (i.e., m) instead of Hz. In addition, as CTSR and STCR are equivalent bistatic configurations when similar antennas and array configurations are used, the simulation results and influence analysis of only the CTSR bistatic radar system are presented here. Simulation results for the STCR bistatic radar system could be obtained by replacing the velocity of the receiving station platform with that of the transmitting station platform.

3.1. Simulation 1: Simulated Space–Time Distribution for Different Headings of the Shipborne Platform

Based on Equation (14), the distribution of the broadening of the first-order sea clutter for a coast–ship bistatic HFSWR when the shipborne platform was anchored is shown in Figure 2. Here, we set the elliptical eccentricity as e = 0.7, frequency as 4.7 MHz, navigation speed v_R = 0 km/h, and wind direction as 90°.

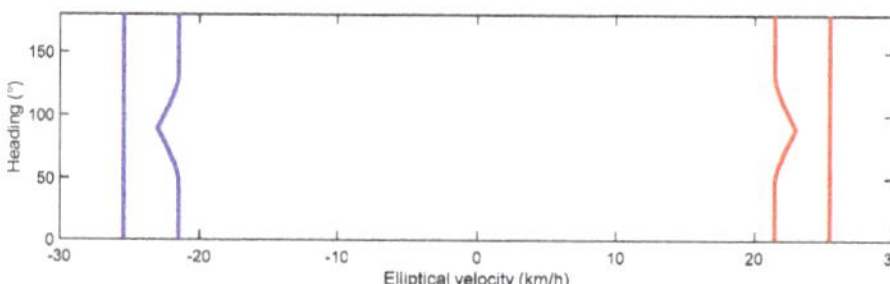

Figure 2. Distribution of the broadening of the first-order sea clutter of a coast–ship bistatic HFSWR when the shipborne platform was anchored.

It can be seen from Figure 2 that the broadening ranges of the right sea clutter spectrum and left sea clutter spectrum were symmetrical for a coast–ship bistatic HFSWR when the shipborne platform was anchored, i.e., the width of the right sea clutter spectrum was equal to that of the left sea clutter spectrum when the heading was given. Here, the right first-order spectrum was selected as an example. The right bound retained the value of $f_{RR} = \sqrt{\frac{g}{\pi\lambda}}$, while the value of the left bound f_{RL} was

$$f_{RL} = \begin{cases} \sqrt{2gk_0cos\left(\frac{\beta_{max}}{2}\right)}, 0° < \varphi_R < \theta_{R1} \\ \sqrt{2gk_0cos\left(\frac{\beta}{2}\right)}, \theta_{R1} < \varphi_R < 90° \end{cases}, \tag{16}$$

where β_{max} is the largest bistatic angle for a given elliptical eccentricity e, i.e., $\beta_{max} = arcsin(e)$, and θ_{R1} is its corresponding direction of arrival. Here, $\beta = 2arctan\left(\frac{sin\varphi_R}{\frac{1}{e}+cos\varphi_R}\right)$.

The space–time distribution of the Doppler frequency shift of the first-order sea clutter and the simulation results of the first-order sea clutter spectrum for a CTSR bistatic HFSWR at three different headings when the shipborne platform was anchored are shown in Figure 3. It can be seen that the simulated space–time distribution of the frequency shift of the sea clutter varied with the direction of arrival. The space–time distribution of the Doppler shift of the first-order sea clutter of a coast–ship bistatic HFSWR system is presented as two nonlinear curves with the cosmic value of the incoming direction, and the two curves are symmetrical along the y-axis.

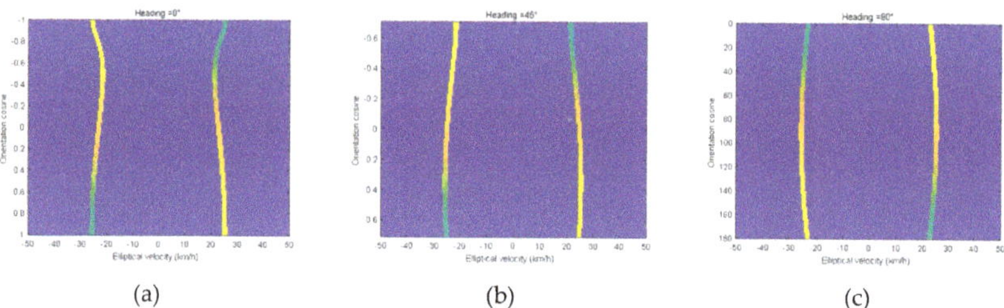

(a)   (b)   (c)

Figure 3. Space–time distribution of the Doppler frequency shift of the first-order sea clutter for a coast-transmit ship-receive (CTSR) bistatic HFSWR at different headings when the shipborne platform was anchored: (**a**) 0°, (**b**) 45°, and (**c**) 90°.

Based on the two-dimensional space–time distribution plots presented in Figure 3, simulation results of the first-order sea clutter spectrum of a coast–ship bistatic HFSWR with different headings on a single channel can be obtained, as shown in Figure 4.

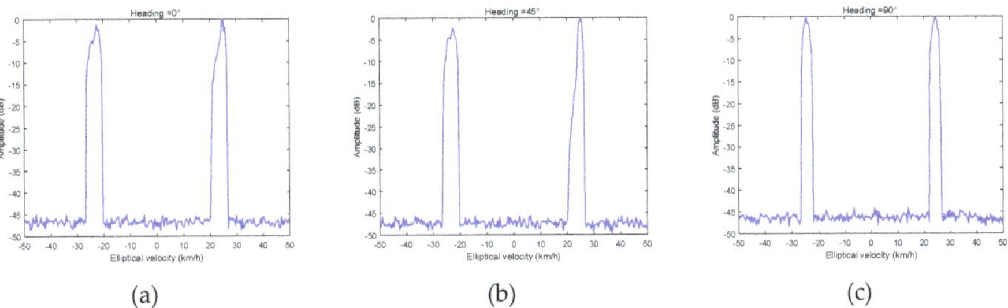

Figure 4. Simulation results of the first-order sea clutter spectrum for a CTSR bistatic HFSWR at different headings: (**a**) 0°, (**b**) 45°, and (**c**) 90°.

It can be seen from Figure 4 that the broadening ranges of the right first-order sea clutter spectrum and left first-order sea clutter spectrum were symmetrical. As $\varphi_R = 45°$, $\theta_{R1} = 46°$, and $\varphi_R < \theta_{R1}$, the widths of both the right sea clutter spectrum and the left sea clutter spectrum at the headings of 0° and 45° were equal. The minimum value of the width of the sea clutter spectrum was obtained when $\varphi_R = 90°$. In addition, owing to the change of heading, the angle of the wind direction relative to the principal axis of the receiving array was equivalent to that change, which induced the amplitude variation of the sea clutter spectrum.

The space–time distribution of the Doppler frequency shift of the first-order sea clutter and the simulation results of the first-order sea clutter spectrum for a CTSR bistatic HFSWR at three different headings when the shipborne platform was navigating with velocity of 11.5 km/h are shown in Figures 5 and 6, respectively. In those cases, the broadening ranges of the right first-order sea clutter spectrum and left first-order sea clutter spectrum were asymmetrical and their widths were not equal.

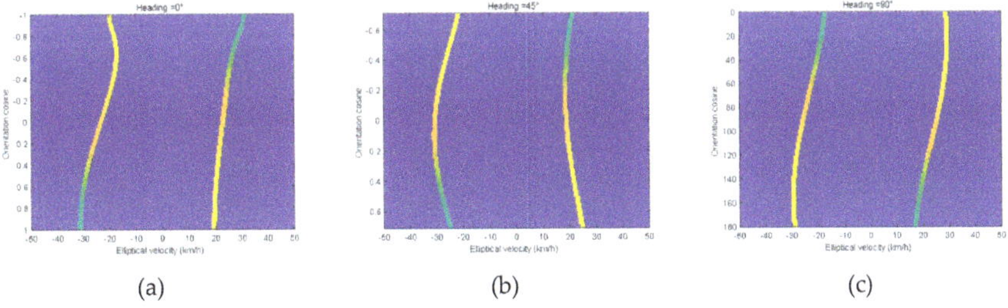

Figure 5. Space–time distribution of the Doppler frequency shift of the first-order sea clutter for a CTSR bistatic HFSWR at different headings when the shipborne platform was navigating: (**a**) 0°, (**b**) 45°, and (**c**) 90°.

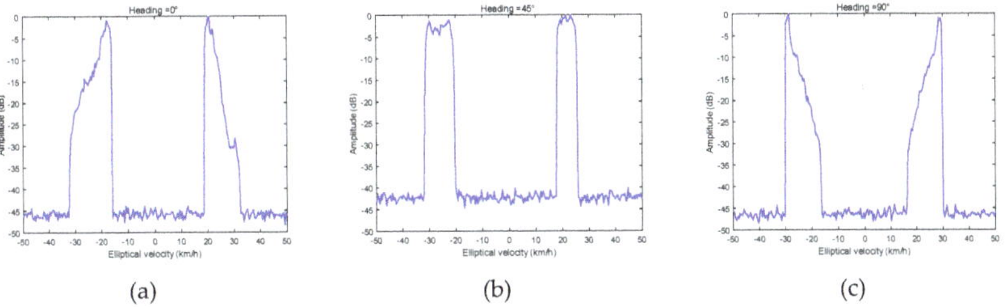

Figure 6. Simulation results of the first-order sea clutter spectrum for a CTSR bistatic HFSWR navigating at different headings: (**a**) 0°, (**b**) 45°, and (**c**) 90°.

3.2. Simulation 2: Simulated Space–Time Distribution for Different Shipborne Platform Velocities

The space–time distribution of the Doppler frequency shift of the first-order sea clutter and the simulation results of the first-order sea clutter spectrum for a CTSR bistatic HFSWR at different velocities are shown in Figures 7 and 8, respectively. Here, we set the elliptical eccentricity as e = 0.22, frequency as 4.7 MHz, heading $\varphi_R = 115°$, and wind direction as 90°. Simulation results of a monostatic shipborne HFSWR with the same operating conditions are presented for comparative analysis in Figure 8.

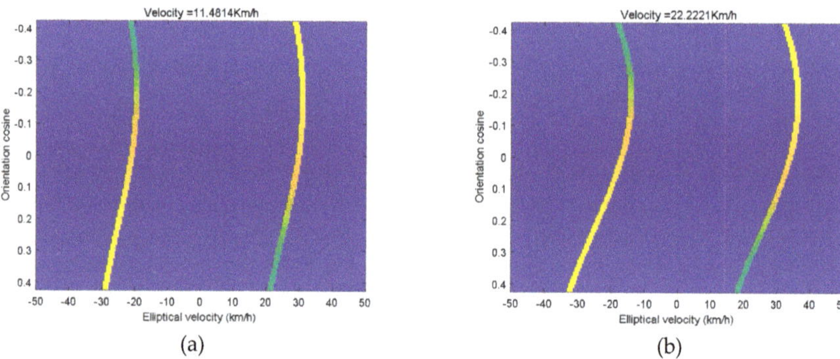

Figure 7. Space–time distribution of the Doppler frequency shift of the first-order sea clutter for a CTSR bistatic HFSWR at different velocities: (**a**) v_R = 11.48 km/h (6.2 knots) and (**b**) v_R = 22.2 km/h (12 knots).

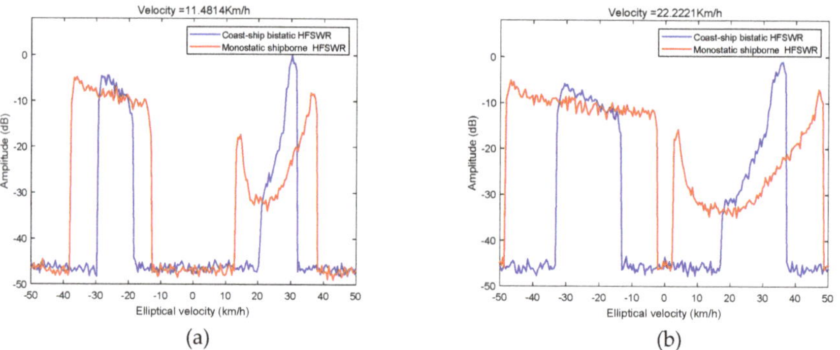

Figure 8. Simulation results of the first-order sea clutter spectrum for a CTSR bistatic HFSWR at different velocities: (**a**) v_R = 11.48 km/h (6.2 knots) and (**b**) v_R = 22.2 km/h (12 knots).

It can be seen from Figure 8a that the range of broadening of the first-order sea clutter spectrum of a coast–ship bistatic HFSWR increased when the platform velocity increased, meaning that the range of the blind area caused by first-order sea clutter was widened, which had a greater effect on moving targets falling within these velocity ranges. This is very disadvantageous to target detection. Conversely, with the increase of platform velocity, the echo amplitude of the first-order sea clutter decreased gradually, which led to a higher signal-to-clutter ratio. This is more conducive to the detection and highlighting of moving targets falling into and becoming submerged in the blind area caused by first-order sea clutter.

As can be seen from Figure 8b, the range of broadening of the first-order sea clutter spectrum for a monostatic shipborne HFSWR was obviously larger than that of the coast–ship bistatic HFSWR under the same platform velocity, and even the left and right sea clutter spectra were almost superimposed when the velocity was greater than 24 km/h. For a CTSR bistatic HFSWR, only the receiving station was on the moving platform, while both the transmitting and the receiving stations were on the moving platform in a monostatic shipborne HFSWR system. Therefore, the Doppler shift of the first-order sea clutter spectrum and the related blind area of a coast–ship CTSR bistatic HFSWR were smaller than those of a monostatic shipborne HFSWR.

3.3. Simulation 3: Simulated Space–Time Distribution for Different Radar Frequencies

The space–time distribution of the Doppler frequency shift of the first-order sea clutter and the simulation results of the first-order sea clutter spectrum for a CTSR bistatic HFSWR at different frequencies are shown in Figures 9 and 10, respectively. Here, we set the elliptical eccentricity as e = 0.134, shipborne platform velocity v_T = 8 km/h (approximately 4.3 knots), heading $\varphi_R = 114°$, and wind direction as 90°.

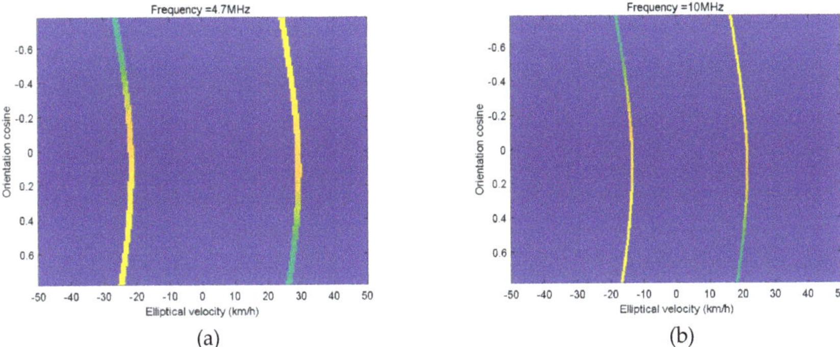

(a) (b)

Figure 9. Space–time distribution of the Doppler frequency shift of the first-order sea clutter for a CTSR bistatic HFSWR at different frequencies: (**a**) f = 4.7 MHz and (**b**) f = 10 MHz.

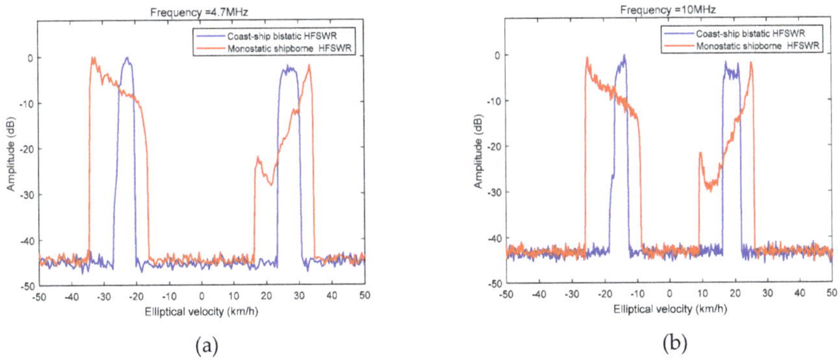

(a) (b)

Figure 10. Simulation results of the first-order sea clutter spectrum for a CTSR bistatic HFSWR at different frequencies: (**a**) f = 4.7 MHz and (**b**) f = 10 MHz.

It can be seen from Figure 10 that the widths of the sea clutter blind area of a monostatic shipborne HFSWR at different frequencies but with a constant platform velocity were similar, while their central positions differed. Conversely, both the range of broadening of the first-order sea clutter spectrum and the width of the sea clutter blind area of a coast–ship bistatic HFSWR changed with radar frequency.

3.4. Simulation 4: Simulated Sea Clutter Spectrum for Different Wind Directions

The simulation results of the first-order sea clutter spectrum at wind directions of 0°, 45°, 90°, 135°, 180°, and 225° are shown in Figure 11a–f. Here, we set the elliptical eccentricity as e = 0.7, frequency as 4.7 MHz, shipborne platform velocity v_R = 8 km/h, and heading φ_R = 0°.

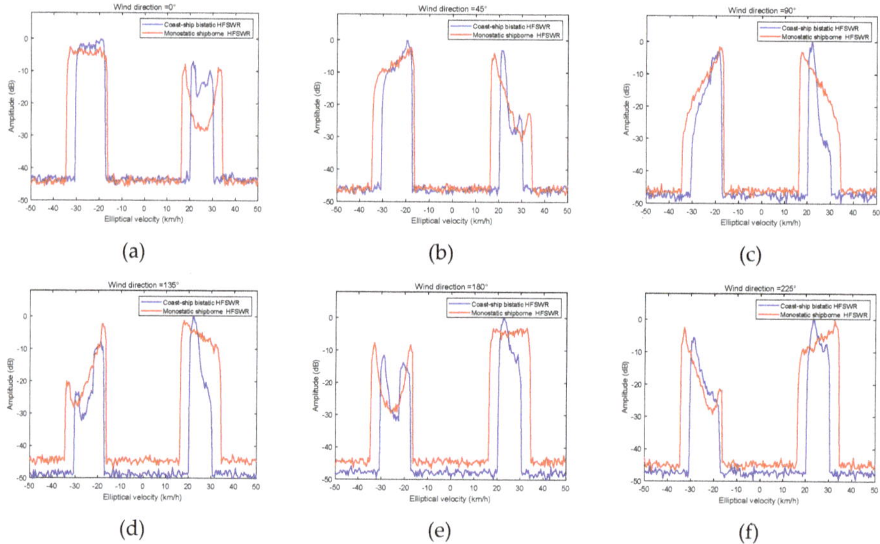

Figure 11. Simulation results of the first-order sea clutter spectrum for a CTSR bistatic HFSWR at different wind directions: (**a**) 0°, (**b**) 45°, (**c**) 90°, (**d**) 135°, (**e**) 180°, and (**f**) 225°.

It can be seen from Figure 11 that the range of broadening of the first-order sea clutter spectrum of a coast–ship bistatic HFSWR remained unchanged under different wind conditions. Moreover, the width of the first-order sea clutter of the coast–ship bistatic HFSWR was always less than that of the monostatic shipborne HFSWR under the same platform velocity and radar configuration. However, the comparative relationship between the amplitude of the left first-order spectrum and right first-order spectrum was changed, and the related blind area had a different influence on target detection.

Under the condition of 180° wind direction, the average amplitude of the right first-order spectrum relative to the underlying noise was 38 dB, while the average amplitude of the left first-order spectrum was 21 dB, i.e., a difference of 17 dB. Thus, for a moving target with amplitude of 30 dB, if its elliptical velocity were within the right first-order spectrum, it would be submerged completely by sea clutter. However, it could be detected easily if its elliptical velocity were within the left first-order spectrum. For the wind condition of 135°, the situation was the reverse of the 45° condition.

3.5. Simulation 5: Simulated Space–Time Distribution for Different Elliptical Eccentricity

The space–time distribution of the Doppler frequency shift of the first-order sea clutter and the simulation results of the first-order sea clutter spectrum for a CTSR bistatic HFSWR at different elliptical eccentricity are shown in Figures 12–15, respectively. Here, we set the frequency as 4.7 MHz, heading φ_R = 48.9°, wind direction as 90°, and velocity v_T = 18.5 km/h (approximately 10 knots) when ship was navigating.

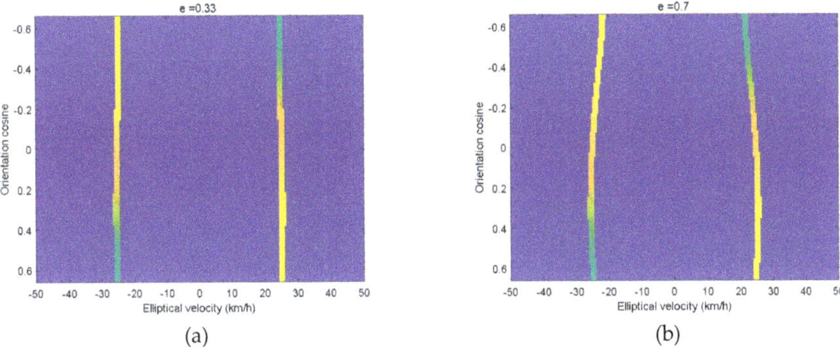

Figure 12. Space–time distribution of the Doppler frequency shift of the first-order sea clutter for a CTSR bistatic HFSWR when the ship was anchored: (**a**) e = 0.33 and (**b**) e = 0.7.

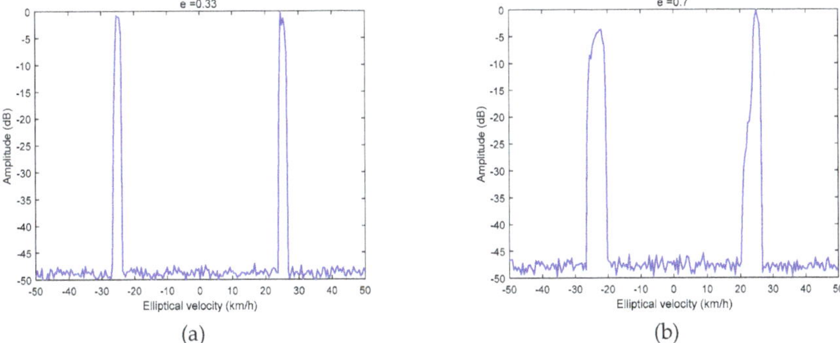

Figure 13. Simulation results of the first-order sea clutter spectrum for a CTSR bistatic HFSWR when the ship was anchored: (**a**) e = 0.33 and (**b**) e = 0.7.

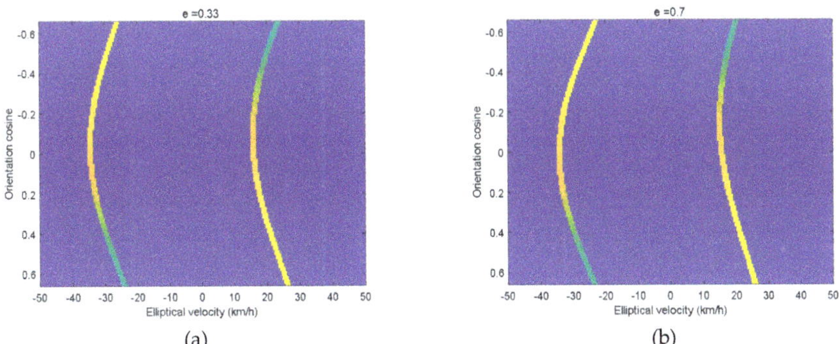

Figure 14. Space–time distribution of the Doppler frequency shift of the first-order sea clutter for a CTSR bistatic HFSWR when the ship was navigating: (**a**) e = 0.33 and (**b**) e = 0.7.

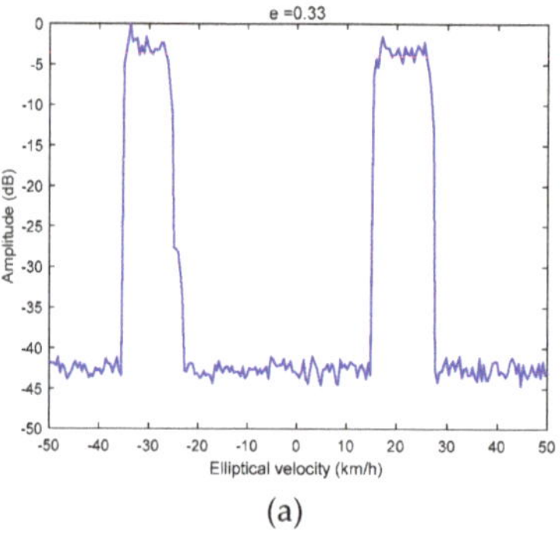
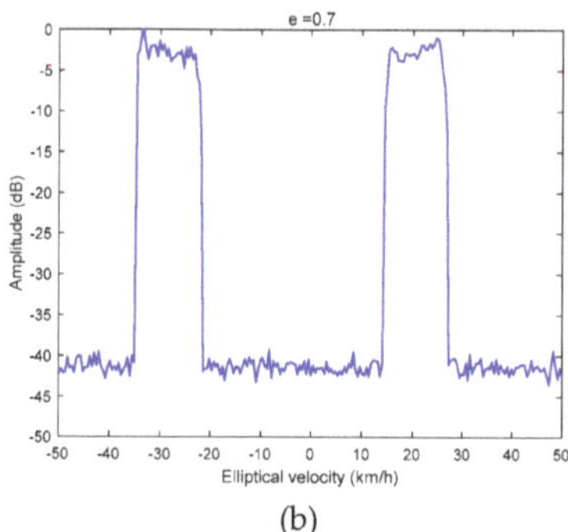

Figure 15. Simulation results of the first-order sea clutter spectrum for a CTSR bistatic HFSWR when the ship was navigating: (**a**) e = 0.33 and (**b**) e = 0.7.

As can be seen from Figures 12–15, the first-order sea clutter spectrum of e = 0.7 was wider than that of e = 0.33 when the heading φ_R = 48.9° in the anchored case, whereas the width of the sea clutter was largely unchanged under different elliptical eccentricity when the ship was navigating with the velocity of 18.5 km/h.

Based on analysis of simulations of the first-order sea clutter spectrum of a coast–ship bistatic HFSWR in navigational state, the range of the detection blind area caused by broadening of the first-order sea clutter spectrum and its influence on target detection can be understood and overcome. The findings could be used to develop appropriate strategies for actual target detection, such as adopting the most suitable navigational speed and heading. For a CTSR bistatic HFSWR system, the principal axis angle of the receiving array and the Doppler frequency shift of the detected target can be changed by adjusting the heading of the platform without changing the platform position. In this way, a moving target signal submerged within each area of sea clutter could be separated from the range of the blind area or adjusted to the side of the sea clutter blind area with less influence. It should be noted that such strategies do not apply to STCR bistatic HFSWR systems.

4. Vessel Target Monitoring Experiment with Coast–Ship Bistatic HFSWR

4.1. Introduction of the Coast–Ship Bistatic HFSWR

A coast–ship HFSWR experiment was conducted in July 2019 using two Compact Over-horizon Radar for Marine Surveillance HFSWR systems (CORMS). One radar was located on the coast near the city of Weihai (Shandong Province, China; Figure 16), and the other was deployed on the M/V *Shun Chang 28*. The experimental area comprised the open-water area off the coast near Weihai. The experiment was divided into two periods. The first period, from 09:18:00 to 10:07:00 local time, is indicated by the green line in Figure 16. During this period, measured experimental data using the CTSR bistatic HFSWR system were obtained. The second period, from 10:50:00 to 14:40:00 local time, is indicated by the red dashed line in Figure 16. During this period, measured experimental data using the STCR bistatic HFSWR system were acquired. The velocity and heading of the shipborne platform during the experiment are shown in Figure 17. Here, the heading is given clockwise relative to north.

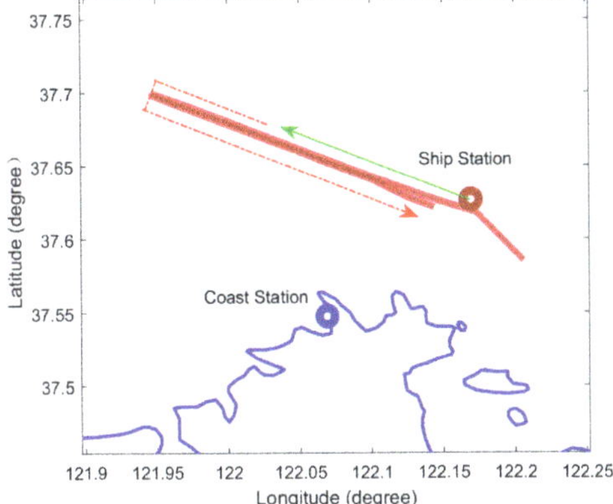

Figure 16. Location of coast-based radar station and navigation route of shipborne platform.

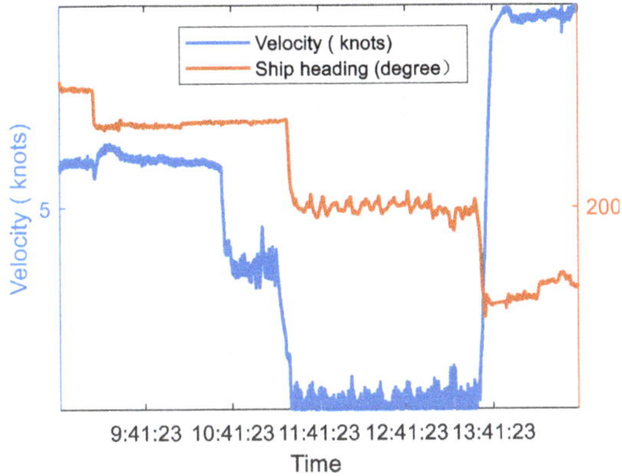

Figure 17. Velocity and heading of shipborne platform.

The CORMS HFSWR system used in the experiment had a solid-state transmitter with maximum peak power of 500 W. The output power of the transmitter could be adjusted continuously. A linear frequency-modulated interrupting continuous wave signal was used, and a double-whip transmitter antenna at the height of 11 m transmitted an omnidirectional pattern. The HF radar receiver was fully digitalized with eight channels, although only five channels were used in the experiment. Each element of the receiving array was a small magnetic cylindrical antenna (length: 0.5 m, diameter: 0.4 m), suitable for shipborne platforms. The radar frequency was 4.7 MHz and the bandwidth was 60 KHz.

In the coast–ship bistatic HFSWR experiment, synchronous transmitter–receiver monostatic HFSWR data were also obtained using a system collocated with the transmitter station. The transmitters and receivers of both radar systems, as well as the two radars, were synchronized using GPS. Both the shore-based receiving array and the shipborne receiving array were composed of five elements but with different element spacing. Owing to the limitation that the length of the shipborne platform was only 88 m, the available array aperture of the shipborne receiving array was only 62 m and its antenna spacing was 15.5 m. Conversely, the antenna spacing of the shore-based receiving array was 29 m, and the array aperture was 116 m, i.e., >100 m. Besides, motion attitude information of the shipborne platform was recorded synchronously using the shipborne inertial navigation system. The measured radar echo spectrum data of the coast–ship bistatic HFSWR system obtained in the experiment are discussed in the following section.

4.2. Interpretation of Experiment Results

Time period 1: CTSR bistatic HFSWR experimental data analysis

A typical two-dimensional range-Doppler (RD) spectrum of the CTSR bistatic HFSWR and that of the coast-based monostatic HFSWR at a specific time is shown in Figure 18a,b, respectively. At the sample time, the baseline distance between the transmitting station and the receiving station was 12.8 km. The shipborne platform was sailing in a straight line with velocity of 11.48 km/h (6.2 knots), the ship heading was 115° (relative to the baseline), and the main axis angle of the shipborne radar receiving station was 205° (relative to the baseline). The broadening range of the simulated sea clutter spectrum with the same parameter configuration as in the experiment is superimposed on the measured radar spectrum, and it is marked with a red box.

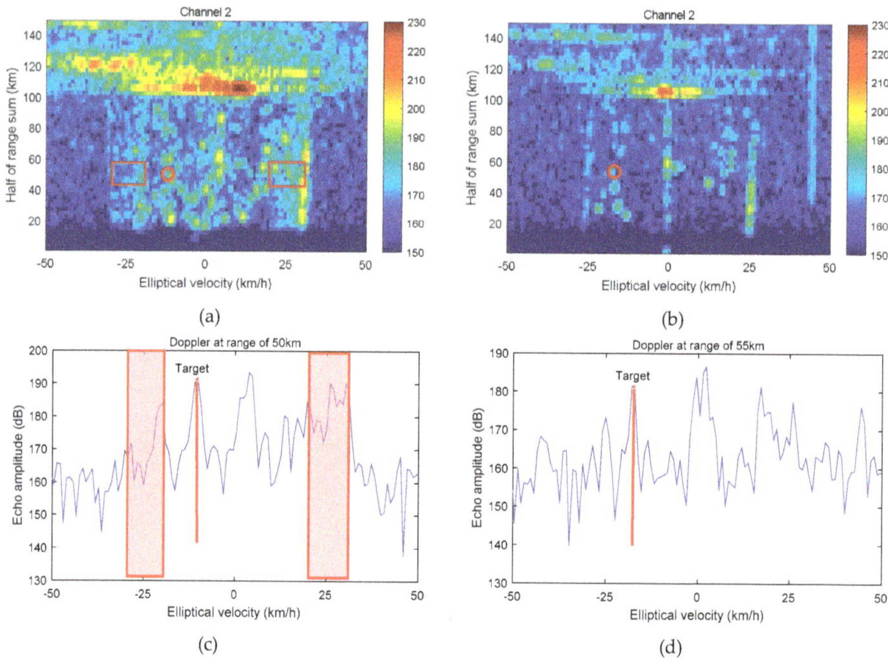

Figure 18. Doppler spectrum of the CTSR bistatic HFSWR and of the coast-based monostatic HFSWR at the velocity of 11.7 km/h (6.2 knots); (**a**) RD spectrum of the CTSR bistatic HFSWR; (**b**) RD spectrum of the coast-based monostatic HFSWR; (**c**) Doppler data of the CTSR bistatic HFSWR at the range of 50 km; (**d**) Doppler data of the coast-based monostatic HFSWR at the range of 55 km.

As can be seen from Figure 18, in comparison with the non-broadened first-order sea clutter spectrum of the coast-based monostatic HFSWR, both the left and right first-order sea clutter RD spectra of the CTSR bistatic HFSWR were broadened at the platform velocity $v_R = 11.48$ km/h (6.2 knots), and the broadening of the right first-order sea clutter spectrum was more obvious than that of the left. The width of the right first-order sea clutter was approximately 11.1 km/h (20.8 to 31.9 km/h), and that of the left first-order sea clutter was 10.3 km/h (−18.5 to −28.8 km/h). According to the simulation results (Figure 8a) discussed in Section 3, the theoretical width of the right first-order sea clutter should be 11.3 km/h (20.2 to 31.5 km/h), and that of the left first-order sea clutter should be 10.5 km/h (−18.9 to −29.4 km/h). It can be seen that the measured width values of both the left and the right first-order sea clutter are consistent with their theoretical values. The small deviation of approximately 0.6 km/h (0.17 m/s) of the location of the sea clutter spectrum might have been caused by the existence of ocean currents. In addition, it can be discerned from Figure 18a that the bending characteristics of the first-order Doppler spectrum caused by the bistatic angle were not obvious. This is mainly attributable to two reasons. The first is that the baseline was small relative to the detection distance. The second is that the shipborne receiving station could only receive on the right side of the ship, limiting the detection range of the radar to a certain extent. Both of these factors could lead to smaller eccentricity and a smaller bistatic angle, resulting in the determined bending characteristics of the first-order sea clutter spectrum.

Compared with the broadened first-order sea clutter spectrum, the echo of a moving target (Maritime Mobile Service Identity No: 413551440) was detected simultaneously by the bistatic HFSWR and the coast-based monostatic HFSWR systems. The positions of the target in the two-dimensional RD spectra and in the one-dimensional Doppler spectra are marked with red circles and straight lines, respectively. Its velocity and heading were 19.96 km/h and 289.4° relative to the north, respectively, i.e., $v = 17.96$ km/h, $\varphi = -167.8529°$, $\frac{\beta}{2} = 6.15°$, $\varphi_R = 115.5°$ relative to the baseline, and $\theta_R = 116.1°$. Based on Equation (5), the theoretical radial velocity was -11.7516 km/h. Its echo appeared in the 20th range unit, i.e., the 74th Doppler unit of the RD spectrum of the CTSR bistatic HFSWR with a signal-to-noise ratio (SNR) value of approximately 30 dB, which means that the half of its range sum was 50.73 km, and the projected elliptical velocity was -11.8127 km/h. It can be seen that the measured velocity of the target is consistent with the theoretical value (-11.7516 km/h). The target's range relative to the monostatic HFSWR was 54.2269 km and the radial velocity was -17.0557 km/h, with an SNR value of approximately 25 dB. The Doppler-time (D-T) distributions over a 10-min period, onto which the elliptical velocity results were projected using Automatic-Identification-System (AIS) real information were superimposed, are shown in Figure 19. It can be seen that the track of the moving target can be detected clearly, and that the Doppler shift of the target echo within the 10-min period is consistent with the AIS track results.

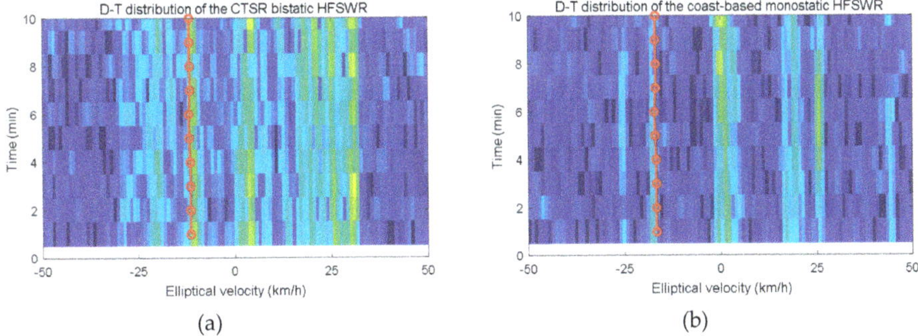

(a) (b)

Figure 19. Doppler-time (D-T) distribution and Automatic-Identification-System (AIS) results for the CTST bistatic HFSWR and coast-based monostatic HFSWR; (**a**) D-T distribution of the CTSR bistatic HFSWR; (**b**) D-T distribution of the coast-based monostatic HFSWR.

Time period 2: STCR bistatic HFSWR experimental data analysis

Measured data of the STCR bistatic HFSRW system with different platform speeds were obtained in the second period. The STCR bistatic HFSWR spectrum and the synchronous shipborne monostatic HFSWR data at the velocity of 0 km/h, 7.96 km/h (4.3 knots), and 18.5 km/h (10.0 knots) are shown in Figures 20–22, respectively. The heading at the three moments was 54°, 141°, and 115°, respectively (relative to the baseline). For the receiving station, the angle of the main axis was fixed at 310°.

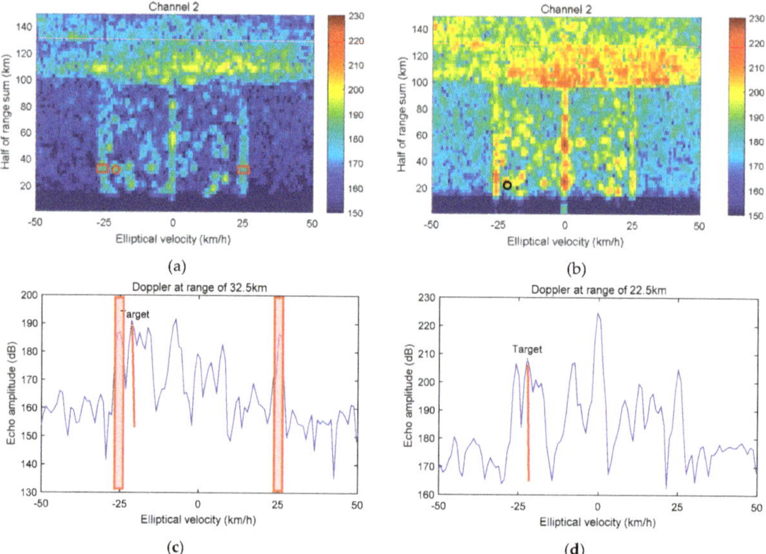

(a)

(b)

(c)

(d)

Figure 20. Doppler spectra of the STCR bistatic HFSWR and ship-based monostatic HFSWR systems at the velocity of 0 knots. (**a**) RD spectrum of the STCR bistatic HFSWR; (**b**) RD spectrum of the ship-based monostatic HFSWR; (**c**) Doppler spectrum of the STCR bistatic HFSWR at the range of 32.5 km; (**d**) Doppler spectrum of the ship-based monostatic HFSWR at the range of 22.5 km.

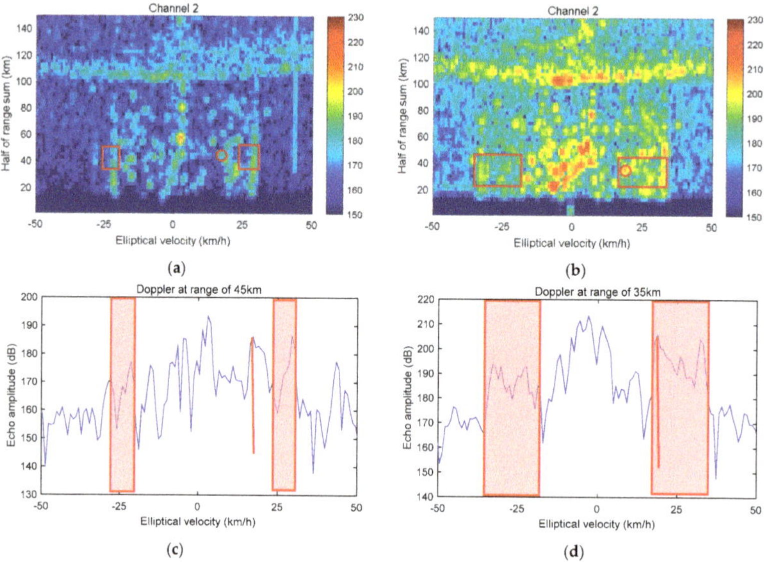

(a)

(b)

(c)

(d)

Figure 21. Doppler spectra of the STCR bistatic HFSWR and ship-based monostatic HFSWR systems at the velocity of approximately 4.3 knots. (**a**) RD spectrum of the STCR bistatic HFSWR; (**b**) RD spectrum of the ship-based monostatic HFSWR; (**c**) Doppler spectrum of the STCR bistatic HFSWR at the range of 45 km; (**d**) Doppler spectrum of the ship-based monostatic HFSWR at the range of 35 km.

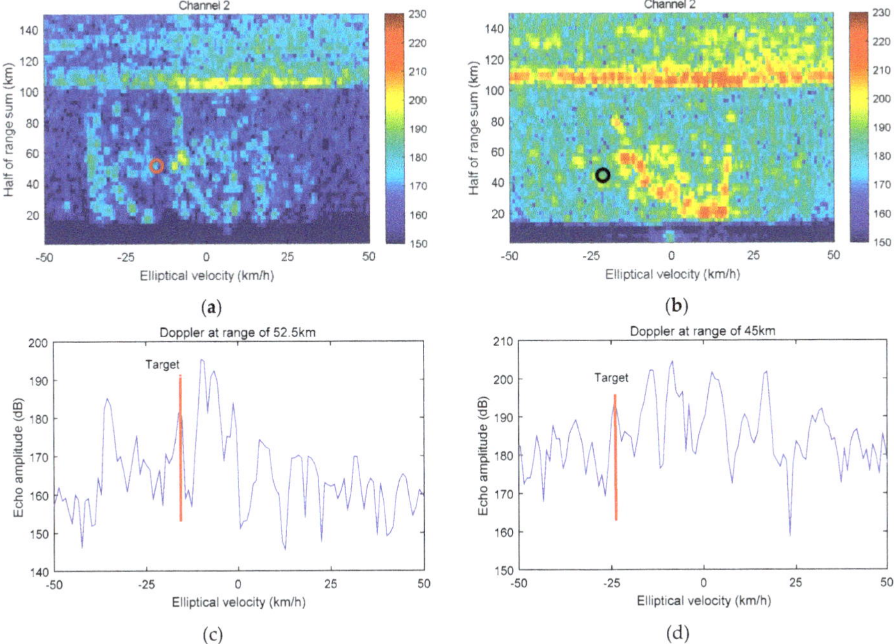

Figure 22. Doppler spectra of the STCR bistatic HFSWR and ship-based monostatic HFSWR systems at the velocity of approximately 10.0 knots. (**a**) RD spectrum of the STCR bistatic HFSWR; (**b**) RD spectrum of the ship-based monostatic HFSWR; (**c**) Doppler spectrum of the STCR bistatic HFSWR at the range of 52.5 km; (**d**) Doppler spectrum of the ship-based monostatic HFSWR at the range of 45 km.

As can be seen from Figures 20–22, the bending characteristics of the first-order Doppler spectrum caused by the bistatic angle can be observed, although they are not obvious. When the platform was anchored, the velocity of the platform was zero (Figure 20). In this case, the radar can be regarded as having a fixed transmitting station and a fixed receiving station. Therefore, the first-order sea clutter of the STCR bistatic HFSWR was not broadened, which was the same outcome as that of the shipborne monostatic HFSWR. The target echo in the RD spectrum of each of the two radar systems was very clear.

When the platform velocity was approximately 4.3 knots, the broadening of the first-order sea clutter in the RD spectrum (Figure 21) of the bistatic HFSWR was not obvious and most targets were easily detected. The width of the left first-order sea clutter was 6.2 km/h (−20.6 to −26.8 km/h), and that of the right first-order sea clutter was 6.1 km/h (24.1 to 30.2 km/h). According to the simulation results (Figure 10a in Section 3), under the same conditions, the theoretical width of the left first-order sea clutter should be 6.1 km/h (−20.6 to −26.7 km/h), and that of the right first-order sea clutter should be 6.5 km/h (23.7 to 30.2 km/h). Thus, the broadening range of the measured sea clutter spectrum is consistent with the theoretical value. In Figure 22c, the SNR value of the target of concern was approximately 17 dB for the STCR bistatic HFSWR. In contrast, the first-order sea clutter in the RD spectrum of the monostatic HFSWR was obviously broadened. Although the range of the first-order spectrum was wide, the obvious outer boundary can be seen coupled with the influence of land clutter, i.e., the moving target marked in the figure can still be detected, but the signal-to-clutter ratio was reduced to approximately 10 dB.

When the platform velocity reached 10 knots (Figure 22), the scene was very different from the previous two examples. For the STCR bistatic HFSWR, as the coast-based receiving station was obscured by surrounding land, the received signal came only from a limited range of the incoming direction. Therefore, the range of broadening of the left first-order sea clutter spectrum and the Doppler shift of the land clutter were obviously smaller than the theoretical values. For the shipborne monostatic HFSWR, as the shipborne platform was on the way back, the main axis angle of the radar receiving station was 205°, and the detection range of the shipborne receiving station was mainly facing land. Consequently, there was no obvious boundary to be found because of the wide range of the first-order sea clutter. However, the presence of land clutter was obvious, varying continuously with range and velocity in the radar echo.

Similar to the above, Figure 23 shows the D-T distributions and AIS results of the STCR bistatic HFSWR and ship-based monostatic HFSWR at the platform velocity of 4.3 knots ($v_T = 7.96$ km/h). At this time, $\varphi = 37.56°$, $\frac{\beta}{2} = 2.9°$, $\varphi_T = 141°$, and $\theta_T = 92°$. The moving target ($v = 18.71$ km/h) was detected at the range of 44.75 km with an elliptical velocity of approximately 17.46 km/h by the STCR bistatic HFSWR based on Equation (6), whereas it was detected at the range of 35 km with a radial velocity of approximately 18.87 km/h by the ship-based monostatic HFSWR. The Doppler shift of the target echo in the measured radar data is largely consistent with the AIS track results.

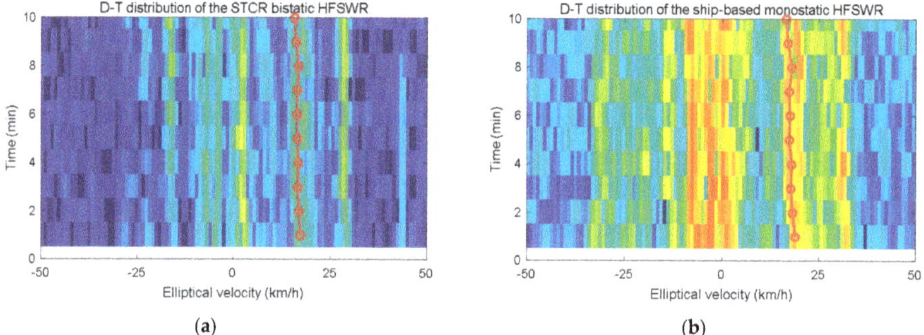

Figure 23. D-T distribution and AIS results for the STCR bistatic HFSWR and ship-based monostatic HFSWR. (**a**) D-T distribution of the STCR bistatic HFSWR; (**b**) D-T distribution of the ship-based monostatic HFSWR.

5. Conclusions

By combining the advantages of coast-based bistatic HFSWR and shipborne monostatic HFSWR, a coast–ship bistatic HFSWR system has potential in anti-electronic interference and expanded detection range by fully exploiting the advantage of the flexibility of a shipborne platform. To investigate the characteristics of the first-order sea clutter spectrum, and the related blind area and its influence on target detection, the theoretical formulas for a coast–ship bistatic HFSWR system were derived. Moreover, simulations of the first-order sea clutter spectrum under different radar parameters and different shipborne platform velocities were analyzed. From the simulation results and its influence on target detection, it can be concluded that the Doppler shift of both the first-order sea clutter and the moving target were affected mainly by the velocity of the moving platform and the bistatic angle in a coast–ship bistatic HFSWR. Broadening of the first-order sea clutter spectrum will cause a blind area for target detection. When the shipborne platform was anchored, the broadening ranges of the right sea clutter spectrum and left sea clutter spectrum were symmetrical, and the widths of the right sea clutter and left sea clutter were equal. When the shipborne platform was navigating, the broadening ranges of the first-order sea clutter spectrum on the left and right were asymmetrical, and their widths were not equal, owing to the combined influence of the bistatic angle and the platform motion. In addition, the broadening range of the first-order sea clutter spectrum of a coast–ship bistatic HFSWR increased with the increase of platform velocity in most cases. At the same platform velocity, both the broadening range of the first-order sea clutter spectrum and the width of the sea clutter blind area of a coast–ship bistatic HFSWR changed with radar frequency, and the width of the first-order sea clutter of a coast–ship bistatic HFSWR was less than that of a shipborne monostatic HFSWR. Under the condition of higher platform velocity, although the detection blind area caused by sea clutter was wider, the echo amplitude of sea clutter was lower, and the lower SNR was beneficial for target detection. Although the comparative relationship between the amplitude of the left first-order spectrum and the right first-order spectrum changed under different wind conditions, the range of broadening of the first-order sea clutter spectrum remained unchanged. The analysis of measured experimental data revealed that the broadening range of the first-order sea clutter and of the frequency shift of a moving target are consistent with theoretical and simulation results. The findings of this study further the understanding of the theoretical formulas and measured spectrum data of the two types of coast-ship bistatic HFSWR.

It should be noted that under the condition of a moving platform, irrespective of whether a monostatic shipborne HFSWR or a coast–ship bistatic HFSWR is used, the first-order sea clutter will be broadened markedly in the channel spectrum. In general, beam data rather than channel data are used in target detection to obtain a better SNR. Therefore, the influence of sea clutter broadening within the beam data of coast–ship bistatic HFSWR should be considered in the process of motion compensation or sea clutter suppression for target detection, and related research will be undertaken in the future.

Author Contributions: Conceptualization, Y.J.; methodology, Y.J. and C.Y. (Chao Yue); validation, W.G., and H.S.; formal analysis, J.L.; investigation, Y.W., C.Y. (Chao Yue) and W.G.; resources, Y.J. and C.Y. (Changjun Yu); Data Curation: W.G. and J.L.; writing—original draft preparation, Y.J.; writing—review and editing, Y.W. and M.L.; visualization, H.S.; supervision, J.Z. and M.L; project administration, Y.J.; funding acquisition, Y.J. All authors have read and agreed to the published version of the manuscript.

Funding: This research was funded by the National Key R & D Program of China (grant number 2017YFC1405202); and the National Natural Science Foundation of China (grant number 61671166).

Acknowledgments: The authors thank the anonymous reviewers for their comments and suggestions that have helped to improve the quality and the readability of this article. We thank James Buxton MSc from Liwen Bianji, Edanz Group China (www.liwenbianji.cn./ac), for editing the English text of this manuscript.

Conflicts of Interest: The authors declare no conflict of interest.

References

1. Ponsford, A.M. Surveillance of the 200 Nautical Mile Exclusive Economic Zone (EEZ) Using High Frequency Surface Wave Radar (HFSWR). *Can. J. Remote Sens.* **2001**, *27*, 354–360. [CrossRef]
2. Liu, Y.T.; Xu, R.Q.; Zhang, N. Progress in HFSWR research at Harbin Institute of Technology. In Proceedings of the International Conference on Radar, Adelaide, Australia, 3–5 September 2003; pp. 522–528.
3. Nicholas, J. *Bistatic Radar*, 2nd ed.; SciTech Publishing Inc.: Troy, NY, USA, 2005; pp. 119–123.
4. Yao, G.W.; Xie, J.H.; Huang, W.M. Ocean Surface Cross Section for Bistatic HF Radar Incorporating a Six DOF Oscillation Motion Model. *Remote Sens.* **2019**, *11*, 2738. [CrossRef]
5. Huang, W.M.; Gill, E.; Wu, X.B.; Li, L. Measurement of Sea Surface Wind Direction Using Bistatic High-Frequency Radar. *IEEE Trans. Geosci. Remote Sens.* **2012**, *50*, 4117–4122. [CrossRef]
6. Gill, E.W. The Scattering of High Frequency Electromagnetic Radiation from the Ocean Surface: An Analysis Based on a Bistatic Ground Wave Radar Configuration. Ph.D. Thesis, Memorial University of Newfoundland, St. John's, NL, Canada, January 1999.
7. Gill, E.W.; Walsh, J. High-frequency bistatic cross sections of the ocean surface. *Radio Sci.* **2001**, *36*, 1459–1475. [CrossRef]
8. Barrick, D.E. First-order Theory and Analysis of MF/HF/VHF Scatter from the Sea. *IEEE Trans. Antennas Propag.* **1972**, *20*, 2–10. [CrossRef]
9. Lipa, B.J.; Barrick, D.E. Extraction of sea state from HF radar sea echo: Mathematical theory and modeling. *Radio Sci.* **1986**, *21*, 81–100. [CrossRef]
10. Xie, J.; Sun, M.; Ji, Z. Space-time Model of the First-Order Sea Clutter in Onshore Bistatic High Frequency Surface Wave Radar. *IET Radar Sonar Navig.* **2015**, *9*, 55–61. [CrossRef]

11. Walsh, J.; Huang, W.; Gill, E.W. The First-Order High Frequency Radar Ocean Surface Cross Section for an Antenna on a Floating Platform. *IEEE Trans. Antennas Propag.* **2010**, *58*, 2994–3003. [CrossRef]
12. Walsh, J.; Huang, W.; Gill, E.W. The second-order high frequency radar ocean surface cross section for an antenna on a floating platform. *IEEE Trans. Antennas Propag.* **2012**, *60*, 4804–4813. [CrossRef]
13. Khoury, J.; Guinvarch, R.; Gillard, R. Sea-echo Doppler spectrum perturbation of the received signals from a floating high-frequency surface wave radar. *IET Radar Sonar Navig.* **2012**, *6*, 165–171. [CrossRef]
14. Xie, J.; Yuan, Y.; Liu, Y.T. Experimental analysis of sea clutter in shipborne HFSWR. *IEE Proc. Radar Sonar Navig.* **2001**, *148*, 67–71. [CrossRef]
15. Xie, J.; Sun, M.; Ji, Z. First-order ocean surface cross-section for shipborne HFSWR. *Electron. Lett.* **2013**, *49*, 1005–1026. [CrossRef]
16. Li, F.L.; Chen, B.X.; Zhang, S.H. Design of Multi-carrier Digital Frequency Synthesizer for Coast-ship Multi-static GroundWave OTH Radar. In Proceedings of the 2006 CIE International Conference on Radar, Shanghai, China, 6–19 October 2006; pp. 1–5.
17. Chen, B.X.; Chen, D.F.; Zhang, S.H. Experimental system and experimental results for coast-ship bi/multistatic ground-wave over-the-horizon radar. In Proceedings of the 2006 CIE International Conference on Radar, Shanghai, China, 6–19 October 2006; pp. 36–40.
18. Chen, B.X.; Chen, D.F. Estimation of Gain and Phase Perturbations for Transmit Array Based on Parameter Transfer in Coast-Ship Bistatic SWOTHR. *Acta Electron. Sin.* **2011**, *39*, 2184–2189.
19. Liu, C.B.; Chen, B.X.; Chen, D.F. Analysis of First-order Sea Clutter in a Shipborne Bistatic High Frequency Surface Wave Radar. In Proceedings of the 2006 CIE International Conference on Radar, Shanghai, China, 6–19 October 2006; pp. 1656–1659.
20. Zhu, Y.; Wei, Y.; Tong, P. First order sea clutter cross section for bistatic shipborne HFSWR. *J. Syst. Eng. Electron.* **2017**, *28*, 681–689.
21. Skolnik, M.I. An Analysis of Bistatic Radar. *Aerosp. Navig. Electron. Ire Trans.* **1961**, *ANE-8*, 19–27. [CrossRef]

 *remote sensing*

Article

Vessel Tracking Using Bistatic Compact HFSWR

Weifeng Sun [1,*], Mengjie Ji [1], Weimin Huang [2], Yonggang Ji [3] and Yongshou Dai [1]

[1] College of Oceanography and Space Informatics, China University of Petroleum (East China),
   Qingdao 266580, China; s17050643@s.upc.edu.cn (M.J.); daiys@upc.edu.cn (Y.D.)
[2] Faculty of Engineering and Applied Science, Memorial University of Newfoundland,
   St. John's, NL A1B 3X5, Canada; weimin@mun.ca
[3] Laboratory of Marine Physics and Remote Sensing, First Institute of Oceanography,
   Ministry of Natural Resources, Qingdao 266061, China; jiyonggang@fio.org.cn
* Correspondence: sunwf@upc.edu.cn; Tel.: +86-532-182-6663-9778

Received: 21 March 2020; Accepted: 15 April 2020; Published: 16 April 2020

Abstract: Bistatic and multi-static high-frequency surface wave radar (HFSWR) is becoming a prospective development trend for sea surface surveillance due to its potential in extending the coverage area, improving the detection accuracy, etc. In this paper, the vessel detection and tracking performance of a newly developed bistatic compact HFSWR system whose transmitting and receiving antennas are not co-located was investigated. Firstly, the representation of the target range and Doppler velocity concerning a bistatic HFSWR was derived and compared with that of a monostatic system. Next, taking the characteristics of target kinematic parameters into account, a target tracking method applicable to a bistatic HFSWR is proposed. The simultaneous target tracking results from both monostatic and bistatic HFSWR field data are presented and compared. The experimental results demonstrate the good performance in target tracking of the bistatic HFSWR and also show that an HFSWR system combining monostatic and bistatic modes has the potential to enhance the target track continuity and improve the detection accuracy.

Keywords: compact HFSWR; bistatic configuration; target detection; target tracking

---

1. Introduction

High-frequency surface wave radar (HFSWR) operated in the 3–30 MHz frequency band has been recognized as an important maritime surveillance tool [1] for both sea state monitoring [2] and hard target detection [3] due to its superiority of over-the-horizon coverage, all-weather and continuous surveillance, high time resolution, low-cost, etc. Existing HFSWRs for hard target detection are monostatic with the transmitter and receiver being co-located. Some HFSWRs employ a linear receiving array with a large aperture size and high transmitting power to achieve high azimuth resolution and long detection range. e.g., the SWR-503 system developed in Canada uses a receiving array with an aperture size of 660 m and a transmitting power of 30 kW. Such a radar system usually requires a large coastline area, which makes it difficult for site selection, system deployment, and maintenance, and thus limits its operational applications. Therefore, system miniaturization has become a new development trend [4]. So far, two kinds of compact HFSWR systems with small aperture size have been developed. One utilizes crossed-loop/monopole antennas, such as the SeaSonde system developed by CODAR [5,6], the OSMAR system developed by Wuhan University [7], etc., while the other still uses phased array antenna but with less antenna array elements (e.g., 3–8 elements), e.g., the WERA-S system seveloped by Helzel MessTechnik [8], the Compact Over-the-horizon Radar for Maritime Surveillance (CORMS) system developed by our team [9–11], etc.

Compared with an HFSWR system that has large aperture size, a small-aperture compact system has the advantages of flexible deployment and maintenance. It can be installed on a small island, or even on a large ship [12], thus increasing the detection flexibility and extending coverage area. However, it should be noted that the direction of arrival estimation performance degrades due to the reduced number of antenna elements. The resultant poor azimuth estimation accuracy may lead to large target positioning errors. In addition, the lower transmitting power and clutter interference, such as the sea clutter, ionospheric clutter, etc., make target detection more challenging.

To improve the target detection performance of compact HFSWRs, on one hand, super-resolution direction-finding methods, such as MUSIC [13], were proposed to improve the azimuth estimation accuracy. On the other hand, distributed multi-radar systems, such as bistatic and multi-static radar [14], MIMO radar [15], etc., were designed to simultaneously monitor the targets in a common area of interest from different perspectives. Also, bistatic or multi-static configuration makes the influence of sea clutter, ionosphere clutter, etc., diverse from different sites. The complementary information obtained from different radars can be associated and fused to produce more precise results. However, the distributed radar systems with multiple radars are inherently more complex and hence tend to be more expensive. Multi-static radar offers a way to obtain good target detection performance but requires extensive research, especially advanced signal processing techniques. Among distributed radar systems, the combination of one monostatic radar and one bistatic radar with a common transmitting station and two separate receiving stations, i.e., T/R-R radar, is a typical configuration that offers a tradeoff between system performance and complexity. In this paper, the target detection and tracking performance of a T/R-R bistatic compact HFSWR system is investigated. It is worth noting that although the T/R-R configuration brings several advantages in target detection, the problems caused by the compact HFSWR system itself, such as low detection precision for both range and azimuth, are inherent.

Although bistatic system is not a new concept, very few bistatic HFSWRs have been developed and limited experiments have been reported. Theoretical studies on bistatic HFSWR have been carried out extensively. For example, various issues related to bistatic HFSWR, such as system configuration, site selection, spectral characteristics, detection performance, etc., were analyzed in [16–18]. Clutter models were established in [19] and analyzed in [20], and interference suppression methods were proposed in [21]. Different ocean surface radar cross section models of bistatic HFSWR were derived and analyzed in [22–28]. However, experimental work is relatively limited. So far, bistatic HFSWRs have been utilized for surface current mapping [29–31], directional ocean spectrum measurement [32], and wind direction measurement [33] over a very short period. Existing research related to bistatic HFSWR mainly focuses on system design and performance analysis, scattering mechanism investigation, clutter suppression, and sea state mapping applications. Most of the research work lies in theoretical analysis and numerical simulation. Compared with the numerous multi-target tracking (MTT) algorithms developed for monostatic HFSWRs [34–36], target detection and tracking using bistatic HFSWR have been much less explored [37], especially with field data.

The primary objective of this paper was to investigate the characteristics of target detection with a bistatic HFSWR and develop an applicable target tracking method accordingly. The field data collected by a newly developed T/R-R compact HFSWR were used to validate the correctness of the derived results and verify the performance of the proposed tracking method. Besides, comparisons were made between the tracking results of monostatic and bistatic HFSWR. The remainder of this paper is organized as follows. Section 2 describes the characteristics of target representation for a bistatic HFSWR, followed by a detailed description of an applicable target tracking method. In Section 3, the experimental results are presented and analyzed. Discussions are provided in Section 4 and conclusions are drawn in Section 5.

2. Target Detection and Tracking with a Bistatic Compact HFSWR

The combined monostatic and bistatic compact HFSWR system, abbreviated as a T/R-R compact HFSWR system in the following description, employs a common transmitter and two receivers. The co-located transmitter and receiver constitute a monostatic radar, while the bistatic radar is composed of the shared transmitter and the other receiver deployed at a considerable distance away from the transmitter. As target detection is the basis of target tracking, the target representation method for a bistatic HFSWR is discussed first.

2.1. Target Representation with a Bistatic Compact HFSWR

Compared with a monostatic HFSWR, a bistatic HFSWR represents a target differently due to different geometry. The target detection geometry of a T/R-R HFSWR system, defined by the position of a transmitter (Tx), a receiver (Rx), and a target using a two-dimensional north-referenced coordinate system, is illustrated in Figure 1.

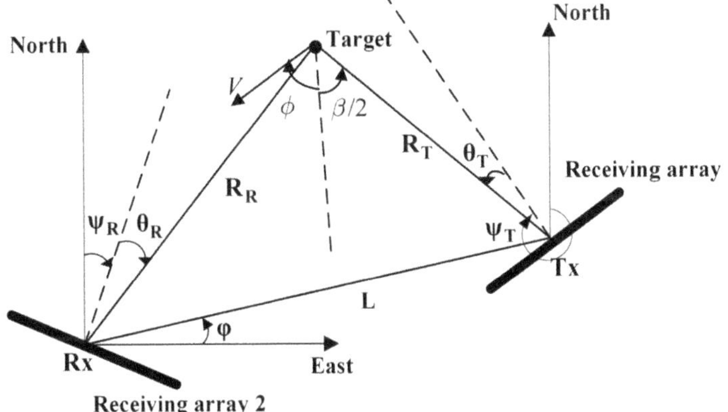

Figure 1. The radar configuration geometry of a T/R-R high-frequency surface wave radar (HFSWR) system. The transmitter and receiving array 1 are deployed at the transmitting station (Tx), while receiving array 2 is deployed at the receiving station (Rx). The transmitter serves the two receiving arrays simultaneously. The distance L between the transmitter and receiver is called the bistatic range or simply the baseline. φ denotes the angle between the baseline and East direction. Denote ψ_T and ψ_R, respectively, as the radar look directions of receiving array 1 and receiving array 2 with respect to true North. β is the bistatic angle.

In Figure 1, the transmitter and receiving array 1 are deployed at the transmitting station (Tx), while receiving array 2 is deployed at the receiving station (Rx). The transmitter serves the two receiving arrays simultaneously. The distance L between the transmitter and receiver is called the bistatic range or simply the baseline. φ denotes the angle between the baseline and East direction. Denote ψ_T and ψ_R, respectively, as the radar look directions of receiving array 1 and receiving array 2 with respect to true North. β is the bistatic angle. For a moving target with a velocity of magnitude V and aspect angle ϕ referenced to the bistatic bisector, the monostatic radar represents a target as a plot with a state vector $[R_T \ \theta_T \ v_{dT}]$ in a polar coordinate, with R_T, θ_T, v_{dT} being the range, azimuth, and Doppler velocity measured at the transmitting station. Similarly, the bistatic radar represents a target as a plot with a state vector $[R_R \ \theta_R \ v_{dR}]$, with R_R, θ_R, v_{dR} being the range, azimuth, and Doppler velocity calculated at the receiving station. As the estimation methods of R_T, θ_T, and v_{dT} for monostatic HFSWR has been investigated [11], only the estimation methods of R_R, θ_R, and v_{dR} for the bistatic HFSWR are discussed here.

Like the monostatic case, a target azimuth θ_R is also estimated using the digital beaming forming (DBF) method for a bistatic HFSWR with a linear phased array as its receiving antenna. θ_R takes negative values on the left side of the radar boresight, and positive values on the other side. However, the estimation methods for range and Doppler velocity are different from those used in a monostatic HFSWR. Firstly, the distance directly measured by a bistatic radar is the sum of R_T and R_R, the total transmitter-to-target-to-receiver scattering path, instead of R_R. Target positions with the same range sum, i.e., the isorange contour, form an ellipse with foci at the transmitter and receiver sites. The bisector of the bistatic angle β is orthogonal to the tangent of the ellipse and passes through a target position. Thus, a target position can be determined by its range sum and estimated azimuth θ_R. As can be seen from Figure 1, either (R_R, θ_R) or (R_T, θ_T) can be exclusively used for target location representation. For monostatic HFSWR, $R_R = R_T$, and $\theta_R = \theta_T$, thus its target tracking methods cannot be directly applied to the bistatic case. Considering that the target azimuth θ_R is directly measured by the receiving antenna array 2 with respect to its normal direction, (R_R, θ_R) instead of (R_T, θ_T) is chosen to specify the target position for bistatic HFSWR. i.e., R_R is used to represent the target range and is derived as follows.

According to the bistatic triangle (i.e., the transmitter–target–receiver triangle shown in Figure 1) relationship, the following relation can be obtained:

$$R_T{}^2 = R_R{}^2 + L^2 - 2R_R L \cos(\frac{\pi}{2} - \psi_R - \theta_R - \varphi). \tag{1}$$

Denote $R = R_T + R_R$, $\theta = \psi_R + \theta_R + \varphi$, then R_R can be calculated as

$$R_R = \frac{R^2 - L^2}{2(R - L \sin \theta)}. \tag{2}$$

Once the radar configuration is set, ψ_R, L, and φ are known. R and θ_R can be determined from the data collected by the linear receiving array 2, then R_R can be calculated by Equation (2). The calculated R_R and the estimated θ_R can also specify the location of the target. However, unlike the monostatic case, the calculated range R_R is a function of the estimated target azimuth θ_R. The coarse azimuth resolution of a compact HFSWR leads to large estimation errors in R_R, which brings greater challenges for target detection and tracking with bistatic compact HFSWR.

Another difference from a monostatic radar is the estimated Doppler velocity. The Doppler velocity measured by a monostatic radar is along the radial direction in the polar coordinate system with origin at the radar site, while the Doppler velocity estimated from a range-Doppler spectrum of a bistatic radar is along the direction of the bistatic bisector. It is the resultant velocity combining the Doppler velocities measured from the transmitting and receiving stations. The estimated elliptical Doppler velocity can be calculated as

$$\begin{aligned}
V_{dR} &= \frac{dR}{dt} = \frac{dR_T}{dt} + \frac{dR_R}{dt} \\
&= V \cos(\phi + \frac{\beta}{2}) + V \cos(\phi - \frac{\beta}{2}) \\
&= 2V \cos \phi \cos(\frac{\beta}{2}).
\end{aligned} \tag{3}$$

It can be concluded that the magnitude of the bistatic Doppler velocity is related to the bistatic angle and is never greater than that of a monostatic radar. In practice, the measured Doppler velocity v_{dR} is obtained from the Doppler shift f_d extracted from a bistatic range-Doppler spectrum by

$$v_{dR} = (f_d \cdot c) / f_o, \tag{4}$$

where f_d denotes the radar operating frequency, c is the light speed.

Sea surface target detection using compact HFSWR is typically affected by either ocean clutter or ionospheric clutter, which can mask the returns from targets at their corresponding Doppler points and make them undetectable. In addition to parameter representations of a target discussed above, the first-order sea clutter should also be considered for target detection with a bistatic HFSWR.

Without the effect of surface current, for stationary transmitting and receiving antennas, the Doppler shift of the first-order sea clutter for a bistatic HFSWR [22] can be written as

$$f_b = \pm\sqrt{\frac{g\cos(\beta/2)}{\pi\lambda}}, \tag{5}$$

where λ denotes the radar wavelength, g is the gravity acceleration. Equation (5) indicates that the first-order Bragg shift of a bistatic HFSWR is a function of the bistatic angle β. It is less than that of a monostatic HFSWR, which can be expressed as $f_b' = \pm\sqrt{\frac{g}{\pi\lambda}}$. For a compact HFSWR, its detection range is limited due to the lower transmitting power; and its beamwidth is wider due to the smaller aperture size. Thus, the bistatic angle β always takes a relatively larger value leading to a spread first-order spectrum, which may mask the targets and increases the challenge for target detection.

2.2. Target Tracking with a Bistatic Compact HFSWR

Once the plot data sequence is consecutively obtained with each plot being denoted by a measured state vector $[R_R^m \ \theta_R^m \ v_{dR}^m]$, a multi-target tracking method is required to produce target tracks. The proposed MTT algorithm was modified from the method presented in [11], which is developed from Converted Measurement Kalman Filter (CMKF) and a data association method based on minimal cost. In this method, data association, state prediction, and state estimation are three key steps, where data association is performed in the polar coordinate with the receiving radar site as its origin, while state prediction and estimation are implemented in a Cartesian coordinate.

The state prediction and estimation using the CMKF method are implemented by a linear Kalman filter, which is based on a specific dynamic model and an observation model. Taking the motion characteristic of large vessels into consideration, the target dynamic model is defined in a Cartesian coordinate as

$$\mathbf{x}_k = \mathbf{F}\mathbf{x}_{k-1} + \boldsymbol{\omega}_k, \tag{6}$$

where $\mathbf{x}_k = [x_k, v_{x_k}, y_k, v_{y_k}]^T$ is the true state vector at time k in the Cartesian coordinate with the boresight of the receiving array 2 as its x-axis, the direction perpendicular to the radar boresight as the y-axis. x_k and y_k denote the true target position components, v_{x_k} and v_{y_k} denote the corresponding true velocity components along x and y directions. $[\cdot]^T$ denotes the transpose operator. $\boldsymbol{\omega}_k$ represents the Gaussian process noise with zero mean and covariance matrix $\mathbf{Q}_k$. $\mathbf{F}$ is the state transition matrix defined as

$$\mathbf{F} = \begin{bmatrix} 1 & T & 0 & 0 \\ 0 & 1 & 0 & 0 \\ 0 & 0 & 1 & T \\ 0 & 0 & 0 & 1 \end{bmatrix},$$

where T denotes the sampling time.

The observation model is also defined in the Cartesian coordinate as

$$\mathbf{z}_k = \mathbf{H}\mathbf{x}_k + \mathbf{v}_k, \tag{7}$$

where $\mathbf{z}_k = [\tilde{x}_k, \tilde{v}_{x_k}, \tilde{y}_k, \tilde{v}_{y_k}]^T$ is a measured state vector at time k, $\tilde{x}_k$ and $\tilde{y}_k$ denote the measured target position components, $\tilde{v}_{x_k}$ and $\tilde{v}_{y_k}$ are the corresponding measured velocity components along x and y directions, respectively. $\mathbf{v}_k$ represents measurement noise following Gaussian distribution with zero mean and covariance matrix $\mathbf{R}_k$. $\mathbf{H}$ is the measurement matrix and it is an identity matrix here.

The difference in the tracking procedure between a monostatic and a bistatic HFSWR lies in that the measured state vector $[R_R^m \ \theta_R^m \ v_{dR}^m]$ instead of $[R_T^m \ \theta_T^m \ v_{dT}^m]$ is used. On one hand, the accuracy of R_R^m, which is related to the coarsely estimated azimuth as shown in Equation (2), is lower than R_T^m. On the other hand, the elliptical Doppler velocity v_{dR}^m that is along the bistatic bisector is used here, and it is assumed that the elliptical Doppler velocity of a target does not change much during a coherent integration time. Thus, the data association procedure is modified by using different target parameters, validation gate thresholds, and association weights. Based on the above analysis, the proposed target tracking procedure for a bistatic compact HFSWR, including three parallel sub-procedures, i.e., track initiation, track maintenance, and track termination, is summarized as follows.

A. Track initiation

Tracks are initiated by the logic method with the M-of-N rule [38]. If a track is successfully initiated with more than M plots connected in the most recent N frames, it will go to the track maintenance procedure; otherwise, it will be dropped.

B. Track maintenance

Step 1: State prediction. For each initiated or maintained track with n plots, denote $\hat{\mathbf{x}}_{k-1} = [\hat{x}_{k-1}, \hat{v}_{x_{k-1}}, \hat{y}_{k-1}, \hat{v}_{y_{k-1}}]^T$ as its previous plot at time $k-1$, its predicted state $\hat{\mathbf{x}}_{k|k-1} = [\hat{x}_{k|k-1}, \hat{v}_{x_{k|k-1}}, \hat{y}_{k|k-1}, \hat{v}_{y_{k|k-1}}]^T$ at time k is obtained by $\hat{\mathbf{x}}_{k|k-1} = \mathbf{F}\hat{\mathbf{x}}_{k-1}$. Meanwhile, the corresponding state prediction covariance $\mathbf{P}_{k|k-1}$ is calculated according to $\mathbf{P}_{k|k-1} = \mathbf{F}\mathbf{P}_{k-1}\mathbf{F} + \mathbf{Q}_k$.

Step 2: Coordinate conversion. The predicted state $\hat{\mathbf{x}}_{k|k-1}$ is converted from Cartesian coordinate to polar coordinate by

$$
\begin{aligned}
R^p_{Rk} &= \sqrt{\hat{x}^2_{k|k-1} + \hat{y}^2_{k|k-1}}, \\
\theta^p_{Rk} &= \arctan\left(\frac{\hat{y}_{k|k-1}}{\hat{x}_{k|k-1}}\right), \\
v^p_{dRk} &= \frac{\hat{x}_{k|k-1}\hat{v}_{x_{k|k-1}} + \hat{y}_{k|k-1}\hat{v}_{y_{k|k-1}}}{\sqrt{\hat{x}^2_{k|k-1} + \hat{y}^2_{k|k-1}}}.
\end{aligned}
\tag{8}
$$

Step 3: Data association. The data association method based on the minimal cost in Equation (9) is utilized to find the most likely measurements $[R^m_{Rk} \quad \theta^m_{Rk} \quad v^m_{dRk}]$ at time k within a predefined validation gate. The minimal cost criterion is defined by

$$
cost = 1 - \left(cost_{v_d} + cost_R + cost_\theta\right),
\tag{9}
$$

where $cost_{v_d}$, $cost_R$, $cost_\theta$ represent the association cost of Doppler velocity, range, and azimuth, respectively, which are defined as

$$
\begin{aligned}
cost_{v_d} &= W_{v_d} * \exp\left(-\left|v^m_{dRk} - v^p_{dRk}\right|^2 \Big/ \sigma^2_{v_d}\right), \\
cost_R &= W_R * \exp\left(-\left|R^m_{Rk} - R^p_{Rk}\right|^2 \Big/ \sigma^2_R\right), \\
cost_\theta &= W_\theta * \exp\left(-\left|\theta^m_{Rk} - \theta^p_{Rk}\right|^2 \Big/ \sigma^2_\theta\right),
\end{aligned}
\tag{10}
$$

where W_{v_d}, W_R, W_θ are the corresponding weights of three parameters, while σ_{v_d}, σ_R, and σ_θ denote their corresponding standard deviations. The candidate with the minimum *cost* value is associated with the current target. If a track can associate a measurement, go to step 4; otherwise, go to step 6.

Step 4: Measurement conversion. The associated measurement $[R^m_{Rk} \quad \theta^m_{Rk} \quad v^m_{dRk}]$ is converted from polar coordinate to Cartesian coordinate to obtain the measured target state $\mathbf{z}_k = [\tilde{x}_k, \tilde{v}_{x_k}, \tilde{y}_k, \tilde{v}_{y_k}]^T$ by

$$
\begin{aligned}
\tilde{x}_k &= R^m_{Rk} \cos\theta^m_{Rk}, \\
\tilde{y}_k &= R^m_{Rk} \sin\theta^m_{Rk}, \\
\tilde{v}_{x_k} &= (\tilde{x}_k - \tilde{x}_{k-1})/T, \\
\tilde{v}_{y_k} &= (\tilde{y}_k - \tilde{y}_{k-1})/T.
\end{aligned}
\tag{11}
$$

In practice, averaging is usually carried out with a longer time interval to make a more robust estimation of $\tilde{v}_{x_k}$ and $\tilde{v}_{y_k}$.

Step 5: State estimation. The target state $\hat{x}_k$ at time k, as well as the state estimation covariance matrix P_k, is updated by

$$K_k = P_{k|k-1} H^T (H P_{k|k-1} H^T + R_k),$$
$$\hat{x}_k = \hat{x}_{k|k-1} + K_k (z_k - H \hat{x}_{k|k-1}),$$
$$P_k = P_{k|k-1} - K_k H P_{k|k-1}, \tag{12}$$

where K_k is the Kalman gain at time k. Then the estimated target state $\hat{x}_k$ is used to update the current track, and $n = n + 1$.

Step 6: Determine if the track termination conditions are satisfied. If the conditions are met, the track will be terminated; otherwise, k is increased by 1 and go to step 1.

C. Track termination

A maintained track will be terminated if one of the following conditions occurs:

(1) There are no associated measurements in the past K frames out of L most recent frames.

(2) The estimated velocity reaches an unrealistic value v_{max}.

3. Experiment Results

To test the target detection and tracking performance of a bistatic compact HFSWR and verify the effectiveness and applicability of the proposed target tracking method, vessel detection and tracking experiments were conducted using field data simultaneously collected by a newly developed T/R-R compact HFSWR system, as shown in Figure 2, in operation at North China Sea from 9:57 a.m. to 13:42 p.m. on 30 April 2019. The monostatic T/R radar is located at Weihai (122.07°E, 37.54°N), while the other independent receiving station is deployed at Yantai (121.49°E, 37.45°N). The baseline distance L is 52 km, and φ is 10.35° for this configuration.

(a)

(b)

Figure 2. Transmitting antenna and one receiving antenna array of the developed T/R-R compact HFSWR system. (a) Transmitting antenna installed at Weihai radar station. (b) One receiving antenna array.

The compact HFSWR system used a solid-state transmitter with a maximum peak power of 2 kW and linear frequency modulated interrupting continuous wave (FMICW) as its transmitted waveform. A 10-meter-high omnidirectional log-periodic antenna, as shown in Figure 2a, was used to transmit electromagnetic waves with a working frequency of 4.7 MHz. Two similar linear receiving antenna arrays with an antenna element height of 4 meters were placed along the coast at Weihai and Yantai, respectively. Here, only the photo of the receiving antenna array at the Yantai station is shown in Figure 2b. Each receiving antenna array consists of eight active whip antenna units with an inter-element distance of 15 m. Thus, the aperture size of each receiving array is 105 m. The maximum detection range is designed as 100 km. As for vessel detection, the coherent integration time is set to be 262.144 s. A moving slide window method with a window length of 266.144 s is used to produce the detection data. The window slides forward with a step of 60 s, thus the data rate is 1 frame/min. The two radars are synchronized using a GPS time reference. Simultaneous automatic identification system (AIS) data were used as ground truth for comparisons and evaluations [39].

From the data collected by the bistatic compact HFSWR at Yantai, the range sum R and the azimuth θ_R of the detected targets were estimated, then their ranges R_R were calculated from Equation (2) and the measured state vectors $[R_R^m \ \theta_R^m \ v_{dR}^m]$ can be obtained. Then the proposed target tracking method was applied to the target detection data to obtain the bistatic target tracks. The target tracking method proposed in [11] was applied to the target plot data sequence measured by the monostatic HFSWR at Weihai to produce the monostatic target tracks. The threshold parameters involved in the tracking algorithm were determined via trial-and-error and they are summarized as follows:

- Track initiation—M is chosen to be equal to 3, and N is 4.
- Data association—The validation gate thresholds of range, azimuth, and Doppler velocity for monostatic HFSWR are 1.5 km, 5°, and 1 km/h, while those for bistatic HFSWR are set to 4 km, 5°, and 1.5 km/h, respectively. The weights W_{w_d}, W_R, and W_θ for monostatic radar are 0.6, 0.3, and 0.1, while those for bistatic radar are set to 0.7, 0.2, and 0.1, respectively.
- Track termination—K is set to be 3 and L is 5. The maximum target velocity v_{max} is set to 70 km/h.

From the obtained tracking results, three typical targets are selected for analysis, as shown in Figure 3.

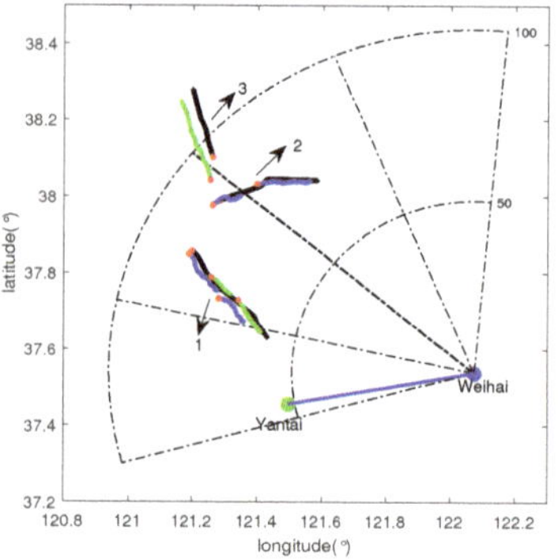

Figure 3. Tracking results of three typical targets. The blue and green dots indicate the location of the monostatic HFSWR at Weihai and the bistatic HFSWR at Yantai, respectively. The black angular sector illustrates the detection region of the monostatic HFSWR at Weihai. The three targets are marked as '1', '2', and '3' and named as target 1, target 2, and target 3, respectively. The tracks in blue, green, and black represent the tracking results from the monostatic radar, bistatic radar, and matched AIS, respectively. The red dot indicates the first plot of a track.

In Figure 3, the blue and green dots indicate the location of the monostatic HFSWR at Weihai and the bistatic HFSWR at Yantai, respectively. The black angular sector illustrates the detection region of the monostatic HFSWR at Weihai. The three targets are marked as '1', '2', and '3' and named as target 1, target 2, and target 3, respectively. The tracks in blue, green, and black represent the tracking results from the monostatic radar, bistatic radar, and matched AIS, respectively. The red dot indicates the first plot of a track. It is shown that target 1 is captured by both the monostatic and bistatic radars simultaneously, target 2 is tracked by the monostatic radar only, and target 3 is only detected by the bistatic radar. The general information of these three targets reported by AIS are listed in Table 1, and their photos are shown in Figure 4.

Table 1. General information for three targets.

	Target 1	**Target 2**	**Target 3**
MMSI	413331140	241491000	412328490
Ship Name	HUI RONG	MARAN HELEN	ZHONGTIEBOHAI 1 HAO
Ship Type	Cargo	Tanker	Passenger
Length (m)	98	274	182
Width (m)	16	46	25
Draught (m)	3.9	9.4	6.0

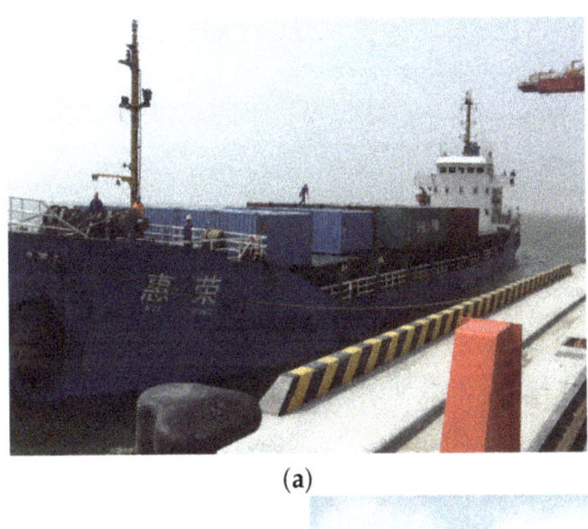

(a)

(b)

(c)

Figure 4. Photos of three targets considered in this paper. (a) Target 1—HUI RONG, the photo is from the website: www.yantaiport.com.cn. (b) Target 2—MARAN HELEN, the photo is from: www.marinetraffic.com. (c) Target 3—ZHONGTIEBOHAI 1 HAO, the photo is from: image.baidu.com.

The tracking results of these three targets are analyzed and compared in detail as follows.
(1) Target tracks obtained by both the monostatic and bistatic radars.
As shown in Figure 3, target 1 moves nearly along the bistatic bisector direction of the bistatic

radar at Yantai, its velocities have significant projection components in the radial direction of the monostatic radar at Weihai. Thus, it is captured by both radars. The tracks in longitudes and latitudes of target 1 are shown in Figure 5.

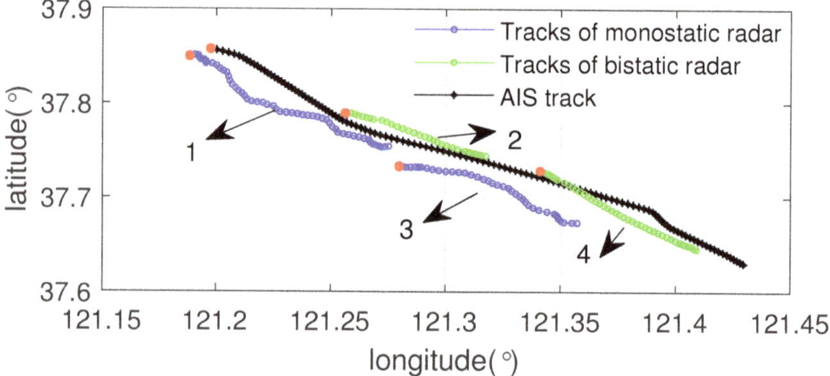

Figure 5. Comparison between radar tracks and automatic identification system (AIS) track of target 1.

It can be seen from Figure 5 that the obtained tracks from both the bistatic radar and monostatic radar are broken into two track segments. The four track segments are marked as '1', '2', '3', and '4', whose durations were 10:38–11:32 a.m. for track segment 1 with 54 plots, 11:13–11:44 a.m. for track segment 2 with 32 plots, 11:37 a.m.–12:16 p.m. for track segment 3 with 40 plots, and 11:54 a.m.–12:37 p.m. for track segment 4 with 44 plots. The duration for the matched AIS track was from 10:38 a.m. to 12:37 p.m. with 120 plots. It is shown that these four track segments together cover the entire AIS track.

In order to quantitatively evaluate the tracking performance, the positions provided by the AIS track in longitudes and latitudes, as well as the velocities, were projected onto the coordinates of the monostatic radar at Weihai and bistatic radar at Yantai, respectively, to obtain the corresponding range data, azimuth data, and Doppler velocity data sequences. The root-mean-square error (RMSE) and statistical error distribution criteria were used for accuracy evaluation.

The range data, azimuth data, and Doppler velocity data sequences corresponding to the four track segments are compared with those obtained by the AIS track projections, the results are shown in Figure 6. Figure 6a,c,e illustrate the range, azimuth, and Doppler velocity comparison results of the monostatic radar, respectively. The corresponding comparison results of the bistatic radar are shown in Figure 6b,d,f, respectively. The Sample number denotes the sequence number of a plot in a track.

As the monostatic radar at Weihai and bistatic radar at Yantai measure targets under different coordinates, the scales of their kinematic parameters are different. According to the velocities reported by AIS, target 1 moves at a nearly constant velocity during the observation period. However, the instability of the instantaneously measured course results in fluctuations in radial velocity projections. It can be observed that the range data, azimuth data, and Doppler velocity data sequences obtained from track segment 1 and track segment 3 of the monostatic compact HFSWR are in good agreement with those of the AIS results, the corresponding RMSEs are 1.24 km, 1.18°, and 1.3 km/h, respectively. By contrast, the range and azimuth data sequences obtained from track segment 2 and track segment 4 of the bistatic compact HFSWR agree well with those of the AIS results, with a relatively larger RMSEs of 3.6 km and 1.94°, respectively. It is worth noting that the agreement between the bistatic Doppler velocity data sequences and the projected results of AIS is fairly good, with a RMSE of 0.55 km/h. For clarity, the results are listed in Table 2.

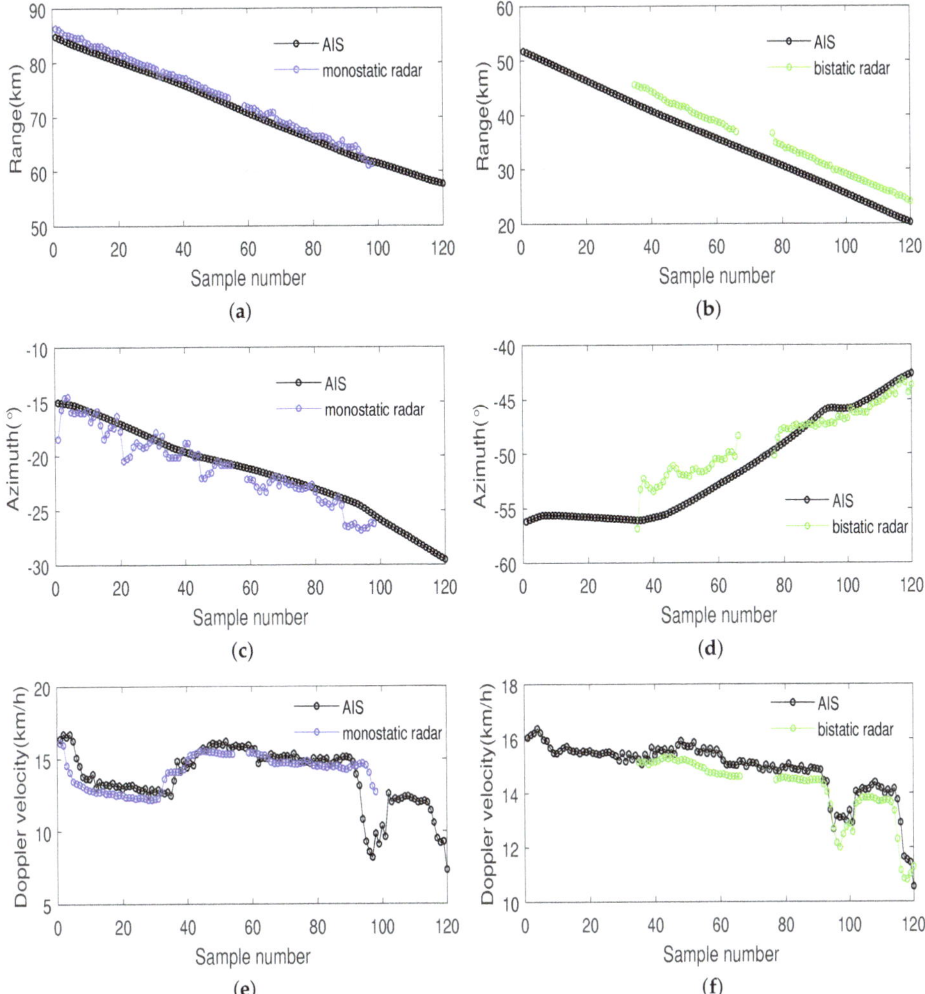

Figure 6. Kinematic parameter comparisons for target 1. (**a**) Range comparison for monostatic radar. (**b**) Range comparison for bistatic radar. (**c**) Azimuth comparison for monostatic radar. (**d**) Azimuth comparison for bistatic radar. (**e**) Doppler velocity comparison for monostatic radar. (**f**) Doppler velocity comparison for bistatic radar.

Table 2. Root-mean-square error (RMSE) of range, azimuth and Doppler velocity for target 1.

	Range (km)	Azimuth (°)	Doppler velocity (km/h)
Monostatic radar	1.24	1.18	1.30
Bistatic radar	3.60	1.94	0.55

The error distributions of the range, azimuth, and Doppler velocity data sequences of target 1 for both monostatic and bistatic HFSWR are illustrated in Figure 7 for more detailed analysis.

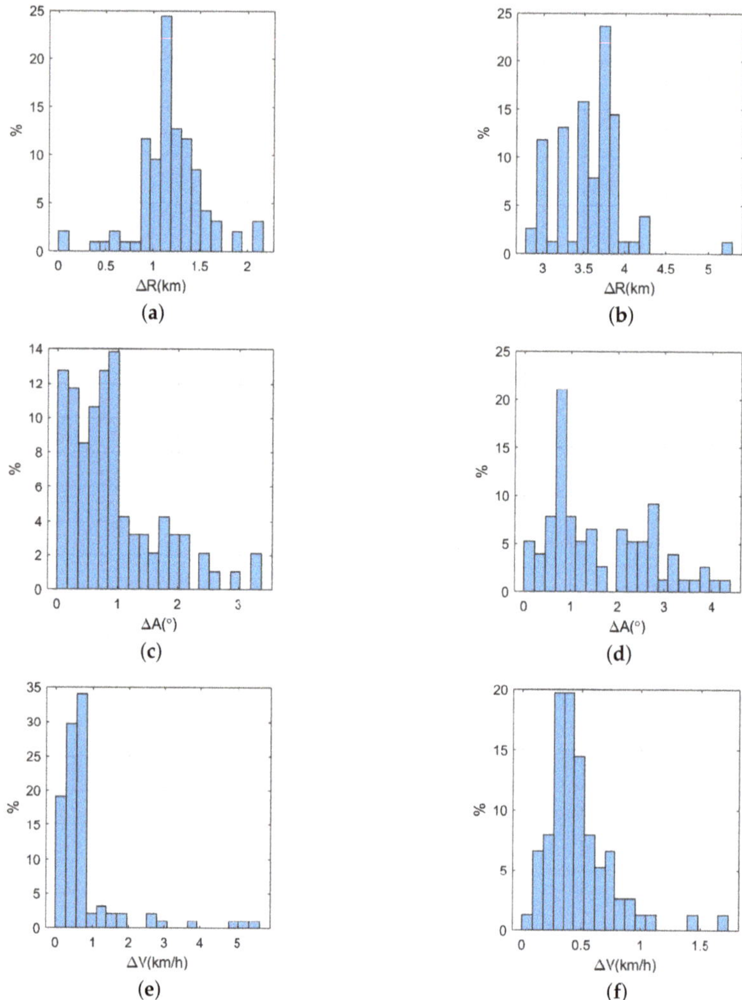

Figure 7. Kinematic parameter error distributions for target 1. (**a**) Range error distribution for monostatic radar. (**b**) Range error distribution for bistatic radar. (**c**) Azimuth error distribution for monostatic radar. (**d**) Azimuth error distribution for bistatic radar. (**e**) Doppler velocity error distribution for monostatic radar. (**f**) Doppler velocity error distribution for bistatic radar.

It can be seen from Figure 7 that the majority of range error, azimuth error, and Doppler velocity error for monostatic HFSWR are less than 1.5 km, 2°, and 1 km/h, respectively, while those for bistatic HFSWR are less than 4 km, 3°, and 1 km/h, respectively.

The above results indicate that the monostatic radar achieves better tracking accuracy than that of the bistatic radar for target 1. However, the RMSE of the Doppler velocity from bistatic HFSWR is lower than that of monostatic HFSWR. The azimuth estimation results from both monostatic radar and bistatic radar display some random fluctuations due to the coarse azimuth resolution caused by reduced aperture size. It is worth mentioning that the range accuracy is different for monostatic radar and bistatic radar. The range resolution of the CORMS is designed to be 2.5 km. The range accuracy of the monostatic radar is much higher than this value as reported in [11], while the range accuracy of the bistatic radar is worse than this design value because the calculated ranges are affected by error in azimuth estimations.

It can also be observed that the three kinematic parameters between track segment 1 and track segment 3, as well as between track segment 2 and track segment 4 are consistent. The track fragmentation is probably due to the missed detections at some sampling time, which may be caused by the weak returned echos due to the relatively smaller ship size, or sea clutter interference, etc. Combining the simultaneous target detections of a monostatic and bistatic radar, the overlapped discontinuous track segments belonging to the same target can be bridged together to obtain a longer track. Thus, there is a potential for this radar configuration to maintain better track consistence.

(2) Target tracks obtained only by the monostatic radar.

As a Doppler radar, HFSWR favors detecting targets that have significant velocity projection components along its radial directions. The track of target 2 can only be produced from the monostatic radar data as it sails nearly along the tangent direction of the isorange ellipse of the bistatic radar. The obtained two track segments, marked as '1' and '2', and the matched AIS track are illustrated in Figure 8. The durations of these two track segments are 11:52 a.m.–12:28 p.m. for track segment 1 with 37 plots, 12:40–1:33 p.m. for track segment 2 with 54 plots. The duration for the matched AIS track is from 11:52 a.m. to 1:33 p.m. with 102 plots.

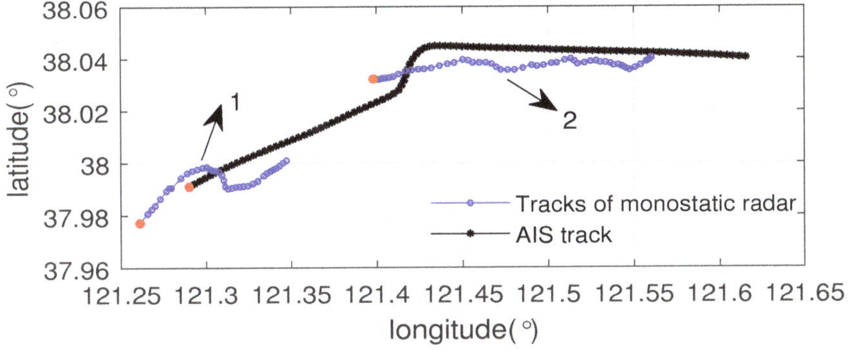

Figure 8. Track comparison between monostatic radar and AIS of target 2.

The comparisons of the corresponding range data, azimuth data, and Doppler velocity data sequences are shown in Figure 9. It is observed that both the range data and Doppler velocity data sequences agree well with those of the AIS projection results with RMSEs of 1.37 km, and 0.36 km/h, respectively, which are similar to the results obtained by the monostatic radar for target 1. However, the accuracy of the azimuth data sequence is a little worse with an RMSE of 2.66° due to the target's longer distance from the radar site, which leads to large target position deviations from its true trajectory. It can be seen from Figure 9c that the Doppler velocity changes from positive values to negative ones, and then becomes positive again around the 40th plots. This is because that target 2 moves nearly along the tangent direction of the Weihai radar. At some locations, its Doppler velocity becomes nearly zero and thus it is difficult to be detected. The missed detections may lead to the track fragmentation. However, the tracking algorithm adopted here has not been considered such situations.

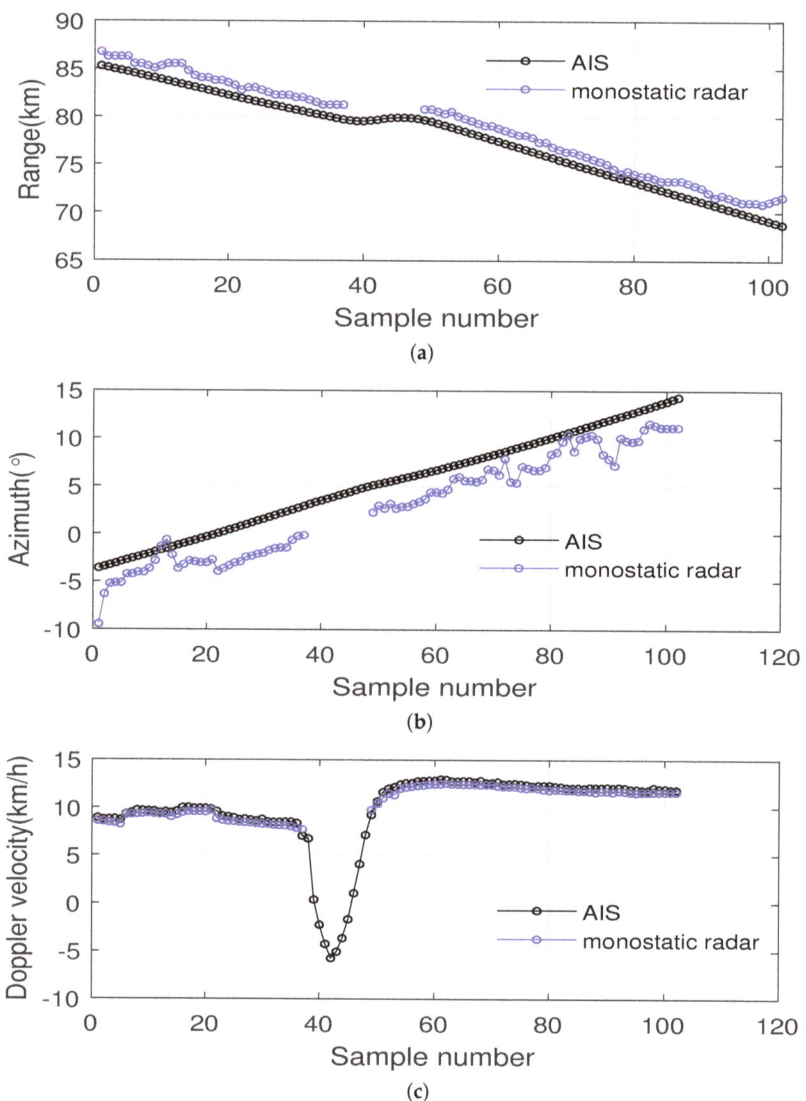

Figure 9. Kinematic parameter comparisons for target 2. (**a**) Range comparison for monostatic radar. (**b**) Azimuth comparison for monostatic radar. (**c**) Doppler velocity comparison for monostatic radar.

The error distributions of the range, azimuth, and Doppler velocity data sequences of target 2 for monostatic HFSWR are illustrated in Figure 10 for more detailed analysis.

It can be seen from Figure 10 that the majority of range error, azimuth error, and Doppler velocity error for monostatic HFSWR are less than 1.5 km, 4°, and 0.6 km/h, respectively.

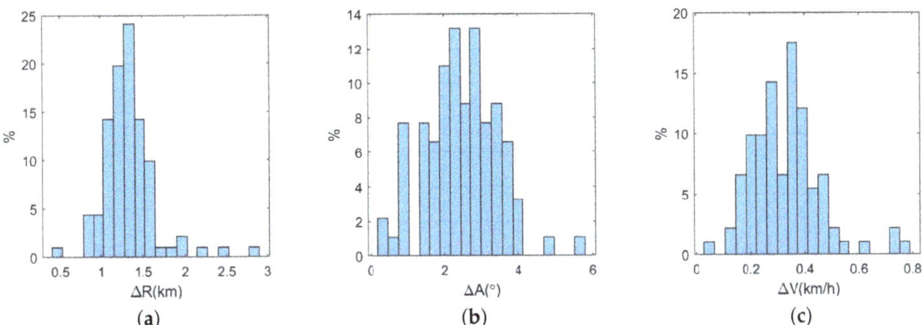

Figure 10. Kinematic parameter error distributions for target 2. (**a**) Range error distribution. (**b**) Azimuth error distribution. (**c**) Doppler velocity error distribution.

(3) Target track obtained only by the bistatic radar.

It can be seen from Figure 3 that target 3 goes beyond the maximum detection range of 100 km of the monostatic radar at Weihai. Fortunately, it is still within the detection range of the bistatic radar at Yantai. Thus, it is only captured by the bistatic radar. From this perspective, the coverage area of the monostatic radar is expanded. The obtained track, as well as its matched AIS track, whose duration are from 10:11 a.m. to 11:08 a.m. with 58 plots, are shown in Figure 11. It is observed that the track obtained by the bistatic radar deviates from its true trajectory but shows a similar course with that of AIS.

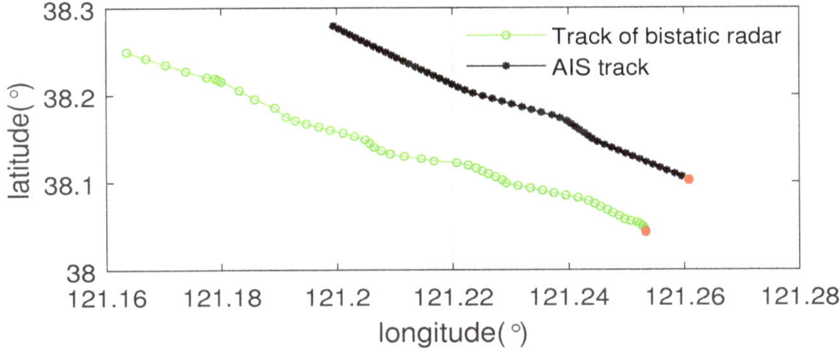

Figure 11. Track comparison between bistatic radar and AIS of target 3.

The kinematic comparisons of the corresponding range data, azimuth data, and Doppler velocity data sequences are shown in Figure 12.

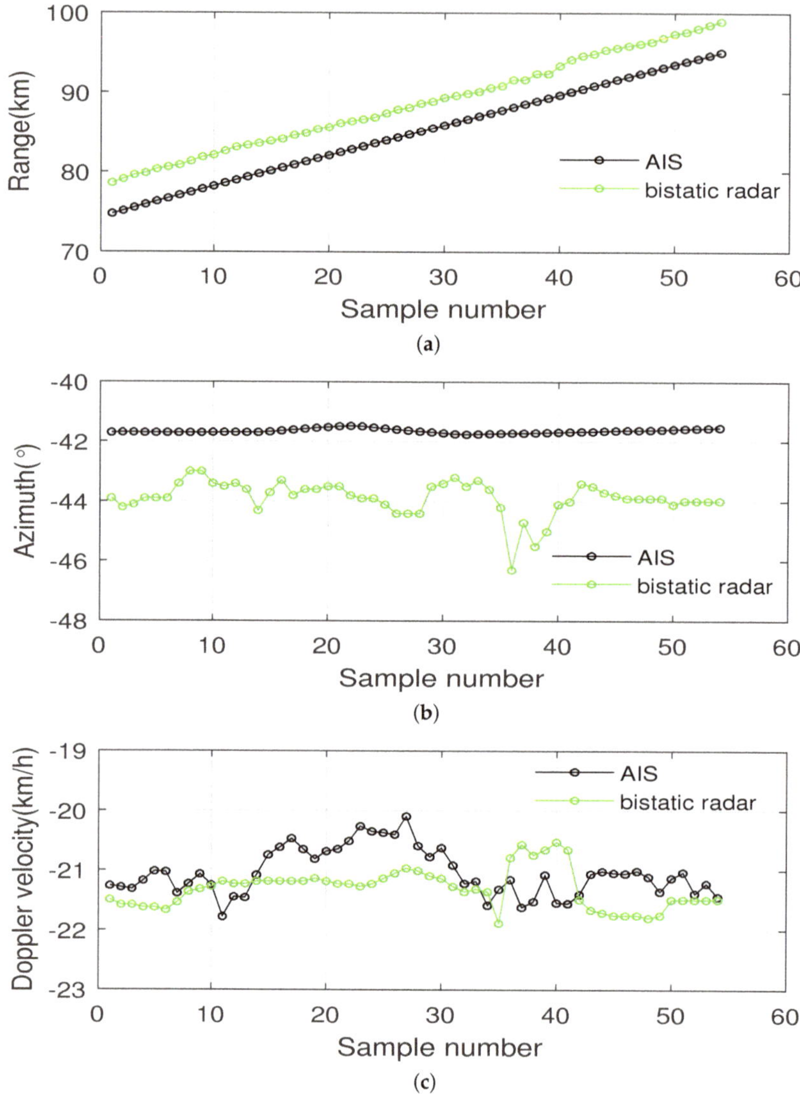

Figure 12. Kinematic parameter comparisons for target 3. (**a**) Range comparison for bistatic radar.
(**b**) Azimuth comparison for bistatic radar. (**c**) Doppler velocity comparison for bistatic radar.

Compared with the results provided by AIS, it can be noted that the range errors are nearly
constant for all the plots with an RMSE of 3.7 km and a standard deviation of 0.3 km. The azimuth
data of AIS keep nearly constant, while the azimuth data sequence of the bistatic radar presents
fluctuations with an RMSE of 2.3° and a maximum deviation of 4.58°. The RMSE of Doppler velocity
is 0.55 km/h and the maximum deviation is 1.05 km/h, indicating again that a bistatic HFSWR can
measure the Doppler velocity of a target with high accuracy.

The error distributions of the range, azimuth, and Doppler velocity data sequences of target 3 for
bistatic HFSWR are illustrated in Figure 13 for more detailed analysis.

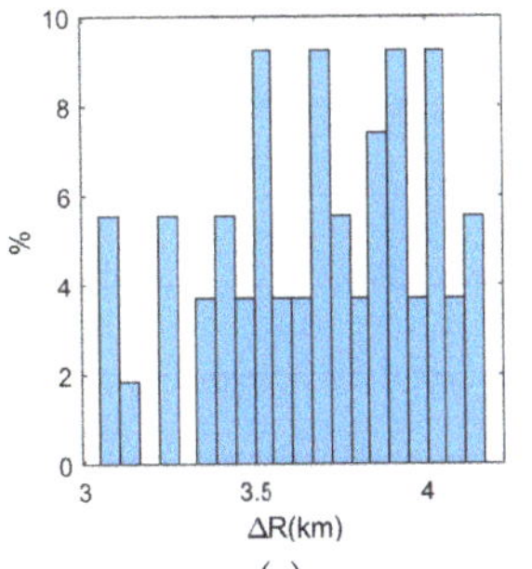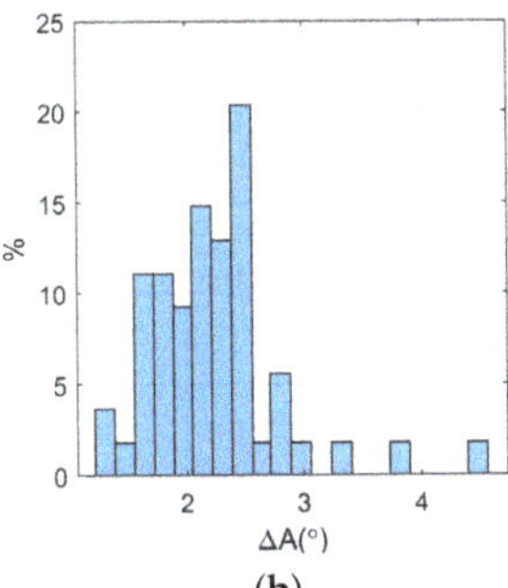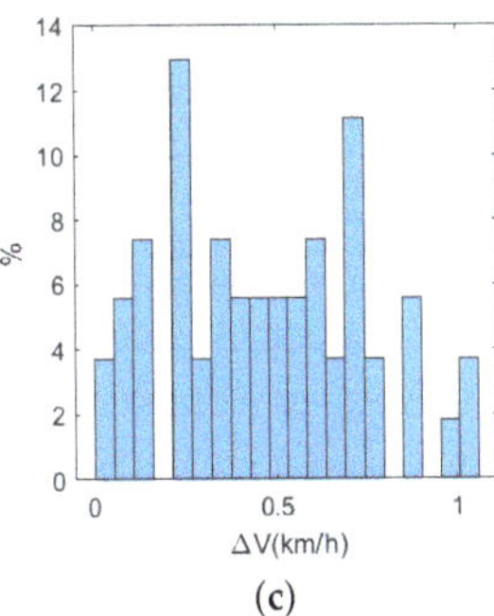

Figure 13. Kinematic parameter error distributions for target 3. (**a**) Range error distribution. (**b**) Azimuth error distribution. (**c**) Doppler velocity error distribution.

It can be seen from Figure 13 that the majority of range error, azimuth error, and Doppler velocity error for monostatic HFSWR are less than 4 km, 3°, and 1 km/h, respectively.

4. Discussion

By analyzing the tracking results from the developed monostatic and bistatic compact HFSWR for three targets considered here, it can be summarized that:

(i) Both monostatic and bistatic compact HFSWR can produce target tracks, simultaneously or complementarily. However, the tracks obtained by compact HFSWR are usually fragmented into short track segments due to the lower detection rate, clutter interference, etc. With the T/R-R configuration, there are potentials that the simultaneously obtained tracks can be fused to improve the tracking accuracy, and the complementary tracks can be associated to enhance the track consistence. Specific track association and fusion methods should be developed to achieve this goal.

(ii) The azimuth and Doppler velocity estimation accuracies of monostatic and bistatic HFSWR are comparable. However, the range estimation accuracy and the resulting tracking accuracy of bistatic HFSWR is relatively lower than those of the monostatic HFSWR. The effect of θ_R on the estimation accuracy of R_R is significant. It is worth developing a method to mitigate the measurement error due to the reduced azimuth resolution of a compact radar. The coupling characteristic between range and azimuth should be fully considered in designing superior tracking algorithms.

(iii) The coverage area of a monostatic HFSWR can be extended by a T/R-R configuration with only a little extra cost.

5. Conclusions

The target detection and tracking performance of a bistatic compact HFSWR was investigated in this paper. An applicable target tracking method for bistatic compact HFSWR was proposed and its performance was verified using the field data collected simultaneously by a monostatic and a bistatic HFSWRs sharing the same transmitter. The experiment results demonstrate that the bistatic HFSWR can produce target tracks with acceptable errors. Moreover, the tracking results of the bistatic HFSWR were compared with those of a monostatic HFSWR. It is found that the range estimation accuracy, thus the tracking accuracy of a bistatic HFSWR is lower than that of a monostatic one based on the data in this work. The combination of a monostatic HFSWR and a bistatic HFSWR provides target observations from different perspectives, thus, the T/R-R configuration can obtain synchronous as well as complementary information for the same target. This configuration may be potentially exploited to increase the target detection probability, improve the target detection and tracking accuracy, and enhance the track continuity. Also, the T/R-R configuration can increase the detection range and extend the coverage area, thus more targets can be monitored.

In the future, more data should be collected by the T/R-R compact HFSWR so that enough statistical sample can be used to determine how long a track should be maintained based on predictions only. According to our experience, improper selection of the termination criteria will lead to track fragmentation. In our experiments, it is found that the fragmentation phenomenon appears more often for the bistatic radar than the monostatic radar. The effect of the parameters involved in the bistatic tracking algorithm on the tracking performance also needs to be studied. Moreover, new track association and fusion methods will be developed to associate and merge the track segments simultaneously obtained by the monostatic radar and bistatic radar into a longer track to improve the continuity as well as the accuracy of target tracking with compact HFSWR.

Author Contributions: Conceptualization, W.S. and W.H.; methodology, W.S. and M.J.; software, W.S.; validation, W.S., W.H., and Y.J.; formal analysis, W.S. and W.H.; investigation, M.J.; resources, Y.J. and Y.D.; data curation, Y.J.; writing—original draft preparation, W.S.; writing—review and editing, W.H.; visualization, M.J.; supervision, W.H. and Y.D.; project administration, W.S.; funding acquisition, W.S. All authors have read and agreed to the published version of the manuscript.

Funding: This research was funded by the National Key R&D Program of China with grant number 2017YFC1405202, National Natural Science Foundation of China with grant numbers 61501520, 61831010, 61671166, the Fundamental Research Funds for the Central Universities with grant numbers 19CX02046A, 17CX02079, and CSC scholarship for 1-year research abroad at Memorial University.

Acknowledgments: The authors would like to thank the anonymous reviewers for their comments and suggestions that helped to improve the quality of this article.

Conflicts of Interest: The authors declare no conflict of interest.

Abbreviations

The following abbreviations are used in this manuscript:

HFSWR	High-Frequency Surface Wave Radar
MUSIC	Multiple SIgnal Classfication
MIMO	Multiple-Input Multiple-Output
MTT	Multi-Target Tracking
DBF	Digital Beam Forming
CMKF	Converted Measurement Kalman Filter
AIS	Automatic Identification System
RMSE	Root Mean Square Error

References

1. Sevgi, L.; Ponsford, A.; Chan, H.C. An integrated maritime surveillance system based on high-frequency surface-wave radars. 1. Theoretical background and numerical simulations. *IEEE Antenn. Propag. Mag.* **2001**, *43*, 28–43. [CrossRef]
2. Green, D.; Gill, E.; Huang, W. An inversion method for extraction of wind speed from high-frequency ground-wave radar. *IEEE Trans. Geosci. Remote Sens.* **2009**, *47*, 3530–3549. [CrossRef]
3. Ponsford, A.M.; Wang, J. A review of high frequency surface wave radar for detection and tracking of ships. *Turk. J. Elect. Eng. Comput. Sci.* **2010**, *18*, 409–428.
4. Park, S.; Cho, C.J.; Ku, B.; Lee, S.; Ko, H. Compact HF surface wave radar data generating simulator for ship detection and tracking. *IEEE Geosci. Remote Sens. Lett.* **2017**, *14*, 969–973. [CrossRef]
5. Roarty, H.J.; Lemus, E.R.; Handel, E.; Glenn, S.M.; Barrick, D.E.; Isaacson, J. Performance evaluation of SeaSonde high-frequency radar for vessel detection. *Mar. Technol. Soc. J.* **2011**, *45*, 14–24. [CrossRef]
6. Smith, M.; Roarty, H.J.; Glenn, S.; Whelan, C.; Barrick, D.E.; Isaacson, J. Methods of associating CODAR SeaSonde vessel detection data into unique tracks. In Proceedings of the MTS/IEEE Oceans, San Diego, CA, USA, 23–27 September 2013; pp. 1–5.
7. Lu, B.; Wen, B.; Tian, Y.; Wang, R. A vessel detection method using compact-array HF radar. *IEEE Geosci. Remote Sens. Lett.* **2017**, *14*, 2017–2021. [CrossRef]

8. Helzel, T.; Hansen, B.; Kniephphoff, M.; Petersen, L.; Valentin, M. Introduction of the compact HF radar WERA-S. In Proceedings of the IEEE/OES Baltic International Symposium, Klaipeda, Lithuania, 8–10 May 2012; pp. 1–3.

9. Ji, Y.; Zhang, J.; Wang, Y.; Sun, W.; Li, M. Target monitoring using small-aperture compact high-frequency surface wave radar. *IEEE Aerosp. Electron. Syst. Mag.* **2018**, *33*, 22–31. [CrossRef]

10. Sun, W.; Huang, W.; Ji, Y.; Dai, Y.; Ren, P.; Zhou, P. Vessel tracking with small-aperture compact high-frequency surface wave radar. In Proceedings of the MTS/IEEE Oceans, Marseille, France, 17–20 June 2019; pp. 1–4.

11. Sun, W.; Huang, W.; Ji, Y.; Dai, Y.; Ren, P.; Zhou, P.; Hao, X. A vessel azimuth and course joint re-estimation method for compact HFSWR. *IEEE Trans. Geosci. Remote Sens.* **2020**, *58*, 1041–1051. [CrossRef]

12. Yao, G.; Xie, J.; Huang, W. First-order ocean surface cross section for shipborne HFSWR incorporating a horizontal oscillation motion model. *IET Radar Sonar Navig.* **2018**, *12*, 973–978. [CrossRef]

13. Chen, Z.; He, C.; Zhao, C.; Xie, F. Enhanced target detection for HFSWR by 2-D MUSIC based on sparse recovery. *IEEE Geosci. Remote Sens. Lett.* **2017**, *14*, 1983–1987. [CrossRef]

14. Barca, P.; Maresca, S.; Grasso, R.; Bryan, K.; Horstmann, J. Maritime surveillance with multiple over-the-horizon HFSW radars: An overview of recent experimentation. *IEEE Aerosp. Electron. Syst. Mag.* **2015**, *30*, 4–18. [CrossRef]

15. Lesturgie, M. Improvement of high-frequency surface waves radar performances by use of multiple-input multiple-output configurations. *IET Radar Sonar Navig.* **2008**, *3*, 49–61. [CrossRef]

16. Anderson, S.J.; Edwards, P.J.; Marrone, P.; Abramovich, Y.I. Investigations with SECAR- a bistatic HF surface wave radar. In Proceedings of the 2003 International Conference on Radar, Adelaide, Australia, 3–5 September 2003; pp. 717–722.

17. Marrone, P.; Edwards, P. The case for bistatic HF surface wave radar. In Proceedings of the 2008 Proceedings of the International Conference on Radar, Adelaide, Australia, 2–5 September 2008; pp. 633–638.

18. Anderson, S.J. Optimizing HF Radar Siting for Surveillance and Remote Sensing in the Strait of Malacca. *IEEE Geosci. Remote Sens. Lett.* **2012**, *51*, 1805–1816. [CrossRef]

19. Xie, J.; Sun, M.; Ji, Z. Space-time model of the first-order sea clutter in onshore bistatic high frequency surface wave radar. *IET Radar Sonar Navig.* **2015**, *9*, 55–61. [CrossRef]

20. Wang, J.; Dizaji, R.; Ponsford, A.M. Analysis of clutter distribution in bistatic high frequency surface wave radar. In Proceedings of the Canadian Conference on Electrical and Computer Engineering, Ontario, QC, Canada, 2–5 May 2004; pp. 1301–1304.

21. Liu, C.; Chen, B.; Zhang, S. Co-channel interference suppression by time and range adaptive processing in bistatic high-frequency surface wave synthesis impulse and aperture radar. *IET Radar Sonar Navig.* **2009**, *3*, 638–645. [CrossRef]

22. Gill, E.; Walsh, J. High frequency bistatic cross section of the ocean surface. *Radio Sci.* **2001**, *36*, 1459–1476. [CrossRef]

23. Gill, E.; Huang, W.; Walsh, J. The effect of the bistatic scattering angle on the high-frequency radar cross sections of the ocean surface. *IEEE Geosci. Remote Sens. Lett.* **2008**, *5*, 143–146. [CrossRef]

24. Ma, Y.; Gill, E.; Huang, W. Bistatic high-frequency radar ocean surface cross section incorporating a dual-frequency platform motion model. *IEEE J. Ocean. Eng.* **2018**, *43*, 205–210. [CrossRef]

25. Ma, Y.; Huang, W.; Gill, E. Bistatic high frequency radar ocean surface cross section for an FMCW source with an antenna on a floating platform. *Int. J. Antennas Propag.* **2016**, *2016*, 8675964. [CrossRef]

26. Ma, Y.; Gill, E.; Huang, W. First-Order Bistatic high-Frequency radar ocean surface cross-section for an antenna on a floating platform. *IET Radar Sonar Navig.* **2016**, *10*, 1136–1144. [CrossRef]

27. Gill, E.; Huang, W.; Walsh, J. On the development of a second-order bistatic radar cross section of the ocean surface—A high-frequency result for a finite scattering patch. *IEEE J. Ocean. Eng.* **2006**, *31*, 740–750. [CrossRef]

28. Ma, Y.; Huang, W; Gill, E. The second-order bistatic high frequency radar ocean surface cross section for an antenna on a floating platform. *Can. J. Remote Sens.* **2016**, *42*, 332–343. [CrossRef]

29. Lipa, B.; Whelan, C.; Rector, B.; Nyden, B. HF radar bistatic measurement of surface current velocities: Drifter comparisons and radar consistency checks. *Remote Sens.* **2009**, *1*, 1190–1211. [CrossRef]

30. Yang, J.; Wen, B.; Zhang, C.; Huang, X.; Yan, Z.; Shen, W. A bistatic HF radar for surface current mapping. *IEICE Electron. Expr.* **2010**, *7*, 1435–1440. [CrossRef]

31. Trizna, D.B. A bistatic HF radar for current mapping and robust ship tracking. In Proceedings of the MTS/IEEE Oceans, Quebec City, QC, Canada, 15–18 September 2008; pp. 1–6.

32. Hardman, R.L.; Wyatt, L.R.; Engleback, C.C. Measuring the directional ocean spectrum from simulated bistatic HF radar data. *Remote Sens.* **2020**, *12*, 313. [CrossRef]

33. Huang, W.; Gill, E.; Wu, X.; Li, L. Measurement of sea surface wind direction using bistatic high-frequency radar. *IEEE Trans. Geosci. Remote Sens.* **2012**, *50*, 4117–4122. [CrossRef]

34. Chan, H.C.; Davies, T.W.; Hall, P. Iceberg tracking using HF surface wave radar. In Proceedings of the MTS/IEEE Oceans, Halifax, NS, Canada, 6–9 October 1997; pp. 1290–1296.

35. Vivone, G.; Braca, P.; Horstmann, P. Knowledge-based multi-target ship tracking for HF surface wave radar systems. *IEEE Trans. Geosci. Remote Sens.* **2015**, *53*, 3931–3949. [CrossRef]

36. Braca, P.; Marano, S.; Matta, V.; Willett, P. Asymptotic efficiency of the PHD in multitarget/multisensor estimation. *IEEE J. Sel. Top. Signal Process.* **2013**, *7*, 553–564. [CrossRef]

37. Zong, H.; Yu, C.; Zhou, G.; Quan, T. Track initiation in monostatic-bistatic composite high frequency surface wave radar network based on NFE model. In Proceedings of the 2010 International Conference on Electronics and Information Engineering, Kyoto, Japan, 1–3 August 2010; pp. 81–85.

38. Bar-Shalom, Y.; Willett, P.; Tian, X. *Tracking and Data Fusion: A Handbook of Algorithms*; YBS Publishing: Storrs, CT, USA, 2011.

39. Nikolic, D.; Stojkovic, N.; Lekic, N. Maritime over the horizon sensor integration: High frequency surface-wave-radar and automatic identification system data integration algorithm. *Sensors* **2018**, *18*, 1147. [CrossRef]

Article

Passive Detection of Moving Aerial Target Based on Multiple Collaborative GPS Satellites

Mingqian Liu [1], Zhiyang Gao [1,*], Yunfei Chen [2], Hao Song [3] and Yuting Li [1] and Fengkui Gong [1]

[1] State Key Laboratory of Integrated Service Networks, Xidian University, Xi'an 710071, China; mqliu@mail.xidian.edu.cn (M.L.); mqliu@xidian.edu.cn (Y.L.); fkgong@xidian.edu.cn (F.G.)

[2] School of Engineering, University of Warwick, Coventry CV4 7AL, UK; yunfei.chen@warwick.ac.uk

[3] Bradley Department of Electrical and Computer Engineering, Virginia Tech, Blacksburg, VA 24060, USA; haosong@vt.edu

* Correspondence: gaozhiyang@stu.xidian.edu.cn

Received: 30 November 2019; Accepted: 9 January 2020; Published: 12 January 2020

Abstract: Passive localization is an important part of intelligent surveillance in security and emergency applications. Nowadays, Global Navigation Satellite Systems (GNSSs) have been widely deployed. As a result, the satellite signal receiver may receive multiple GPS signals simultaneously, incurring echo signal detection failure. Therefore, in this paper, a passive method leveraging signals from multiple GPS satellites is proposed for moving aerial target detection. In passive detection, the first challenge is the interference caused by multiple GPS signals transmitted upon the same spectrum resources. To address this issue, successive interference cancellation (SIC) is utilized to separate and reconstruct multiple GPS signals on the reference channel. Moreover, on the monitoring channel, direct wave and multi-path interference are eliminated by extensive cancellation algorithm (ECA). After interference from multiple GPS signals is suppressed, the cycle cross ambiguity function (CCAF) of the signal on the monitoring channel is calculated and coordinate transformation method is adopted to map multiple groups of different time delay-Doppler spectrum into the distance–velocity spectrum. The detection statistics are calculated by the superposition of multiple groups of distance-velocity spectrum. Finally, the echo signal is detected based on a properly defined adaptive detection threshold. Simulation results demonstrate the effectiveness of our proposed method. They show that the detection probability of our proposed method can reach 99%, when the echo signal signal-to-noise ratio (SNR) is only −64 dB. Moreover, our proposed method can achieve 5 dB improvement over the detection method using a single GPS satellite.

Keywords: cyclic cross ambiguity function; data fusion; GPS; multiple satellites collaboration; passive detection

---

1. Introduction

With the development of space technologies, Global Navigation Satellite Systems (GNSSs) have been widely applied in various applications and have been playing an extremely important role in many fields [1–3]. In a GNSS, GNSS navigation satellites need to be widely distributed and used as radiation sources. Among existing GNSSs [4–7], Global Positioning System (GPS) has been broadly recognized and acknowledged as an advanced and mature technology in target detection due to its wider coverage and shorter observation time compared to other GNSSs, such as the Beidou satellite system [8] and Golbal navigation satellite system (GLONASS). Generally, target detection is conducted using a single GPS as the radiation source [9]. However, due to the widely deployed GNSSs, a GPS satellite signal receiver will inevitably receive multiple GPS signals, causing signal contamination and the failure of echo signal detection. Hence, instead of relying on only a single GPS satellite radiation source, effective weak echo signal detection methods using multiple GPS satellite radiation sources need to be studied.

Due to the benefits of global coverage, 24-h operation, and easy-to-access signal sources, GPS navigation satellite signals will be employed as third-party illumination sources to study GPS-based external radiation source target detection methods in this paper. For weak echoes of GPS satellites, many researchers have made preliminary explorations [10–14]. These works focus on studying the detection of a single GPS satellite signal. In fact, different GPS satellites in the zenith may share the same frequency bands to send GPS signals. On the other hand, a near-Earth orbit target may be simultaneously covered by multiple satellite beams. As a result, reference signals transmitted on reference channels may be contaminated by other unexpected GPS signals [15], which makes the estimation of reference signals very difficult. In addition, interference caused by other GPS signals may degrade the performance of direct-path interference (DPI)/multi-path interference (MPI) suppression and the subsequent echo detection on an echo channel [16]. For effective echo detection, multiple GPS signals received on the reference channels and the DPI/MPI on monitoring channels should be purified and suppressed.

On the other hand, the received power of GPS signals is very low due to a long-distance propagation [17,18]. Furthermore, the signals after the target reflection, also referred to as the target echo, would be even weaker. Even under the condition that the clutter and interference suppression could be realized, the extremely weak GPS echo still needs a long coherent accumulation time to be detected. In order to improve the detection probability of the target echo, the existing methods improved the detection performance by constructing a multi-station joint detection system and merging the detection results of multiple radiation sources [19–21]. Unfortunately, the signal processing methods introduced in these works are not suitable for GPS signals. Therefore, effective use of the received multiple GPS signals to construct a joint detection system with high signal-to-noise ratio (SNR) is an important technical problem that has to be tackled.

The main contributions of this paper can be summarized as follows. A GPS weak echo signal detection method is proposed based on multi-star data fusion. To be specific, under the condition of multiple satellite sources, multiple reference signals will be mixed into a reference channel, resulting in degrading the DPI and MPI suppression effects on a monitoring channel. Therefore, firstly, the proposed method separates and reconstructs multiple GPS reference signals on the reference channel, based on which the Extensive Cancellation Algorithm (ECA) is used to monitor and suppress DPI and MPI. Then, to address the problem of the weak target reflection echo, which is very difficult to be detected, a coordinate conversion algorithm is applied to fuse detection statistics of multiple GPS satellites and obtain a final detection statistic. By this way, the peak value of weak echo detections and the probability of weak echo detections could be improved. Finally, by defining an adaptive detection threshold, the weak echo could be adaptively detected.

The reminder of this paper is organized as follows. In Section 2, the system model under multiple GPS satellite radiation sources is presented. A novel method of joint target detection is proposed and the corresponding technical details are described in the third Section 3, including the separation and reconstruction of direct wave in a reference channel, the suppression of direct and multi-path interference in a monitoring channel, and the construction of detection quantity and the design of detector. In the Section 4, the extensive simulation studies are conducted.

2. System Model

The system model of the echo signal detection and reception system based on data fusion of multiple GPS satellites is shown in Figure 1, where R_t is the distance from the satellite to a target, L is the distance from a satellite to a receiver, θ is the arrival angle of the echo, φ is the arrival angle of the direct wave, and R_r is the distance from a target to a receiver.

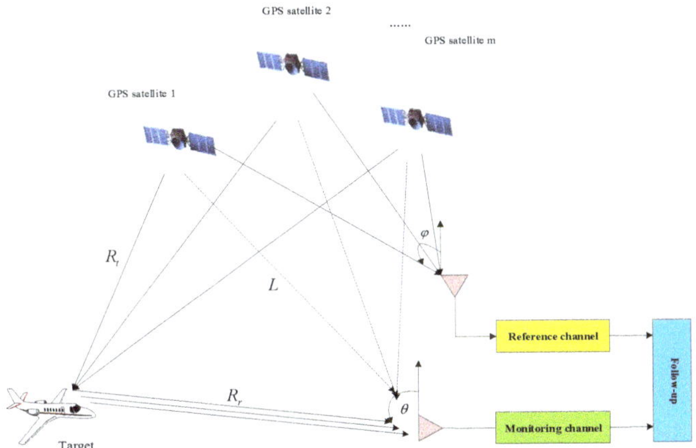

Figure 1. Passive detection based on multiple collaborative GPS satellites.

As shown in Figure 2, a standard GPS receiver exists in the reference channel, which is vertically pointed to the zenith in order to receive the reference signals. Then, the received reference signals will be used for the DPI and MPI suppression in the monitoring channel. Through the monitoring channel, the receiver is able to realize the self-positioning of the detection system, the baseline measurement, and the tracking of the current satellite, obtaining the ephemeris position information of the zenith at the current moment. These measurements will facilitate subsequent offline signal processing.

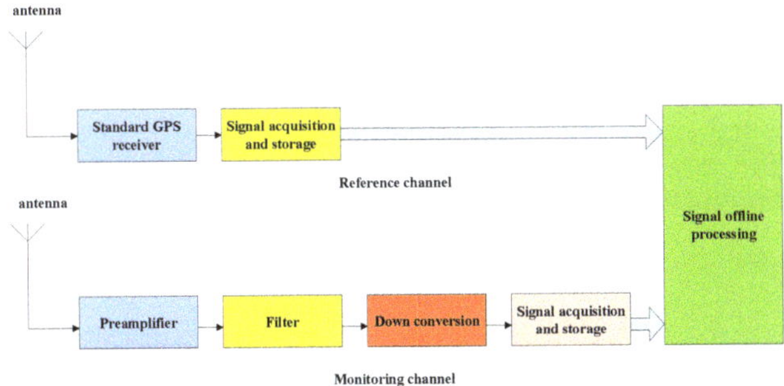

Figure 2. The diagram of the receiving signal channels.

The monitoring channel for reflected echo receptions is mainly composed of a GPS receiving antenna, an amplifier, a filter, a down conversion circuit, and a digital storage oscilloscope. In the GPS receiving antenna, a left-handed circularly polarized receiving antenna is deployed, which is tilted towards the target. The amplifier amplifies the GPS echo signal to achieve a 20 dB to 30 dB amplification. The filter is employed for interference cancellation of clutter signals outside the GPS L-band, reducing the influence of out-of-band clutter on subsequent detection processes. The down-conversion module is applied to down-convert the GPS signal from the L-band to the intermediate frequency to reduce the complexity of the processing. The digital storage oscilloscope is used for the rapid sampling and storage of data to enable following offline processing.

In the monitoring channel, it is assumed that there are M GPS satellites as the radiation source. In addition to the reflected echoes received by the plurality of GPS satellite signals from the target, on the monitoring channel, direct wave signals and multipath interference can be received, which are produced by the reflection of GPS direct wave signals from close ground objects. Accordingly, the received signals of the monitoring channel can be expressed as

$$x_s(t) = \alpha_1 x_1(t - \tau_1)\exp(j2\pi f_{d_1}) + \alpha_2 x_2(t - \tau_2)\exp(j2\pi f_{d_2}) + \ldots + \alpha_M x_M(t - \tau_m)\exp(j2\pi f_{d_M})$$
$$+ \sum_{i=0}^{W_1} \omega_{1_i} x_1(t - \tau_{1_i}) + \sum_{i=0}^{W_2} \omega_{2_i} x_2(t - \tau_{2_i}) + \ldots\ldots + \sum_{i=0}^{W_M} \omega_{M_i} x_M(t - \tau_{M_i}) + n_s(t), \tag{1}$$

where $\alpha_M x_M(t - \tau_m)\exp\left(j2\pi f_{d_M}\right)$ represents the echo signal of the Mth GPS satellite, α_M represents the amplitude of the echo signal, and $x_M(t)$ is the Mth GPS satellite signal. In addition, τ_M and f_{d_M} are the delay and frequency offset of that signal, respectively, $n_s(t)$ is the noise of the monitoring channel, $\sum_{i=0}^{W_M} \omega_{M_i} x_M\left(t - \tau_{M_i}\right)$ stands for the multipath of the Mth GPS satellite affected by multipath interference, i represents the subscript of the ith path in the multipath, W_M is the number of multipath components of the Mth GPS signal, ω_{M_i} is the gain of the ith path in the Mth GPS satellite signal, and τ_{M_i} is the delay of the ith path in the Mth GPS satellite signal.

In the reference channel, GPS satellites share and reuse the same frequency band due to the characteristics of GPS satellite system distribution and Code Division Multiple Access (CDMA) modulation. Thus, the ground receiver is likely to receive more than four frequency-overlapped GPS signals. The received signal by the GPS receiver could be given by

$$x_r(t) = \sum_{k=1}^{M} x_k\left(t - \tau_k\right) + n_r(t), \tag{2}$$

where $x_k\left(t - \tau_k\right)$ is the direct wave signal of the kth GPS satellite, and $n_r(t)$ represents the noise of the reference channel.

3. Interference Suppression

Since GPS signals may be transmitted on the same frequency bands and the zenith can simultaneously have multiple GPS satellites, a near-earth orbit target can be simultaneously illuminated by multiple satellite beams. Therefore, multiple different GPS signals may be received on the reference channel, contaminating desired reference signals and degrading the suppression of DPI and MPI.

3.1. Influence of Reference Channel Interference on DPI and MPI Suppression

To suppress DPI and MPI, an adaptive filtering algorithm was adopted [22–24]. The direct wave signal received on the reference channels is used as reference signals to cancel the DPI and MPI on the monitoring channels. The specific suppression principle is shown in Figure 3, where $X_{\text{ref}}(n)$ is the direct wave signal of the reference channel, $X_s(n)$ is the mixed signal received by the monitoring channel, and $W(n)$ is the coefficient of the filter. The algorithm is able to adjust the filter coefficients adaptively to minimize the output error $e(n)$ of the filter, and $e(n)$ also gives the signal for the monitoring channel after interference suppression, which is obtained by

$$e(n) = X_s(n) - W^H(n - 1)X_{ref}(n). \tag{3}$$

From Figure 3, reference signals are required in this method. The reference signal is used to cancel the DPI and MPI in the monitoring channel, and is also used as a reference signal for time-frequency two-dimensional correlation with the echo signal in the monitoring channel. Therefore, the reference signal is very important throughout the process. This section analyzes the influence of reference channel noise and interference signals on the direct wave multipath suppression. In Figure 4, the DPI and MPI suppression are performed by using the algorithm in Figure 3, and the monitoring channel signal and the reference signal after suppression are used as fuzzy functions, and the DPI and MPI inhibition effects are judged by observing whether or not there is a peak corresponding to the echo on the time delay-Doppler spectrum.

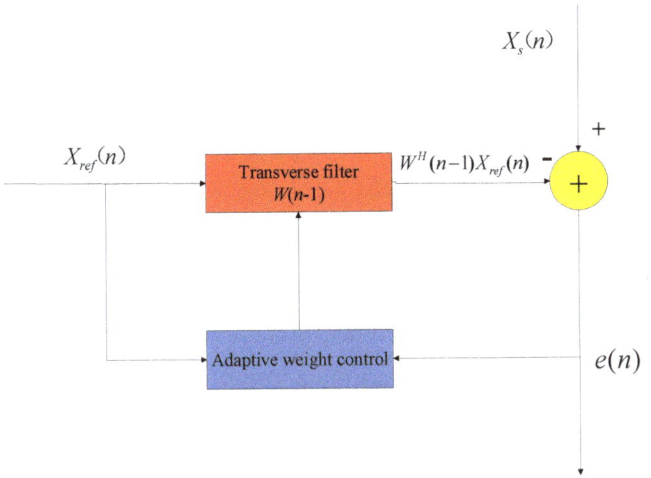

Figure 3. The suppression methods of DPI and MPI.

First, set the monitoring channel to include both the target echo and DPI and MPI. The reference channel only has the reference signal and noise corresponding to the target echo. There are no other GPS interference signals. The parameter settings are shown in Table 1. The DPI and MPI suppression effects are assessed by observing whether there is a peak corresponding to the echo on the delay-Doppler spectrum, which are shown in Figures 4 and 5. As can be seen from Figure 4, when the SNR of the reference signal is -15 dB, after using this reference signal to perform DPI and MPI suppression on the monitoring channel, the interference peak in the delay-Doppler spectrum is still strong, making the peak corresponding to the target echo invisible. When the SNR of the reference signal is 5 dB, after the suppression of the direct wave, it can be seen from Figure 5 that, although the interference peaks of DPI and MPI still exist, the peak of the echo can be seen in the delay-Doppler spectrum. If the SNR is gradually increased, DPI and MPI can be completely suppressed. Therefore, it can be concluded that the noise of the reference channel has a great influence on the suppression process of DPI and MPI.

Table 1. Parameter setting.

	Doppler Shift (Hz)	Delay (us)	Power (dBm)
The noise of the reference channel	–	–	$-105/-95$
The reference signal in the reference channel	0	0	-110
The target echo of the monitoring channel	500	5	-150

Second, set the echo channel to include the target echo plus DPI and MPI. In this case, only the reference signal and other GPS interference signals are in the reference channel. The parameter settings are shown in Table 2.

Table 2. Parameter setting.

	Doppler Shift (Hz)	Delay (us)	Power (dBm)
The reference signal in reference channel	0	0	-100
The interference signal 1 and 2 in reference channel	0	0	$-110/-140$
The noise of reference channel	–	–	–
The target echo of monitoring channel	500	5	-140

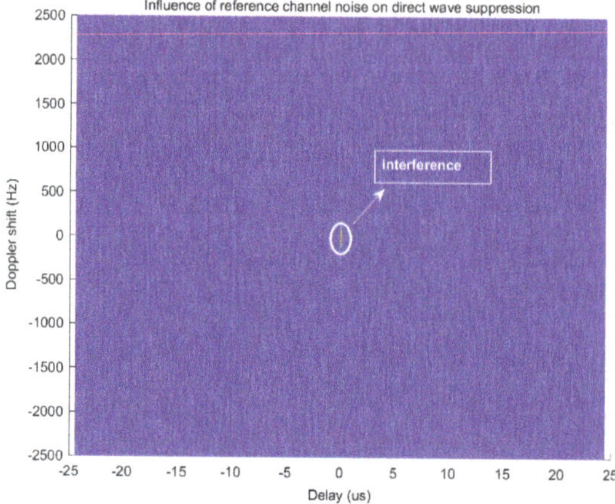

Figure 4. The influence of reference channel SNR on interference suppression process with SNR = −15 dB.

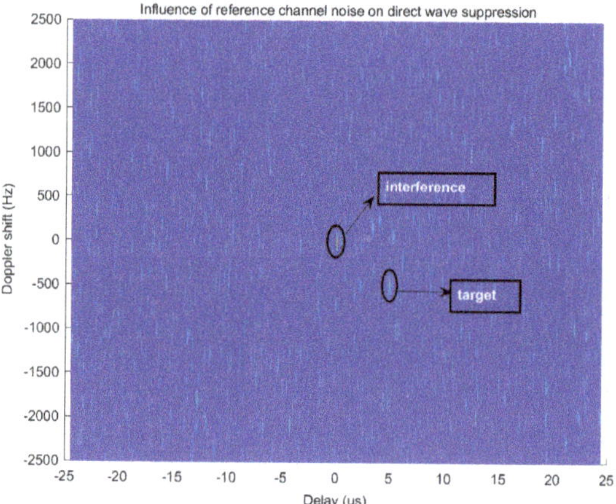

Figure 5. The influence of reference channel SNR on interference suppression process with SNR = 5 dB.

As can be seen from Figure 6, when the power of the interference signal in the reference channel is −110 dBm, that is, the signal-to-interference ratio (SIR) is 10 dB, the peak corresponding to the target is still invisible after interference suppression, and the DPI and MPI suppression methods fail in this scenario. As can be seen from Figure 7, when the power of the interference signal in the reference channel is reduced to −140 dBm and the SIR is 30 dB, the peak corresponding to the target can be seen after interference suppression, but the interference is still not eliminated. Therefore, it can be concluded that, under normal circumstances, the reference channel inevitably receives signals from multiple GPS satellites as co-channel interference, and it has a great influence on the suppression process of DPI and MPI.

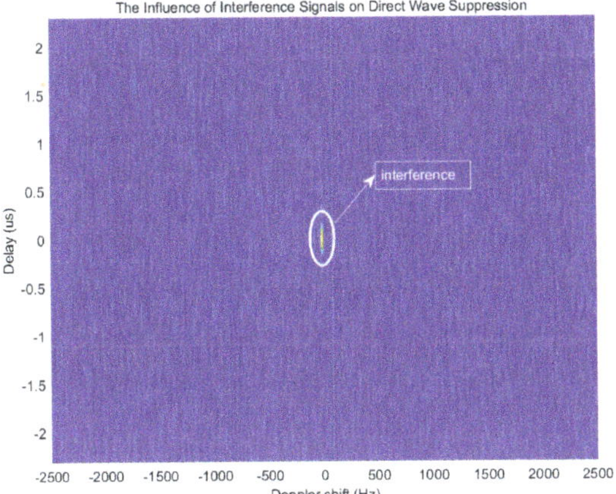

Figure 6. The influence of reference channel SIR on DPI and MPI suppression with SNR = 10 dB.

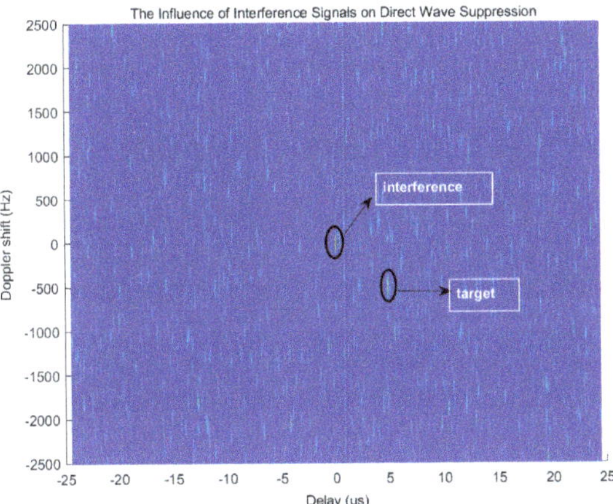

Figure 7. The influence of reference channel SIR on DPI and MPI suppression with SNR = 30 dB.

Based on the above analysis, the noise and interference signals in the reference channel have a great influence on DPI and MPI suppression. The traditional external source detection system does not process the reference channel signal and uses it directly as a reference signal for the DPI and MPI interference suppression algorithms. In order to correctly detect the echo signal, the GPS reference signal of the reference channel must be purified and separated.

3.2. Multiple GPS Signals Separation and Reconstruction

The GPS signal uses C/A code $C(t)$ and P code $P(t)$ to spread the data code $D(t)$. This paper only considers the GPS satellite signal modulated by C/A code [25,26]. The signal received by the reference channel can be expressed as

$$x_r(t) = \sum_{k=1}^{M} \sqrt{P_k} \cdot C_k\left(t - \tau_k\right) \cdot D_k\left(t - \tau_k\right) \cdot e^{j2\cdot\pi kt} + n(t), \tag{4}$$

where P_k is the power of the transmitted signal of the kth GPS satellite, $C_k(t)$ represents the C/A code of the kth GPS satellite, $D_k(t)$ represents the navigation data of the kth GPS satellite, f_k is the carrier frequency of the received signal, and τ_k is the delay of the received signal. By using the CDMA principle of the GPS system and the characteristics of the C/A code disclosure, a successive interference canceller (SIC) can be used to effectively separate and reconstruct multiple GPS signals $x_r(t)$ by the reference channel.

SIC is implemented in multiple steps, where each step requires signal acquisition to reconstruct the signal; then, the interference signal is removed from the received signal, and the "detection-reconstruction-cancel" step is repeated until all GPS signals are recovered. The steps of the method are as follows:

Step 1: Using a GPS acquisition algorithm to detect the GPS signal in the received signal of the reference channel, and obtain its corresponding spreading code information $C_l(t)$, amplitude estimation value P_l , phase offset value τ_l, and frequency offset value f_l;

Step 2: Demodulate and reconstruct the signal $x_l(t)$; the received signal is down-converted by using the frequency offset information f_l, and then the phase offset τ_l is obtained by the lth local C/A code $C_l(t)$ to obtain $C_l(t - \tau_l)$, and, according to the orthogonality of the C/A codes of different satellites, the information $D_l'(t)$ of the satellite is de-spreaded. In order to correctly recover the navigation data, the de-spreaded data are processed by the envelope averaging method, and finally the navigation data $D_l(t)$ are determined. The process is shown in Figure 8.

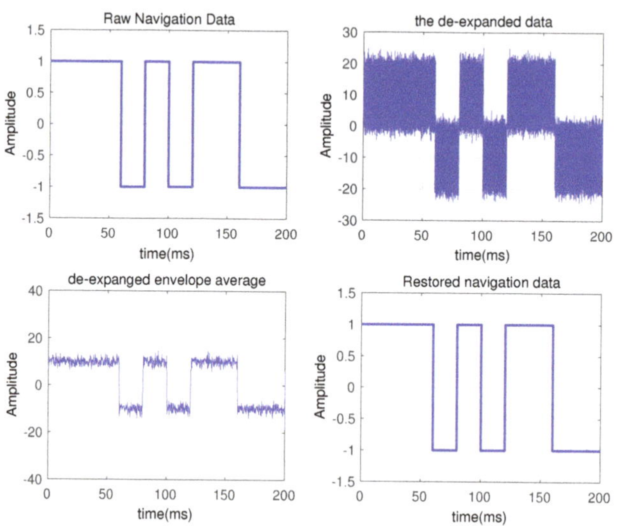

Figure 8. Reference signal demodulation process.

The recovered navigation data $D_l(t)$ is modulated using the phase-synchronized local C/A code, and the reconstructed reference signal $x_l(t)$ is obtained by using the amplitude P_l of the signal and then up-converting, which is expressed as

$$\hat{x}_l(t) = \sqrt{P_l} \cdot C_l(t - \tau_l) D_l(t - \tau_l) e^{j2\pi f_l t}. \tag{5}$$

The correlation coefficient between the original signal and the reconstructed signal obtained by simulation calculation is 0.99. As shown in Figure 9, it shows that the original signal is well reconstructed.

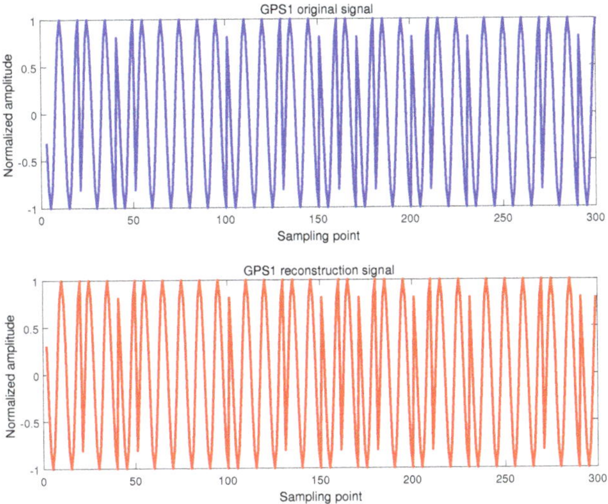

Figure 9. Compare the reconstructed signal with the original signal.

Step 3: Subtract the signal recovered in step 2 from the received signal of the reference channel to reduce the interference when reconstructing the next signal. In this paper, the adaptive filtering method is used to eliminate the interference signal, that is, the input end of the reference signal in Figure 3 is replaced by the reconstructed signal $\hat{x}_l(t)$, and the input end of the monitoring channel signal is replaced by $x_r(t)$, so that the signal after strong interference cancellation can be obtained

$$x_r^{(1)}(n) = \varepsilon(n) = x_r(n) - w^T(n-1)\hat{x}_l(n). \tag{6}$$

Step 4: Repeat the "capture-reconstruction-cancellation" process from step one to step three for the signal $x_r^{(1)}(t)$ of the output signal of step three until the GPS signal is not detected in the reference channel. Finally, a plurality of reconstructed GPS reference signals can be obtained. At this time, the reference signals $\hat{x}_1(t), \hat{x}_2(t) \ldots \hat{x}_M(t)$ have been separated and there is no noise, so there is no other GPS signals and noise interfering with the reference signals. The process not only eliminates the noise in the reference signal, but also separates multiple GPS signals in the reference channel, providing a good reference signal for the DPI and MPI suppression processes of multiple GPS satellites in the monitoring channel. At the same time, it also provides a useful reference signal for the joint detection of multiple GPS weak echoes. In addition, the GPS signal that interferes with the original reference channel is converted into a reference signal that is advantageous to the system.

3.3. DPI and MPI Suppression Based on ECA

After the reference signal separation and reconstruction, the DPI and MPI of multiple satellites can be suppressed. Different from the DPI and MPI suppression processes of a single GPS satellite, this section suppresses the DPI and MPI brought by multiple GPSs in the surveillance channel based on the extensive cancellation algorithm (ECA).

Firstly, the ECA uses multiple GPS reference signals $\hat{x}_1(t), \hat{x}_2(t) \ldots \hat{x}_M(t)$ to construct a delay spread matrix $\mathbf{X}^{\text{ref}}$ of multi-satellite signals, which can be expressed as

$$\mathbf{X}^{\text{ref}} = \begin{bmatrix} \mathbf{X}_1^{\text{ref}} & \mathbf{X}_2^{\text{ref}} & \ldots \mathbf{X}_i^{\text{ref}} \ldots & \mathbf{X}_M^{\text{ref}} \end{bmatrix}, \tag{7}$$

where $X_i^{\text{ref}}(i \in (1, M))$ is the reconstructed matrix of the reconstructed ith reference signal $\hat{x}_i(t)$ through different delays, and the extension matrix X_i^{ref} is an element in the matrix X^{ref} and can be expressed as

$$
X_i^{\text{ref}} = \begin{bmatrix} \hat{x}_i(0) & \hat{x}_i(N-1) & \cdots & \hat{x}_i(N-K) \\ \hat{x}_i(1) & \hat{x}_i(0) & \cdots & \hat{x}_i(N-K+1) \\ \cdots & \cdots & \cdots & \cdots \\ \hat{x}_i(N-1) & \hat{x}_i(N-2) & \cdots & \hat{x}_i(N-K-1) \end{bmatrix},
\tag{8}
$$

where N is the number of sampling points, and K is the maximum delay, which can be obtained by dividing the maximum detection distance by the speed of light($K = \frac{R_{\max}}{c}$). DPI and MPI can be expressed as

$$
DMPI = \sum_{i=0}^{W_1} \omega_{1_i} x_1(t - \tau_{1_i}) + \sum_{i=0}^{W_2} \omega_{2_i} x_2(t - \tau_{2_i}) + \ldots\ldots + \sum_{i=0}^{W_M} \omega_{M_i} x_M(t - \tau_{M_i}).
\tag{9}
$$

Then, adjust the value of $\varepsilon = [\varepsilon_0, \varepsilon_1 \ldots \varepsilon_{N-1}]^T$ to make $\varepsilon X^{\text{ref}}$ approach the direct wave and multipath interference. The problem is transformed into the following problem:

$$
\min \left\| x_s(t) - \varepsilon X^{\text{ref}} \right\|^2.
\tag{10}
$$

The solution of Equation (10) uses the least squares criterion [27–29]; then, Equation (10) is equivalent to the following:

$$
\frac{\partial \left(\left\| x_s(t) - \varepsilon X^{\text{ref}} \right\|^2 \right)}{\partial(\varepsilon)} = 0.
\tag{11}
$$

Thus, $\varepsilon = \left(\left(X^{\text{ref}} \right)^H X^{\text{ref}} \right)^{-1} X^{\text{ref}} x_s(t)$ is obtained, where $\left(X^{\text{ref}} \right)^H$ is the transpose of X^{ref}, and the signal in the monitoring channel after interference suppression is expressed as

$$
x_s'(t) = x_s(t) - DMPI = x_s(t) - \varepsilon X^{\text{ref}} = x_s(t) - X^{\text{ref}} \left(\left(X^{\text{ref}} \right)^H X^{\text{ref}} \right)^{-1} X^{\text{ref}} x_s(t),
\tag{12}
$$

where $x_s'(t)$ only contains the echo signal and noise $n_s(t)$.

The proposed algorithm does not need to know the gain value, and the DMPI can be directly solved by the proposed algorithm. In order to verify the DPI and MPI suppression algorithms based on signal separation and reconstruction proposed in this paper, the specific parameter settings are shown in Table 3 as follows: the reference channel contains noise, five GPS signals; the monitoring channel contains noise, three GPS echoes, and DPIs corresponding to five reference channel GPS signals. Firstly, the reference signal separation and reconstruction algorithms are used to separate and reconstruct the reference signals 1, 2, 3, 4, and 5 of the reference channel. Then, the purified reference signal is used together with the suppression algorithm proposed in this section to perform DPI suppression on the signal of the monitoring channel. Finally, the signal of the monitoring channel after the suppression and the reference signal are subjected to cross ambiguity function (CAF) processing [30–32], and the DPI suppression effect is judged by observing whether there is a peak corresponding to the echo on the delay-Doppler spectrum.

Figure 10 shows the time-frequency two-dimensional correlation of un-suppressed DPI of multiple GPS satellites. It can be seen that the echo generated by the target is completely submerged in the peak generated by DPI. Figure 11 shows the comparison of the monitoring channel signals before and after the DPI and MPI suppression methods proposed in this paper. It can be seen that the amplitude of the monitoring channel signal decreases after interference suppression, which proves that strong DPI interference has been effectively suppressed.

Table 3. Parameter setting.

	Doppler Shift (Hz)	Delay (us)	Power (dBm)
The noise of reference channel	–	–	-95
Reference channels reference signal (1, 2, 3, 4, 5)	0	0/2/3/5/6	$-100/-105/-110/-112/-106$
The noise of monitoring channel	–	–	-110
Monitoring channel's target echo (1, 2, 3)	500/650/850	5/5.2/10	$-140/-145/-150$
Monitoring channel DPI	0	0/2/3/5/6	$-100/-105/-110/-112/-106$

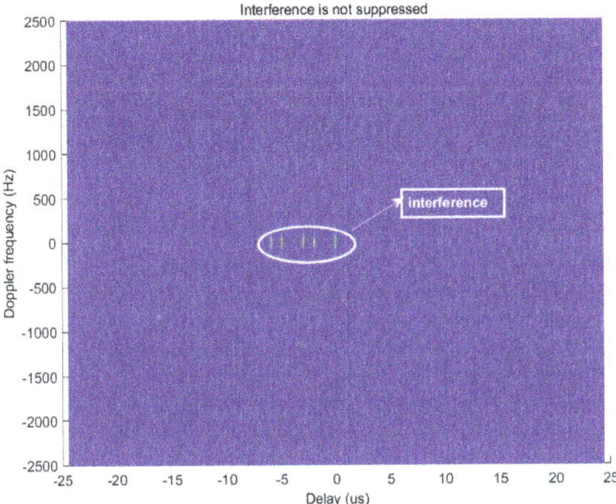

Figure 10. Time-frequency two-dimensional correlation graph with unsuppressed interference.

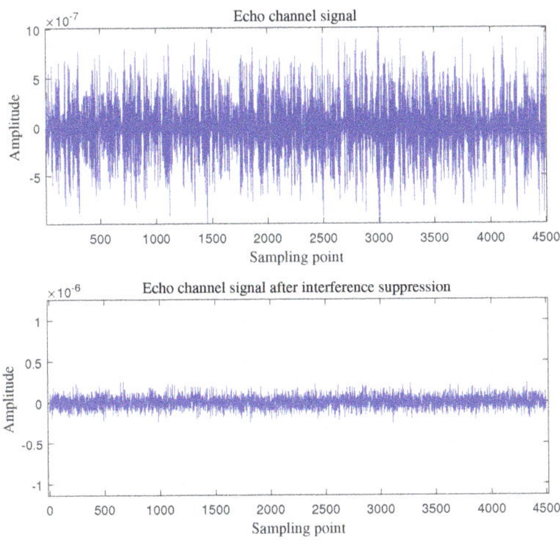

Figure 11. Monitoring channel signal before and after interference suppression.

Figure 12 shows the CCF of the reference and echo signals after DPI interference suppression. It can be seen that the peak of the echo signal is clearly highlighted after the DPI interference suppression. Figure 12 proves that the interference suppression scheme proposed in this paper can effectively suppress the DPI of the monitoring channel when the reference channel SNR is as low as -15 dB, and the reference channel has multiple GPS signals with similar power.

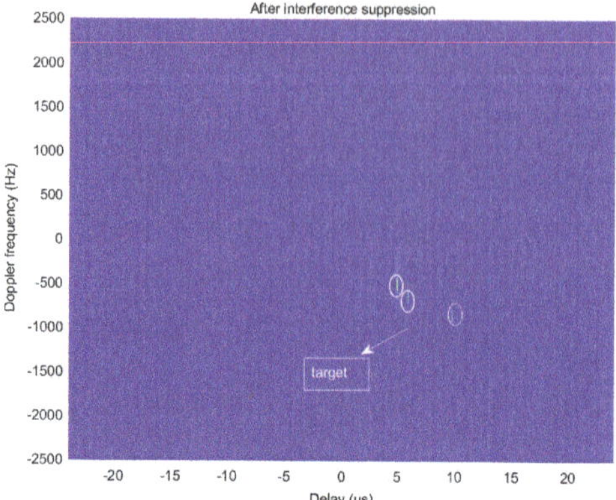

Figure 12. Time-frequency two-dimensional graph after direct wave multipath interference suppression.

This section uses the multiple GPS reference signals separated and reconstructed to construct the delay spread matrix $\mathbf{X}^{\text{ref}}$ of the multiple satellite signals, and then finds the optimal weight ε based on the least squares criterion. Then, let $\varepsilon \mathbf{X}^{\text{ref}}$ approach multipath and then subtract $\varepsilon \mathbf{X}^{\text{ref}}$ from the monitoring channel to get the monitoring channel signal after DPI and MPI suppression.

4. Detection Statistics Construction

Due to the characteristics of the external radiation source detection system and the weak power of the GPS itself, the target echo power is relatively low. The traditional CAF often needs to increase the coherence time of the direct wave and the echo to accumulate the energy of the weak echo [33]. However, long-term coherent accumulation leads to an increase in computational complexity, and, since the range of detection is limited and the target moves faster, the accumulation time is greatly limited. In order to effectively enhance the signal-to-noise ratio of the detected quantity under the condition of equivalent coherence time, in this section, the anti-jamming properties of cycle cross ambiguity function (CCAF) are used to construct the detection statistics of multiple GPS weak echoes, and the coordinate detection algorithm is used to fuse multiple detection statistics to obtain the final detection statistics. This process enhances the signal-to-noise ratio of the detection statistic from the two aspects of detection structure and multi-star data fusion, and improves the detection probability of weak echo. In order to judge the performance of CCAF in the multi-satellite system compared with the traditional CAF detection structure, in this paper, the CAF-based detection statistic and the CCAF-based detection metric are constructed respectively, and the theoretical analysis and simulation performance verification are carried out, respectively.

4.1. Detection Statistic Construction with a Single GPS Satellite

(1)   Detection statistics based on CAF

The signal $\hat{x}_i(t)$ of the reconstructed reference signal $\hat{x}_1(t), \hat{x}_2(t) \ldots \hat{x}_M(t)$ is selected, and the carrier frequency information f_i of the locally known reference signal $\hat{x}_i(t)$ is used for down-conversion processing to obtain the down-converted reference signal $\hat{x}_i^{IF}(t)$. Then, the same down-conversion process is performed on the monitoring channel signal $x_s'(t)$ after DPI and MPI suppression, and the down-converted reference signal $x_s^{IF}(t)$ is obtained. In addition, calculate the CAF of $\hat{x}_i^{IF}(t)$ and $x_s^{IF}(t)$ to obtain the Doppler-time delay spectrum of the i-th GPS satellite, which is expressed as

$$S_i(\tau, f) = \int_{-T/2}^{T/2} \hat{x}_i^{IF}(t) x_s^{IF}(t - \tau) e^{j2\pi f t} dt. \tag{13}$$

The discretization of Equation (13) is expressed as

$$S_i(\tau, f) = \sum_{n=0}^{N-1} \hat{x}_i^{IF}(nT_s) \, x_s^{IF}(nT_s - \tau) \, e^{j2\pi f nT_s},\tag{14}$$

where τ is the delay, f is the Doppler shift, T represents the accumulation time, T_s stands for the sampling period, and N is the number of sampling points.

(2) Detection statistics based on CCAF

To construct a detection using CCAF, we first need to do a cyclic autocorrelation of the reference signal $\hat{x}_i^{IF}(nT_s)$ as

$$R_{r_i r_i}^{\alpha}(\tau) = \frac{1}{N} \sum_{n=0}^{N-1} \hat{x}_i^{IF}(nT_s + \tau/2) \, \hat{x}_i^{IF}(nT_s - \tau/2)^* \, e^{-j2\pi\alpha nT_s}.\tag{15}$$

Then, the cyclic cross-correlation of the reference signal $\hat{x}_i^{IF}(nT_s)$ and the echo signal $\hat{x}_s^{IF}(nT_s)$ is

$$R_{r,s}^{\alpha-f}(\tau) = \frac{1}{N} \sum_{n=0}^{N-1} x_s^{IF}(nT_s + \tau/2) \, \hat{x}_i^{IF}(nT_s - \tau/2)^* \, e^{-j2\pi\alpha nT_s},\tag{16}$$

where τ is the delay, α represents the cyclic frequency, and N stands for the number of sampling points.

The vectors at the cyclic frequencies α' and $\alpha' - f'$ corresponding to the maximum peak values of $R_{r_i r_i}^{\alpha}(\tau)$ and $R_{r_i s}^{\alpha}(\tau)$ are respectively extracted, and are recorded as $R_{r_i r_i}^{\alpha'}(\tau)$ and $R_{r_i s}^{\alpha'-f'}(\tau)$. The mutual fuzzy function processing is performed on these two vectors to obtain

$$\Psi_i(u, f) = \sum_{\tau=0}^{N-1} R_{r_i s}^{\alpha'-f'}(\tau) R_{r_i r_i}^{\alpha'}(\tau - u)^* e^{j\pi f\tau},\tag{17}$$

where $\Psi_i(u, f)$ represents the CCAF between the monitoring channel signal and the reference signal, u is the delay, and f is the Doppler shift.

4.2. Detection Statistics Construction with Multiple GPS Satellites

Due to the different distribution positions of different GPS satellites, the peak coordinates of multiple GPS satellite echo detections $S_i(\tau, f)$ and $\Psi_i(u, f)$ are also different. Thus, it is impossible to add a plurality of detection amounts to the fusion structure detection statistic. Aiming at this problem, this section unifies the detection peak coordinates of different satellites by coordinate transformation, which can superimpose the echo detection spectrum of several different GPS satellites to achieve the purpose of non-correlated cumulative enhanced signal-to-noise ratio. Thereby, a detection statistic with a higher SNR is constructed.

Figure 13 shows the geometry of the receiving system, where θ is the signal arrival angle, δ is the angle between the bistatic angle bisector and the speed v of the aircraft, and β is the bistatic angle. It can be seen from the figure that the positions of different GPS satellites are different, so the delay τ and the Doppler shift f_d corresponding to the peak values of the two-dimensional correlation between the different satellite reference signals and the monitoring channel echo signals are different. However, the common edge R_r and the velocity v corresponding to different peak coordinates τ and f_d are the same, so the detection spectrum can be converted from the delay-Doppler dimension to the distance velocity dimension so that the coordinate peaks are the same. Thereby, it is possible to accumulate different detection amounts. The relationship between τ and R_r and f_d and v is

$$\begin{cases} R_r + R_t = L + c\tau, \\ R_t^2 = R_r^2 + L^2 - 2R_r L \cos\theta. \end{cases}\tag{18}$$

Solving Equation (18) can obtain:

$$R_r = \frac{c^2\tau^2 + 2Lc\tau}{2(L + c\tau - L\cos\theta)} = f(\tau),\tag{19}$$

$$f_d = \frac{2v}{\lambda}\cos\delta\cos\frac{\beta}{2} = \frac{2v'}{\lambda}\cos\frac{\beta}{2} = g(v),\tag{20}$$

where $v' = v\cos\delta$, and v' represents the speed v of the target on the bisecting angle bisector, and β can be obtained from

$$\sin(\beta) = \frac{2R_r\sin\theta\,(R_r + c\tau - R_r\cos\theta)}{R_r^2 - 2R_r\,(R_r + c\tau)\cos R_r + (L + c\tau)^2}.\tag{21}$$

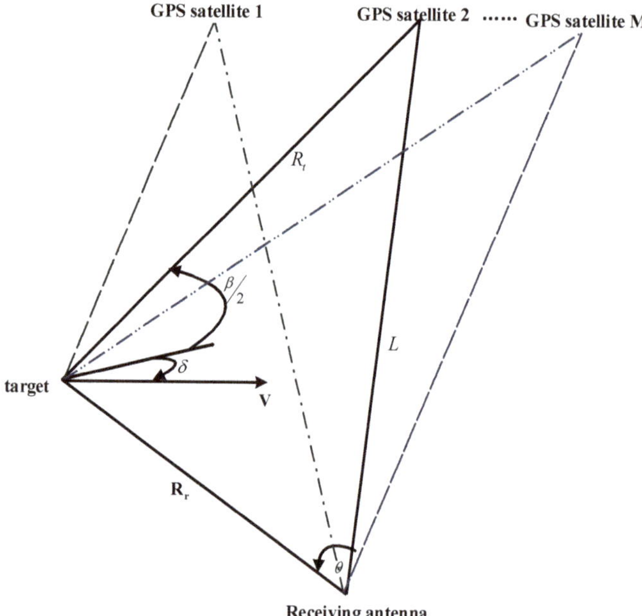

Figure 13. The geometry of the receiving system.

The detection quantities $S_i(\tau, f)$ and $\Psi_i(u, f)$ obtained by different methods are transformed by Equations (19) and (21) to obtain distance-velocity spectra $S_i\left(f^{-1}(R_r), g(v)\right)$ and $\Psi_i\left(f^{-1}(R_r), g(v)\right)$, which are expressed as:

$$S_i\left(f^{-1}(R_r), g(v)\right) = \sum_{n=0}^{N} x_s^{IF}(nT_s)\,\hat{x}_i^{IF}\left(nT_s - f^{-1}(R_r)\right)^* e^{j2\pi g(v)nT_s},\tag{22}$$

$$\Psi_i\left(f^{-1}(R_r), g(v)\right) = \sum_{n=0}^{N} R_{r_is}^{\alpha'-f'}(nT_s)\,R_{r_ir_i}^{\alpha'}\left(nT_s - f^{-1}(R_r)\right)^* e^{j2\pi g(v)nT_s}.\tag{23}$$

At this time, the detection amount $S_i\left(f^{-1}(R_r), g(v)\right)$ or $\Psi_i\left(f^{-1}(R_r), g(v)\right)$ of the plurality of GPS satellites can be non-coherently superimposed in the distance-speed domain. The final superimposed detection statistics obtained by the two methods as

$$\Lambda(R_r, V) = \sum_{i=1}^{M} S_i\left(f^{-1}(R_r), g(v)\right),\tag{24}$$

$$\Omega\left(R_r, V\right) = \sum_{i=1}^{M} \Psi_i\left(f^{-1}\left(R_r\right), g(v)\right), \tag{25}$$

where R_r represents the distance from the target to the receiver and V represents the speed of the target. It can be seen from the above process that this paper uses the multiple GPS reference signals separated and reconstructed to calculate the detection quantity of different GPS satellites as the illumination source, which provides more favorable information for target detection. However, due to the different peak values of the detected quantities obtained by multiple GPSs, these detection quantities cannot be effectively fused, so this paper uses the coordinate fusion algorithm to combine them to obtain the final detection amount after peak enhancement.

5. Moving Aerial Target Detection Performance Analysis

In order to better design the decision threshold of the detector and evaluate the detection performance of the two detection quantities, this section analyzes the probability distributions of $\Lambda(L, V)$ and $\Omega(L, V)$, respectively.

For convenience, the reference baseband signal $\hat{x}_i^{IF}(t)$ of the GPS satellite is first discretized and expressed as

$$\hat{x}_i^{IF}\left(nT_s\right) = \beta_i p_i\left(nT_s\right), \tag{26}$$

where β_i is the amplitude of the baseband reference signal, $p_i\left(nT_s\right)$ is the baseband reference signal after the amplitude normalized, and T_s stands for the sampling period.

The signal $x_s^{IF}(t)$ of the monitoring channel is discretized and then expressed as

$$x_s^{IF}\left(nT_s\right) = \sum_{j=1}^{M} \alpha_j p_j\left(nT_s - \tau_j\right) \exp\left(j2\pi f_{d_j} nT_s\right) + w\left(nT_s\right), \tag{27}$$

where α_j is the amplitude of the echo signal, $p_j\left(nT_s - \tau_j\right) \exp\left(j2\pi f_{d_j} nT_s\right)$ is the amplitude-normalized echo signal, $w(n)$ is the complex Gaussian noise obeying the $N\left(0, \sigma_w^2\right)$ distribution, and $N(\cdot)$ is the Gaussian distribution.

The binary hypothesis of echo signal detection is: assuming that H_1 is the target existence, the signal $x_s'''\left(nT_s\right)$ of the monitoring channel contains the echo signal $\alpha_i p_i\left(nT_s - \tau_i\right) \exp\left(j2\pi f_{d_i} nT_s\right)$ corresponding to a certain reference signal $\hat{x}_i^{IF}\left(nT_s\right)$. Assuming that H_0 indicates that the target does not exist, the signal $x_s'''\left(nT_s\right)$ of the monitoring channel does not contain any echo signal corresponding to $\hat{x}_i^{IF}\left(nT_s\right)$, which is expressed as follows:

$$\begin{cases} H_0 : x_s^{IF}\left(nT_s\right) = w\left(nT_s\right) \\ H_1 : x_s^{IF}\left(nT_s\right) = \sum_{\substack{j=1 \\ \exists j = i}}^{M} \alpha_j p_j\left(nT_s - \tau_j\right) \exp\left(j2\pi f_{d_i} nT_s\right) + w\left(nT_s\right). \end{cases} \tag{28}$$

5.1. Performance Analysis of Detection Based on CAF

Lemma 1. *The distribution of the detection statistic under the H_1 hypothesis is*

$$\left(S_i(\tau, f) | H_1\right) \sim CN\left(\left\{\alpha_i^* \beta_i \chi_{pp}\left(\tau - n_\tau, f - f_d\right)\right\}, \ 2N\beta_i^2 \sigma_w^2\right), \tag{29}$$

where $CN(.)$ represents the complex Gaussian process.

Proof. See Appendix A.1. □

As with the analysis under the H_1 hypothesis, it can be concluded that the detection statistic distribution under the H_0 hypothesis is

$$(S_i(\tau, f)|H_0) \sim CN\left(0, 2N\beta^2\sigma_\omega^2\right). \tag{30}$$

Using the data fusion technique of multiple GPS satellites, according to the cumulative nature of the Gaussian distribution, the distribution of the final detection statistic $\Lambda\left(R_r, V\right)$ is

$$(\Lambda|H_0) \sim CN\left(0, \sum_{i=1}^{M} 2N\beta_i^2\sigma_\omega^2\right), \tag{31}$$

$$(\Lambda|H_1) \sim CN\left(\sum_{i=1}^{M}\left\{\alpha_i^*\beta_i\chi_{pp}\left(\tau - n_\tau, f - f_d\right)\right\}, \sum_{i=1}^{M} 2N\beta_i^2\sigma_\omega^2\right). \tag{32}$$

The false alarm probability of detecting weak echo signals from different GPS satellite sources by Equations (28) and (31) is given by

$$P_{FA} = \int_\lambda^\infty f\left(\Lambda|H_0\right) d\Lambda = \exp\left(-\frac{\lambda}{\sum_{i=1}^{M} 2N\beta_i^2\sigma_\omega^2}\right), \tag{33}$$

where $f\left(\Lambda|H_0\right)$ is the probability density function of $(\Lambda|H_0)$.

From Equation (33), it can be concluded that the adaptive detection threshold λ is

$$\lambda = -\sum_{i=1}^{M} 2N\beta_i^2\sigma_\omega^2 \ln\left(P_{FA}\right). \tag{34}$$

According to Equationss (A2), (32), and (34), the detection probability can be obtained by using signal detection theory:

$$P_D = \int_\lambda^\infty f(\Lambda|H_1) d\Lambda = Q_m\left(\sqrt{\frac{\left|\sum_{i=1}^{M}\left(\beta_i\alpha_i N\right)\right|^2}{\sum_{i=1}^{M} N2\beta_i^2\sigma_\omega^2}}, \sqrt{\frac{\lambda}{\sum_{i=1}^{M} N2\beta_i^2\sigma_\omega^2}}\right), \tag{35}$$

where $Q_m(\cdot, \cdot)$ is the Marcum Q function, and $f\left(\Lambda|H_1\right)$ is the probability density function of $(\Lambda|H_1)$.

From Equation (35), it can be seen that the theoretical detection probability of the multi-star weak echo joint detection based on the CCA detection quantity construction method is related to the parameters such as the monitoring channel noise, the number of sampling points N, the number of satellites, and the false alarm probability. It can be seen that the detection probability is proportional to the number of satellites M, that is, the detection probability increases with the number of satellites. It is theoretically proved that the weak echo combined detection of multiple GPS satellites has a higher detection probability than the weak echo detection of a single GPS satellite, but the method does not reflect the noise suppression performance.

5.2. Performance Analysis of Detection Based on CCAF

Lemma 2. *The probability distribution of $\Psi_i(u, f)$ under H_1 hypothesis is*

$$\Psi_i\left((u, f)|H_1\right) \sim CN\left(\Psi_{R_iR_i}(u, f), \frac{\sigma_\omega^2\beta_i^6}{N}\right). \tag{36}$$

Proof. See Appendix A.2. □

Using the same analysis method to analyze the H_0 hypothesis, the distribution of the detection quantity $\Psi_i(u, f)$ of a single GPS satellite under the H_0 hypothesis is given by

$$\Psi_i\left((u, f)|H_0\right) \sim CN\left(0, \frac{\sigma_\omega^2 \beta_i^6}{N}\right). \tag{37}$$

Using the data fusion technique of multiple GPS satellites, according to the cumulative nature of the Gaussian distribution, the distribution of the final detection statistic $\Omega(R_r, V)$ is

$$(\Omega(R_r, V)|H_0) \sim CN\left(0, \sum_{i=1}^{M} \frac{\sigma_\omega^2 \beta_i^6}{N}\right) \tag{38}$$

and

$$(\Omega(R_r, V)|H_1) \sim CN\left(\sum_{i=1}^{M} \{\Psi_{R,R_i}(u, f)\}, \sum_{i=1}^{M} \frac{\sigma_\omega^2 \beta_i^6}{N}\right). \tag{39}$$

According to Equation (38), the false alarm probability can finally obtain:

$$P_{FA} = \int_\lambda^\infty f(\Omega|H_0)d\Omega = \exp\left(-\frac{\lambda}{\sum\limits_{i=1}^{M} \frac{\sigma_\omega^2 \beta_i^6}{N}}\right), \tag{40}$$

where $f(\Omega|H_0)$ is the probability density function of $(\Omega|H_0)$.

The detection threshold λ in the solution Equation (40) is

$$\lambda = -\ln(P_{FA}) \cdot \sum_{i=1}^{M} \frac{\sigma_\omega^2 \beta_i^6}{N}. \tag{41}$$

According to Equations (A14), (39), and (41), the detection probability can be obtained by using the signal detection theory

$$P_D = Q_m\left(\sqrt{\frac{\left|\sum\limits_{i=1}^{M} \beta_i^3 \alpha_i\right|^2}{\sum\limits_{i=1}^{M} \frac{\sigma_\omega^2 \beta_i^6}{N}}}, \sqrt{\frac{\lambda}{\sum\limits_{i=1}^{M} \frac{\sigma_\omega^2 \beta_i^6}{N}}}\right), \tag{42}$$

where $Q_m(\cdot, \cdot)$ is the Marcum Q function.

It can be seen from Equation (42) that the theoretical detection probability of the GPS weak echo signal detection method based on multi-star data fusion is related to the parameters, such as the monitoring channel noise, the number of sampling points N, the number of satellites, and the false alarm probability. It can be seen that the detection probability is proportional to the number of sampling points N and the number of satellites, that is, the detection probability increases with the number of sampling points N and the number of satellites. It is theoretically proved that the GPS weak echo signal detection method based on multi-star data fusion can improve the detection performance from the number of sampling points and the number of satellites. In addition, by comparing Equation (32) and Equation (39), it can be found that the power of the noise of the detection statistic constructed based on the CCAF algorithm decreases as the number of sampling points increases. However, the detection statistic based on the CAF algorithm does not have this property, so the CCAF-based detection statistic construction method has good noise immunity in the proposed algorithm.

6. Numerical Results and Discussion

In order to analyze and verify the effectiveness of the proposed detection algorithm and the influence of various factors on detection, this section presents several simulation experiments using MATLAB (9.5.0.944444 (R2018b), MathWorks Company, Natick, MA, USA) and simulation parameter setting according to [34,35].

Experiment 1: In order to compare the detection performance of CAF and CCAF, this experiment fixes the false alarm probability, the power difference between the reference signal and the echo signal, the number of sampling points, the number of satellites, etc. The delay and Doppler shift of the echo relative to the direct wave are set as 1 us and 500 Hz, respectively. The SNR of the echo is in the range from −90 dB to −30 dB. For different detection quantity construction methods, 2000 Monte Carlo simulation experiments are carried out on the GPS weak echo detection method based on multiple satellites data fusion. The parameters are substituted into the theoretical detection probability Equations (35) and (42) to compare them with the simulation. The specific simulation parameters are shown in Table 4.

Table 4. Parameter setting of experiment 1.

Sampling Frequency (Hz)	Duration (ms)	False Alarm Probability	Number of Satellites	Direct Wave Power (dBm)	Echo Power (dBm)	SNR (dB)
10.23 MHz	1000 ms	0.0001	1	−99.3	−170	−90~−30

It can be seen from Figure 14 that, under the same conditions, the simulated detection probability could reach 99% at a SNR of −55 dB. When the SNR of the CCAF algorithm is −64 dB, the simulated detection probability also reaches 99%. Therefore, the CCAF algorithm is about 9 dB better than the CAF algorithm, and the difference between simulation results and theoretical results is 2 dB, which verifies the effectiveness of the method. The above simulation results show that CCAF has a certain noise suppression capability compared with CAF because the noise does not have cyclostationarity. Moreover, according to Equation (A19), it can be seen that the noise power in the CCAF-based detection decreases as the number of points increases, while CAF does not have this property. This experiment verifies the validity of the theory, showing the noise resistance of the CCAF.

Experiment 2: This verifies the detection performance of the GPS weak echo signal detection method based on multi-star data fusion for different satellite number conditions, the false alarm probability, GPS direct wave power, and sampling points. Assuming that the target is 10 km away from the receiver, the speed is 600 m/s, and the arrival angles θ of the three echoes are 45°, 50°, 60°, respectively. It is concluded that the time delay and Doppler shift of the echo relative to the direct wave are 10 us, 3157 Hz, 11 us, 2669 Hz, 16 us, 2467 Hz, and the echo power is 70 dB different from the direct wave power. The SNR range of the echo is from −90 dB to −30 dB. For different satellite numbers, 2000 Monte Carlo experiments are carried out on the detection method of the GPS weak echo signal based on data fusion of multiple satellites. Finally, the parameters are substituted into Equations (35) and (42), and the theoretical detection probability is compared with the simulation. The specific simulation parameters are shown in Table 5.

Table 5. Parameter setting of experiment 2.

Sampling Frequency (Hz)	Duration (ms)	False Alarm Probability	Number of Satellites	Direct Wave Power (dBm)	Echo Power (dBm)	SNR (dB)
10.23 MHz	1000 ms	0.0001	1/2/3	−99.3	−170	−90~−30

From Figure 15, it can be seen that, under the same conditions, the detection probability of a satellite is 99% when the echo SNR is −65 dB, and the detection probability of two satellites is 99% when the echo SNR is −71 dB, and the detection probability of three satellites is 99% when the echo SNR is −74 dB, and the simulation and the theoretical detection performances are only about 2 dB in difference. The validity of the method is verified. The above simulation results show that, compared with the traditional single satellite detection, the fusion of multiple satellite detections can effectively enhance the detection performance. On the one hand, the algorithm can effectively convert multiple GPS delay–Doppler detection peak coordinates (10 us, 3157 Hz), (11 us, 2669 Hz), (16 us, 2467 Hz) to distance–speed detection peak (10 km, 600 m/s). Furthermore, a plurality of echo peaks can be superimposed, thereby enhancing the peak value of the echo peak. On the other hand, the method uses the CCAF algorithm to construct the detection amount, which suppresses the noise caused by the cross terms. Therefore, the method enhances the detection performance from two aspects: the cumulative effect of the multi-star detection fusion on the echo peak and the de-noise of the CCAF detection structure. However, it is necessary to consider that there are only 7–8 GPS satellites in the zenith at the same time, and detection performance improves as the number of satellites increases.

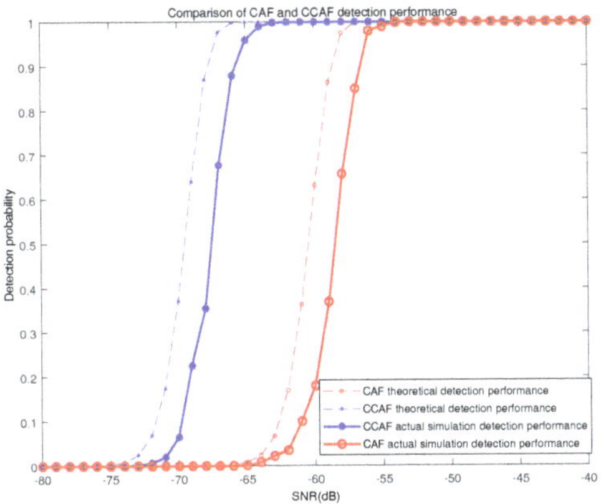

Figure 14. Comparison of detection performance between CCAF and CAF.

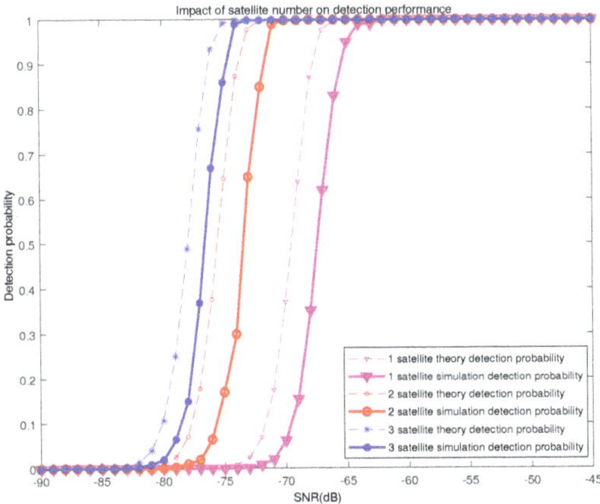

Figure 15. Comparison of detection performance between CCAF and CAF.

Experiment 3: In order to verify the influence of the accumulated time of the signal on the detection performance, we set the parameters of the false alarm probability, the direct wave power of GPS, the number of satellites, etc. The echo parameters are set according to Experiment 2. The SNR of the echo ranges from -90 to -30 dB. The Monte Carlo experiment was performed on the proposed algorithm under the condition that the number of sampling points changed from 10^6 to 10^9 points. Finally, the parameters are substituted into Equation (35) and Equation (42), and the theoretical detection probability is compared with the actual simulation. The specific simulation parameters are shown in Table 6.

Table 6. Parameter setting of experiment 3.

Sampling Frequency (Hz)	Duration (ms)	False Alarm Probability	Number of Satellites	Direct Wave Power (dBm)	Echo Power (dBm)	SNR (dB)
10.23 MHz	0.1 s/1 s/10 s	0.0001	3	−110	−170	−90~−30

From the results in Figure 16, it can be seen that, under the same conditions, in the case of echo SNR of −60 dB, the detection probability of the cumulative time of 0.1 s reaches 99%; in the case of echo SNR of −71 dB, the detection probability of the cumulative time of 1 s reaches 99%, in the case of echo SNR of −78 dB, the detection probability of the cumulative time of 10 s reaches 99%, and the simulated and theoretical detection performances are only 2 dB in difference, which verifies the effectiveness of the method. From the simulation results, it can be found that, as the accumulation time increases, the detection probability of the algorithm will increase. This is because CCAF is two-dimensional correlation. Therefore, the longer the correlation accumulation time, the larger the energy accumulated by the target echo will be. At the same time, as the accumulation time increases, the noise floor of the detection amount is also suppressed. Therefore, the number of sampling points N is one of the key factors affecting the detection performance of the proposed method. It is possible to increase the detection probability by using a longer accumulation time as much as possible, but the accumulation time cannot be increased indefinitely due to the limited detection area.

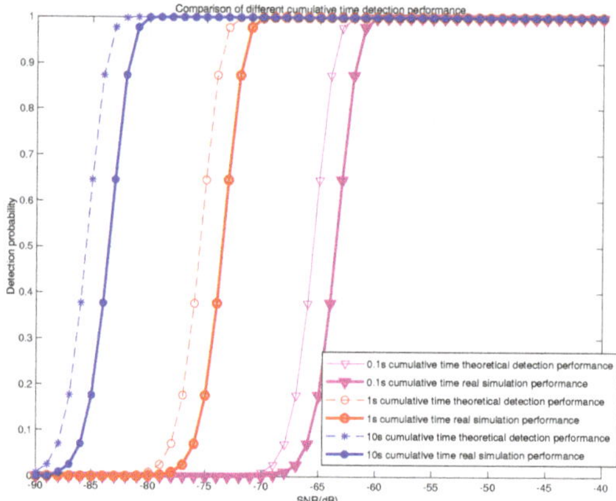

Figure 16. Influence of different accumulation time on detection performance.

Experiment 4: When the detection system detects the target of the specified area, the detection performance is mainly affected by the radar cross section (RCS) area of the target due to the fixed detection distance, and the change of the RCS size causes the change of the direct wave and the echo power ratio (SDR). In order to verify the influence of the attenuation of GPS signal caused by RCS on the detection performance of the proposed algorithm, the experimental fixed false alarm probability, reference signal power, noise power, number of satellites, and sampling points. The echo parameters are set according to the second experiment. Under the condition that the SDR variation range is 105 dB to 60 dB, the RCS corresponding variation range is 0.01 m^2 to 96.3 m^2, and 2000 Monte Carlo experiments are performed on the proposed algorithm. Finally, these parameters are substituted into Equations (35) and (42) to compare with the actual simulation. The specific simulation parameters are shown in Table 7.

Table 7. Parameter setting of experiment 4.

Sampling Frequency (Hz)	Duration (ms)	False Alarm Probability	Number of Satellites	Direct Wave Power (dBm)	Noise Power (dBm)	SNR (dB)	Detection Distance (Km)
10.23 MHz	10 s	0.0001	3	−100	−110	−90∼−60	10

From the results in Figure 17, it can be seen that the detection probability can reach 99% when the difference between the direct wave and echo power reaches 80 dB. At the same time, when the detection distance is about 10 km, the weak echo detection algorithm based on the multiple satellites data fusion proposed in this paper can effectively detect the weak GPS signal with an 80 dB attenuation. Both theoretical analysis and simulation verify that the proposed algorithm can effectively detect close-range targets.

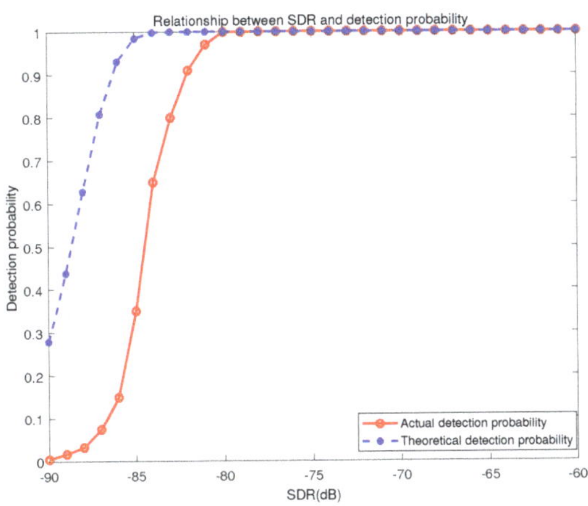

Figure 17. Relationship between SDR and detection probability.

7. Conclusions

In this paper, a GPS weak echo signal detection method has been proposed based on data fusion of multiple satellites, in which the SIC algorithm is utilized to separate and reconstruct multiple GPS reference signals on a reference channel. Then, the detection statistic of the weak echo of multiple GPS satellites has been constructed by using the anti-interference property of CCAF. Through the coordinate transformation algorithm, multiple detection statistics have been superimposed to obtain the final detection statistic. Finally, based on the theoretical analysis of the detection statistics, the relationships between the detection performance and a series of key parameters have been studied. Numerical results have shown that the proposed method is proved to be able to effectively detect the weak echo of the target in multiple GPS scenarios and significantly outperforms the existing methods.

Author Contributions: M.L. conceptualized and performed the algorithm and wrote the paper; Z.G. and Y.L. analyzed the experiment data; Y.C. helped modify the language; H.S. provided technical assistance to the research; F.G. is the research supervisor. The manuscript was discussed by all co-authors. All authors have read and agreed to the published version of the manuscript.

Funding: This research was funded by the National Natural Science Foundation of China (61501348 and 61801363), the Shaanxi Provincial Key Research and Development Program (2019GY-043), the Joint Fund of Ministry of Education of the People's Republic of China (6141A02022338), the China Postdoctoral Science Foundation (2017M611912), the 111 Project (B08038), and the China Scholarship Council (201806965031).

Conflicts of Interest: The authors declare no conflict of interest.

Appendix A

Appendix A.1. Proof of Lemma 1

Substituting Equations (26) and (27) into Equation (14), we can obtain

$$
\begin{aligned}
S_i(\tau, f) &= \sum_{n=0}^{N} x_s^{IF}(nT_s)\, \hat{x}_i^{IF}(nT_s - \tau)^*\, e^{j2\pi f nT_s} \\
&= \sum_{n=0}^{N} \mathrm{Re}\left\{ x_s^{IF}(nT_s)\, \hat{x}_i^{IF}(nT_s - \tau)^*\, e^{j2\pi f nT_s} \right\} \\
&\quad + j \sum_{n=0}^{N} \mathrm{Im}\left\{ x_s^{IF}(nT_s)\, \hat{x}_i^{IF}(nT_s - \tau)^*\, e^{j2\pi f nT_s} \right\}.
\end{aligned}
\tag{A1}
$$

Analyzing the probability distribution of the real part in Equation (A1), and the real part in Equation (A1) is given by

$$
\begin{aligned}
\mathrm{Re}\left(S_i(\tau, f)\right) &= \sum_{n=0}^{N} \mathrm{Re}\left\{ x_s^{IF}(nT_s)\hat{x}_i^{IF}(nT_s - \tau)^* e^{j2\pi f nT_s} \right\} \\
&= \sum_{n=0}^{N} \mathrm{Re}\left\{ \alpha_i^* \beta_i p_i(nT_s - \tau_i) e^{-j(2\pi f_{d_i} nT_s)} p_i^*(nT_s - \tau) e^{j2\pi f nT_s} \right\} \\
&\quad + \sum_{n=0}^{N} \mathrm{Re}\left\{ \beta_i p_i(nT_s - \tau_i) e^{-j(2\pi f_{d_i} nT_s)} \left(\sum_{\substack{j=1 \\ \forall i \neq j}}^{M} \alpha_j p_j(nT_s - \tau_j) \right)^* e^{j2\pi f nT_s} \right\} \\
&\quad + \sum_{n=0}^{N} \mathrm{Re}\left\{ \beta_i p_i^*(nT_s - \tau) e^{j2\pi f nT_s} \omega(nT_s) \right\} \\
&\approx \sum_{n=0}^{N} \mathrm{Re}\left\{ \alpha_i^* \beta_i p_i(nT_s - \tau_i) e^{-j(2\pi f_{d_i} nT_s)} p_i^*(nT_s - \tau) e^{j2\pi f nT_s} \right\} \\
&\quad + \sum_{n=0}^{N} \mathrm{Re}\left\{ \beta_i p_i^*(nT_s - \tau) e^{j2\pi f nT_s} \omega(nT_s) \right\} \\
&= S_{r_i s_i}(\tau, f) + S_{r_i \omega}(\tau, f).
\end{aligned}
\tag{A2}
$$

The reason for the sign being approximately equal in Equation (A2) is that the baseband signals of different GPS satellites are not correlated with each other, and the power of the echo signals is much smaller than the noise. The first term $S_{r_i s_i}(\tau, f)$ in Equation (A2) represents the real part of the mutual ambiguity function between the reference signal and its corresponding echo signal, and belongs to the determined detection amount, which can be expressed as

$$
\begin{aligned}
S_{r_i s_i}(\tau, f) &= \sum_{n=0}^{N} \mathrm{Re}\left\{ \alpha_i^* \beta_i p_i(nT_s - \tau_i)\, e^{-j\left(2\pi f_{d_i} nT_s\right)} p_i^*(nT_s - \tau) e^{j2\pi f_{d_i} nT_s} \right\} \\
&= \mathrm{Re}\left\{ \alpha^* \beta \chi_{pp}(\tau - n_\tau T_s, f - f_d) \right\},
\end{aligned}
\tag{A3}
$$

where n_τ stands for the time delay of the echo relative to the direct wave, f_d is the frequency offset of the echo relative to the direct wave, and $\chi_{pp}(\tau, f)$ is the self-fuzzy function of the amplitude-normalized baseband signal $p_i(nT_s)$, which can be expressed as

$$
\chi_{pp}(\tau, f) = \sum_{n=0}^{N-1} p_i(nT_s)\, p_i^*(nT_s - \tau)\, e^{j2\pi f nT_s},
\tag{A4}
$$

where $|p_i(nT_s)| = 1$, $E[p_i(nT_s)] = 0$. When $\tau = 0$, $f_d = 0$, the maximum value reached by Equation (A4) is $\max\left\{ |\chi_{pp}(\tau, f_d)| \right\} = N$.

The second term $S_{r_i \omega}(\tau, f)$ in Equation (A2) represents the mutual blur function between the monitoring channel noise and the reference signal, which can be regarded as the product of the

Gaussian noise and the amplitude scaling factor $p_i(nT_s)$, and still obeys the Gaussian distribution. Therefore, the mean and variance of the term can represent the probability distribution of the term. The mean of $S_{r_i\omega}(\tau, f)$ is given by

$$E\{S_{r_i\omega}(\tau, f)\} = 0 \tag{A5}$$

and the variance of $S_{r_i\omega}(\tau, f)$ is given by

$$
\begin{aligned}
&Var\left[S_{r_i\omega}(\tau, f)\right]\\
&= E\left[\left(\sum_{n=0}^{N} \mathrm{Re}\left\{\beta_i\omega(nT_s)p_i^*(nT_s - \tau)e^{j2\pi fnT_s}\right\}\right)^2\right] - E\left[\sum_{n=0}^{N} \mathrm{Re}\left\{\beta_i\omega(nT_s)p_i^*(nT_s - \tau)e^{j2\pi fnT_s}\right\}\right]^2\\
&= E\left[\left(\sum_{n=0}^{N} \mathrm{Re}\left\{\beta_i\omega(nT_s)p_i^*(nT_s - \tau)e^{j2\pi fnT_s}\right\}\right)^2\right]\\
&= \sum_{n=0}^{N}\sum_{m=0}^{N} E\left[\mathrm{Re}\left\{\beta_i\omega(nT_s)p_i^*(nT_s - \tau)e^{j2\pi fnT_s} \cdot \beta_i\omega(mT_s)^* p_i(mT_s - \tau)e^{-j2\pi fmT_s}\right\}\right]\\
&= \beta_i^2 \sum_{n=0}^{N}\sum_{m=0}^{N} R_{\omega\omega}(nT_s - mT_s) \cdot R_{p_i p_i}(nT_s - mT_s)e^{j2\pi f(n-m)T_s}\\
&= \beta_i^2 \sum_{n=0}^{N} \sigma_\omega^2 \delta(0)R_{pp}(0)\\
&= N\beta_i^2\sigma_\omega^2,
\end{aligned}
\tag{A6}
$$

where $R_{\omega\omega}(nT_s - mT_s)$ is the autocorrelation of the noise, and $R_{\omega\omega}(nT_s - mT_s) = \sigma_\omega^2 \delta(nT_s - mT_s)$, $\delta(nT_s)$ is the sequence of unit samples, and $R_{p_i p_i}(nT_s - mT_s)$ is the autocorrelation of the normalized signal. According to Equations (A4)–(A6), under the assumption of H_1, the distribution of the real part of $S_i(\tau, f)$ can be expressed as

$$\mathrm{Re}\left(S_i(\tau, f)\right) \sim N\left(\mathrm{Re}\left\{\alpha_i^* \beta_i \chi_{pp}(\tau - n_\tau, f - f_d)\right\}, N\beta_i^2\sigma_\omega^2\right). \tag{A7}$$

The same can be obtained. Under the H_1 hypothesis, the probability distribution of the imaginary part of $S_i(\tau, f_d)$ can be expressed as

$$\mathrm{Im}\left(S_i(\tau, f)\right) \sim N\left(\mathrm{Im}\left\{\alpha_i^* \beta_i \chi_{pp}(\tau - n_\tau, f - f_d)\right\}, N\beta_i^2\sigma_\omega^2\right). \tag{A8}$$

Therefore, the distribution of the detection statistic under the H_1 hypothesis is

$$\left(S_i(\tau, f)|H_1\right) \sim CN\left(\left\{\alpha_i^* \beta_i \chi_{pp}(\tau - n_\tau, f - f_d)\right\}, \ \ 2N\beta_i^2\sigma_\omega^2\right), \tag{A9}$$

where $CN(.)$ represents the complex Gaussian process.

Appendix A.2. Proof of Lemma 2

First, the reference signal $\hat{x}_i^{IF}(nT_s)$ is cyclically autocorrelated by

$$
\begin{aligned}
R_{r_i r_i}^\alpha(\tau) &= \frac{1}{N}\sum_{n=0}^{N-1} \hat{x}_i^{IF}(nT_s + \tau/2)\hat{x}_i^{IF}(nT_s - \tau/2)^* e^{-j2\pi\alpha nT_s}\\
&= \beta_i^2 R_{p_i p_i}^\alpha(\tau),
\end{aligned}
\tag{A10}
$$

where $R_{r_i r_i}^\alpha(\tau)$ is the cyclic autocorrelation of two reference signals, α is the cyclic frequency, and $R_{p_i p_i}^\alpha(\tau)$ stands for the cyclic autocorrelation normalized by the amplitudes of the two reference signals, which can be expressed as

$$R_{p_i p_i}^\alpha(\tau) = \frac{1}{N}\sum_{n=0}^{N-1}\left[C_i(nT_s + \tau/2) \cdot D_i(nT_s + \tau/2)\right] \cdot \left[C_i(nT_s - \tau/2) \cdot D_i(nT_s - \tau/2)\right]^* e^{-j2\pi\alpha nT_s}. \tag{A11}$$

Then, the reference signal $\hat{x}_i^{IF}(nT_s)$ and the echo signal $x_s^{IF}(nT_s)$ are subjected to cyclic cross-correlation processing under the assumption of H_1. In addition, using the uncorrelated properties of the baseband GPS reference signal pseudo-random code to obtain

$$
\begin{aligned}
R_{r_i s}^{\alpha-f}(\tau) &= \frac{1}{N} \sum_{n=0}^{N-1} \hat{x}_i^{IF}(nT_s + \tau/2) x_s^{IF}(nT_s - \tau/2)^* e^{-j2\pi(\alpha-f)nT_s} \\
&= \frac{1}{N} \sum_{n=0}^{N-1} [\beta_i p_i(nT_s)] \cdot \left[\sum_{\substack{j=1 \\ i\in j}}^{M} \alpha_j p_j(nT_s - \tau_j) e^{j2\pi f_{d_j} nT_s} + \omega(nT_s) \right]^* e^{-j2\pi(\alpha-f)nT_s} \\
&= \frac{1}{N} \sum_{n=0}^{N-1} [\beta_i p_i(nT_s)] \cdot \left[\alpha_i p_i(nT_s - \tau_i) e^{j2\pi f_{d_i} nT_s} + \omega(nT_s) \right]^* e^{-j2\pi(\alpha-f)nT_s} \\
&= \beta_i \alpha_i e^{-j\pi f_{d_i}\tau} e^{-j\pi(\alpha-f+f_{d_i})\tau_i} R_{p_i p_i}^{\alpha-f+f_{d_i}}(\tau - \tau_i) + N^\alpha(\tau),
\end{aligned}
\tag{A12}
$$

where τ_i is the time delay of the echo relative to the direct wave, f_{d_i} is the Doppler shift of the echo relative to the direct wave, and $N^\alpha(\tau)$ is the cyclic cross-correlation of the reference signal $\hat{x}_i^{IF}(nT_s)$ and the monitoring channel noise $\omega(nT_s)$, which can be expressed as follows:

$$
N^\alpha(\tau) = \frac{1}{N} \sum_{n=0}^{N-1} \hat{x}_i^{IF}(nT_s + \tau/2)\, \omega(nT_s - \tau/2)^* \, e^{-j2\pi(\alpha-f)nT_s}.
\tag{A13}
$$

Equation (A13) obeys a Gaussian distribution with a variance of $\frac{1}{N}\beta_i^2 \sigma_\omega^2$ and a mean of 0, which is denoted here by $N^\alpha(\tau)$. It can be seen that $R_{r_i s}^{\alpha-f}(\tau)$ is obtained by delay and frequency offset of $R_{r_i r_i}^\alpha(\tau)$, and the delay and frequency offset of $R_{r_i s}^{\alpha-f}(\tau)$ with respect to $R_{r_i r_i}^\alpha(\tau)$ is just the delay and frequency offset of the echo. Next, the column vectors corresponding to the cyclic frequencies of the maximum peaks of $R_{r_i r_i}^\alpha(\tau)$ and $R_{r_i s}^{\alpha-f}(\tau)$ are extracted, denoted respectively as $R_{r_i s}^{\alpha'-f'}(\tau)$ and $R_{r_i r_i}^{\alpha'}(\tau)$, and subjected to mutual blur function processing to obtain

$$
\begin{aligned}
\Psi_i(u, f) &= \sum_{\tau=0}^{N} R_{r_i s}^{\alpha'-f'}(\tau) R_{r_i r_i}^{\alpha'}(\tau - u)^* e^{j2\pi f\tau} \\
&= \beta_i^3 \alpha_i e^{-j\pi\alpha'\tau_i} \sum_{\tau=0}^{N} R_{p_i p_i}^{\alpha'-f+f_{d_i}}(\tau - \tau_i) R_{p_i p_i}^{\alpha'}(\tau - u)^* e^{j2\pi(f-f_{d_i})\tau} \\
&\quad + \beta_i^2 \sum_{\tau=0}^{N} N^{\alpha'}(\tau) R_{p_i p_i}^{\alpha'}(\tau - u)^* e^{j2\pi f\tau} \\
&= \Psi_{R_i R_i}(u, f) + \Psi_{NR_i}(u, f),
\end{aligned}
\tag{A14}
$$

where $\Psi_{R,R_i}(u, f) \leq \beta_i^3 \alpha_i \int \left| R_{p_i p_i}^{\alpha'}(\tau) \right|^2 d\tau$. When $u = \tau_i$ and $f = f_{d_i}$,

$$
\Psi_{R_i R_i}(u, f) = \beta_i^3 \alpha_i \int \left| R_{p_i p_i}^{\alpha'}(\tau) \right|^2 d\tau,
\tag{A15}
$$

where $\Psi_{NR_i}(u, f)$ is the noise term. According to the principle of constant false alarm detection, the size of the detection threshold is related to the false alarm probability. Therefore, the probability distribution of the detection quantity $\Psi_i(u, f)$ needs to be analyzed under the assumption of H_1. The first term of Equation (A14) belongs to the determined detection amount and is expressed as

$$
\Psi_{R_i R_i}(u, f) = \beta_i^3 \alpha_i e^{-j\pi\alpha'\tau_i} \sum_{\tau=0}^{N} R_{p_i p_i}^{\alpha'-f+f_{d_i}}(\tau - \tau_i) R_{p_i p_i}^{\alpha'}(\tau - u)^* e^{j2\pi(f-f_{d_i})\tau}.
\tag{A16}
$$

For the second term of Equation (A14), this term represents the mutual fuzzy function of the cyclic autocorrelation of the reference channel and the noise of the echo channel. If the cycle frequency is α', the term can be expressed as

$$\Psi_{NR_i}(u,f) = \beta_i^2 \sum_{\tau=0}^{N} N^{\alpha'}(\tau) R_{p_i p_i}^{\alpha'}(\tau - u)^* e^{j2\pi f\tau}, \tag{A17}$$

where $N^{\alpha'}(\tau)$ obeys a Gaussian distribution with a mean of 0 and a variance of $\frac{1}{N}\beta_i^2\sigma_\omega^2$, and $R_{p_i p_i}^{\alpha'}$ is a cyclic autocorrelation of the signal. Since the term is obtained by linear integral operation on noise, the term is still subject to Gaussian distribution, so the mean and variance can be used to characterize the probability distribution. The mean and variance are given by

$$E\left\{\Psi_{NR_i}(u,f)\right\} = 0, \tag{A18}$$

$$
\begin{aligned}
Var\{\psi_{NR_i}(u,f)\} &= E\left\{\left|\psi_{NR_i}(u,f)\right|^2\right\} - E\{\psi_{NR_i}(u,f)\}^2 \\
&= E\left\{\left[\beta_i^2\sum_{\tau_1=0}^{N-1} N^{\alpha'}(\tau_1 - u) R_{p_i p_i}^{\alpha'}(\tau_1) e^{j\pi f\tau_1}\right] * \left[\beta_i^2\sum_{\tau_2=0}^{N-1} N^{\alpha'}(\tau_2 - u) R_{p_i p_i}^{\alpha'}(\tau_2) e^{j\pi f\tau_2}\right]^*\right\} \\
&= \beta_i^4\sum_{\tau_1=0}^{N-1}\sum_{\tau_2=0}^{N-1} E\left\{N^{\alpha'}(\tau_1 - u) R_{p_i p_i}^{\alpha'}(\tau_1) e^{j\pi f\tau_1} * \left(\sum_{\tau_2=0}^{N-1} N^{\alpha'}(\tau_2 - u) R_{p_i p_i}^{\alpha'}(\tau_2) e^{j\pi f\tau_2}\right)^*\right\} \\
&= \beta_i^4\sum_{\tau_1=0}^{N-1}\sum_{\tau_2=0}^{N-1} R_{NN}(\tau_1 - \tau_2) E\left\{R_{p_i p_i}^{\alpha'}(\tau_1) R_{p_i p_i}^{\alpha'*}(\tau_2) e^{j\pi f(\tau_1-\tau_2)}\right\} \\
&= \beta_i^4\sum_{\tau_1=0}^{N-1}\sum_{\tau_2=0}^{N-1} \frac{1}{N}\beta_i^2\sigma_\omega^2\delta(\tau_1 - \tau_2) E\left\{R_{p_i p_i}^{\alpha'}(\tau_1) R_{p_i p_i}^{\alpha'*}(\tau_2) e^{j\pi f(\tau_1-\tau_2)}\right\} \\
&= \frac{\sigma_\omega^2\beta_i^6}{N}.
\end{aligned}
\tag{A19}
$$

Therefore, the distribution of $\Psi_{NR_i}(u,f)$ can be expressed as $\Psi_{NR_i}(u,f) \sim CN\left(0, \frac{\sigma_\omega^2\beta_i^6}{N}\right)$. It can be seen that the variance of the noise of the item is inversely proportional to the number of sampling points. Obviously, as the number of sampling points increases, the variance of the noise decreases, showing good noise suppression performance. Thus, the distribution of $\Psi_i(u,f)$ under the assumption of H_1 is given by

$$\Psi_i\left((u,f)|H_1\right) \sim CN\left(\Psi_{R_iR_i}(u,f), \frac{\sigma_\omega^2\beta_i^6}{N}\right). \tag{A20}$$

References

1. Foged, L.J.; Saporetti, M.; Scialacqua, L. Measurement and Simulation Comparison Using Measured Source Antenna Representation of GNSS Antenna on Sentinel Satellite. In Proceedings of the 2017 IEEE International Symposium on Antennas and Propagation USNC/URSI National Radio Science Meeting, San Diego, CA, USA, 9–15 July 2017; pp. 1933–1934.
2. Powell, S.J.; Akos, D.M. GNSS Reflectometry Using the L5 and E5a Signals for Remote Sensing Applications. In Proceedings of the 2013 US National Committee of URSI National Radio Science Meeting, Boulder, CO, USA, 9–12 January 2013; p. 1.
3. Giangregorio, G.; Bisceglie, M.; Addabbo, P.; Beltramonte, T.; D'Addio, S.; Galdi, C. Stochastic modeling and simulation of Delay-Doppler Maps in GNSS-R over the ocean. *IEEE Trans. Geosci. Remote Sens.* **2016**, *54*, 2056–2069. [CrossRef]
4. Fernez-Prades, C.; Arribas, J.; Majoral, M.; Ramos, A.; Vilá-Valls, J.; Giordano, P. A Software-Defined Spaceborne GNSS Receiver. In Proceedings of the 2018 9th ESA Workshop on Satellite NavigationTechnologies and European Workshop on GNSS Signals and Signal Processing, Noordwijk, The Netherlands, 5–7 December 2018; pp. 1–9.

5. Imam, R.; Pini, M.; Marucco, G.; Dominici, F.; Dovis, F. Data from GNSS-Based Passive Radar to Support Flood Monitoring Operations. In Proceedings of the 2019 International Conference on Localization and GNSS, Nuremberg, Germany, 4–6 June 2019; pp. 1–7.

6. Khosravi, F.; Moghadas, H.; Mousavi, P. A GNSS Antenna with a Polarization Selective Surface for the Mitigation of Low-Angle Multipath Interference. *IEEE Trans. Antennas Propag.* **2015**, *63*, 5287–5295. [CrossRef]

7. Liu, F.; Antoniou, M.; Zeng, Z.; Cherniakov, M. Coherent change detection using passive GNSS-based BSAR: Experimental proof of concept. *IEEE Trans. Geosci. Remote Sens.* **2013**, *51*, 4544–4555. [CrossRef]

8. Qiao, J.; Chen, W.; Ji, S.; Weng, D. Accurate and Rapid Broadcast Ephemerides for Beidou-Maneuvered Satellites. *Remote Sens.* **2019**, *11*, 787. [CrossRef]

9. Yan, S.; Chen, N.; Gong, J.; Zhang, X. Sea State Sensing with Single Frequency GPS Receiver. In Proceedings of the 2011 IEEE International Geoscience and Remote Sensing Symposium, Vancouver, BC, Canada, 24–29 July 2011; pp. 2073–2076.

10. Clarizia, M.P.; Chotiros, N.P.; Vaccaro, M.G. A GPS-Reflectometry Simulator for Target Detection Over Oceans. In Proceedings of the IGARSS 2018—2018 IEEE International Geoscience and Remote Sensing Symposium, Valencia, Spain, 22–27 July 2018; pp. 450–451.

11. Garvanov, I.; Kabakchiev, C.; Behar, V.; Garvanova, M. Target Detection Using a GPS Forward-Scattering Radar. In Proceedings of the 2015 International Conference on Engineering and Telecommunication, Mosocw, Russia, 18–19 November 2015; pp. 29–33.

12. Clarizia, M.; Braca, P.; Ruf, C.S.; Willett, P. Target Detection Using GPS Signals of Opportunity. In Proceedings of the 2015 18th International Conference on Information Fusion, Washington, DC, USA, 6–9 July 2015; pp. 1429–1436.

13. Gronowski, K.; Samczyński, P.; Stasiak, K.; Kulpa, K. First, Results of Air Target Detection Using Single Channel Passive Radar Utilizing GPS Illumination. In Proceedings of the 2019 IEEE Radar Conference (RadarConf), Boston, MA, USA, 22–26 April 2019; pp. 1–6.

14. Mojarrabi, B.; Homer, J.; Kubik, K. Power Budget Study for Passive Target Detection and Imaging Using Secondary Applications of GPS Signals in Bistatic Radar Systems. In Proceedings of the IEEE International Geoscience and Remote Sensing Symposium, Toronto, ON, Canada, 24–28 June 2002; pp. 449–451.

15. AlJewari, Y.H.; Ahmad, R.B.; AlRawi, A.A. Impact of multipath interference and change of velocity on the reliability and precision of GPS. In Proceedings of the 2014 2nd International Conference on Electronic Design (ICED), Penang, Malaysia, 19–21 August 2014; pp. 427–430.

16. Wu, X.; Gong, P.; Zhou, J.; Liu, Z. The applied research on anti-multipath interference GPS signal based on narrow-related. In Proceedings of the 2014 IEEE 5th International Conference on Software Engineering and Service Science, Beijing, China, 27–29 June 2014; pp. 771–774.

17. Suberviola, I.; Mayordomo, I.; Mendizabal, J. Experimental Results of Air Target Detection with a GPS Forward-Scattering Radar. *IEEE Geosci. Remote Sens. Lett.* **2011**, *9*, 47–51. [CrossRef]

18. Lashley, M.; Bevly, D.; Hung, J. Performance Analysis of Vector Tracking Algorithms for Weak GPS Signals in High Dynamics. *IEEE J. Sel. Top. Signal Process.* **2009**, *3*, 661–673. [CrossRef]

19. Wang, L.; Wang, J.; Xiao, L. Passive Location and Precision Analysis Based on Multiple CDMA Base Stations. In Proceedings of the 2009 IET International Radar Conference, Guilin, China, 20–22 April 2009; pp. 1–4.

20. Michael, E.; Alexander, S.; Fabienne, M. Design and Performance Evaluation of a MatureFM/DAB/DVB-T Multi-illuminator Passive Radar System. *IET Radar Sonar Navig.* **2014**, *8*, 114–122.

21. Stephen, D.; Songsri, S. Passive Radar Detection Using Multiple Ransmitters. In Proceedings of the 2013 Asilomar Conference on Signals Systems and Computers, Pacific Grove, CA, USA, 3–6 November 2013; pp. 945–948.

22. Zhang, Y.; Xi, S. Application of New LMS Adaptive Filtering Algorithm with Variable Step Size in Adaptive Echo Cancellation. In Proceedings of the 2017 IEEE 17th International Conference on Communication Technology, Chengdu, China, 27–30 October 2017; pp. 1715–1719.

23. Niranjan, D.; Ashwini, B. Noise Cancellation in Musical Signals Using Adaptive Filtering Algorithms. In Proceedings of the 2017 International Conference on Innovative Mechanisms for Industry Applications, Bangalore, India, 21–23 February 2017; pp. 82–86.

24. Mugdha, A.C.; Rawnaque, F.S.; Ahmed, M.U. A Study of Recursive Least Squares (RLS) Adaptive Filter Algorithm in Noise Removal from ECG Signals. In Proceedings of the 2015 International Conference on Informatics, Electronics Vision, Fukuoka, Japan, 15–18 June 2015; pp. 1–6.
25. Liu, J.; Liu, L. GPS C/A Code Signal Simulation Based on MATLAB. In Proceedings of the 2011 First, International Conference on Instrumentation, Measurement, Computer, Communication and Control, Beijing, China, 21–23 October 2011; pp. 4–6.
26. Beldjilali, B.; Benadda, B. Real Time Software Based L1 C/A GPS receiver. In Proceedings of the 2017 Seminar on Detection Systems Architectures and Technologies, Algiers, Algeria, 20–22 February 2017; pp. 1–8.
27. Ke, Z.; Ying, S. Steady-state Kalman Fusion Filter Based on Improved Multi-innovation Least Squares Algorithm. In Proceedings of the 2018 37th Chinese Control Conference, Wuhan, China, 25–27 July 2018; pp. 4434–4437.
28. Yang, Y.H.; Shi, Y. Iterative Least Squares Method Based Fusion Kalman Filter for Unknown Model Parameters System. In Proceedings of the 2018 37th Chinese Control Conference, Wuhan, China, 25–27 July 2018; pp. 1628–1631.
29. Ballal, T.; Suliman, M.A.; Al-Naffouri, T.Y. Bounded Perturbation Regularization for Linear Least Squares Estimation. *IEEE Access* **2017**, 5, 27551–27562. [CrossRef]
30. He, H.; Stoica, P.; Li, J. On Synthesizing Cross Ambiguity Functions. In Proceedings of the 2011 IEEE International Conference on Acoustics, Speech and Signal Processing, Prague, Czech Republic, 22–27 May 2011; pp. 3536–3539.
31. Huang, L.; Zhang, Y.; Li, Q.; Pan, C.; Song, J. Fast Calculation for Cross Ambiguity Function in Passive Detection Systems. In Proceedings of the 2018 International Conference on Sensing, Diagnostics, Prognostics, and Control, Xi'an, China, 15–17 August 2018; pp. 690–694.
32. Li, J.; He, Y.; Song, J. The Algorithms and Performance Analysis of Cross Ambiguity Function. In Proceedings of the 2009 IET International Radar Conference, Guilin, China, 20–22 April 2009; pp. 1–4.
33. Zeng, H.; Chen, J.; Wang, P.; Yang, W.; Liu, W. 2D Coherent Integration Processing and Detecting of Aircrafts Using GNSS-Based Passive Radar. *Remote Sens.* **2018**, *10*, 1164. [CrossRef]
34. Liu, M.; Yi, F.; Liu, P.; Li, B. Cramer-Rao Lower Bounds of TDOA and FDOA Estimation Based on Satellite Signals. In Proceedings of the 2018 14th IEEE International Conference on Signal Processing (ICSP), Beijing, China, 12–16 August 2018; pp. 1–4.
35. Liu, M.; Zhang, J.; Tang, J.; Jiang, F.; Liu, P.; Gong, F.; Zhao, N. 2D DOA Robust Estimation of Echo Signals Based on Multiple Satellites Passive Radar in the Presence of Alpha Stable Distribution Noise. *IEEE Access* **2019**, *7*, 2169–3536.

Letter

Continuous Tracking of Targets for Stereoscopic HFSWR Based on IMM Filtering Combined with ELM

Ling Zhang [1], Dongwei Mao [1], Jiong Niu [1,*], Q. M. Jonathan Wu [2] and Yonggang Ji [3]

[1]   College of Engineering, Ocean University of China, Qingdao 266100, China; zhangling@ouc.edu.cn (L.Z.); maodongwei@stu.ouc.edu.cn (D.M.)
[2]   Department of Electrical and Computer Engineering, University of Windsor, Windsor, ON N9B3P4, Canada; jwu@uwindsor.ca
[3]   Laboratory of Marine Physics and Remote Sensing, First Institute of Oceanography, Ministry of Natural Resources, Qingdao 266061, China; Jiyonggang@fio.org.cn
*   Correspondence: niujiong@ouc.edu.cn

Received: 28 November 2019; Accepted: 11 January 2020; Published: 14 January 2020

Abstract: High frequency surface wave radar (HFSWR) plays an important role in marine surveillance on account of its ability to provide wide-range early warning detection. However, vessel target track breakages are common in large-scale marine monitoring, which limits the continuous tracking ability of HFSWR. The following are the possible reasons for track fracture: highly maneuverable vessels, dense channels, target occlusion, strong clutter/interference, long sampling intervals, and low detection probabilities. To solve this problem, we propose a long-term continuous tracking method for multiple targets with stereoscopic HFSWR based on an interacting multiple model extended Kalman filter (IMMEKF) combined with an extreme learning machine (ELM). When the trajectory obtained by IMMEKF breaks, a new section of the track will start on the basis of the observation data. For multiple-target tracking, a number of broken tracks can be obtained by IMMEKF tracking. Then the ELM classifies the segments from the same vessel by extracting different features including average velocity, average curvature, ratio of the arc length to the chord length, and wavelet coefficient. Both the simulation and the field experiment results validate the method presented here, showing that this method can achieve long-term continuous tracking for multiple vessels, with an average correct track segment association rate of over 91.2%, which is better than the tracking performance of conventional algorithms, especially when the vessels are in dense channels and strong clutter/interference area.

Keywords: HFSWR; target tracking; interacting multiple model; extended Kalman filter; track association; extreme learning machine

---

1. Introduction

High-frequency surface wave radar (HFSWR) has become the primary technical means for maritime-state monitoring [1,2]. However, track breakages frequently occur in large-scale marine surveillance due to highly maneuverable targets, dense channels, target occlusion, strong clutter/interference, long sampling intervals, and low detection probabilities, which substantially degrade the overall tracking performance and adversely affect situation assessment [3,4]. Hence, continuous tracking of vessels is one of the key problems to be solved in the field of target tracking in marine surveillance.

At present, algorithm research studies for long-term vessel tracking mainly center on segment association, which can be divided into two categories: one is based on statistics [5–9]; the other is based on fuzzy mathematics [10–13]. The former takes the difference of the state estimation as the statistic, establishes the hypothesis, and then uses the given probability to accept or reject the hypothesis to determine whether the track is associated or not. The latter selects the membership degree of association, and calculates the membership value of two tracks to determine whether the track is associated or not. Bar-Shalom et al. [5] applied the fixed distance metric in the weighted statistical method. Yeom et al. [6] presented a track segment association algorithm on the basis of discrete optimization. Zhang et al. [7] stitched broken tracks by using an interacting multiple model (IMM) estimator. Aybars et al. [8] calculated an association cost in order to associate tracks on the basis of Mahalanobis distance. Zhu et al. [9] proposed a mixed integer nonlinear programming (MINLP) model in the maximum likelihood rule. Ashraf et al. [10] presented a fuzzy correlation approach on the basis of the fuzzy clustering means algorithm. Stubberud et al. [11] proposed a straightforward fuzzy-logic-based association method based on the chi 2 metric. Shao et al. [12] used fuzzy k-nearest neighbors and fuzzy C-means clustering to achieve track segment association. Hong et al. [13] presented a track segment association algorithm by calculating the fuzzy membership matrix and clustering methodology. All of these methods compare the distances of track segments point by point without considering the relevance of the track segments' features, which can cause invalid track association when a vessel is in a complex situation, such as in a dense channel or strong clutter interference region.

These track segment association (TSA) algorithms are feasible in theory, however, the backward prediction error of the present track is often large due to the noise of the system and measurement, which results in poor accuracy of association. Even worse, the performance of TSA algorithms drops suddenly due to track crossing and bifurcation when affected by surrounding large vessels or strong clutters. In order to solve the track breakage problem effectively and track targets steadily and continuously, we consider combining the conventional tracking algorithm with the machine learning method, which can achieve track segment association through studying the training dataset. The back propagation (BP) network has a strong nonlinear mapping ability, although it easily becomes trapped in local minimums during its training, which limits the stability and accuracy of classification [14,15]. Although the support vector machine (SVM) is able to avoid a local optimal solution, it has some shortcomings, such as training slowly and having a low efficiency [16,17]. The extreme learning machine (ELM) is a single hidden layer neural network with low computational complexity and good general performance [18]. Compared with the traditional neural network, ELM can randomly initialize the input weight and offset without adjusting during the training process, which make it simple and fast with the guarantee of accuracy [19]. Moreover, the ELM is more suited to handle small sample classification problems than deep learning [20,21].

We propose a long-term continuous tracking method based on an interacting multiple model extended Kalman filter (IMMEKF) combined with an extreme learning machine (ELM). When using the IMMEKF tracking method alone, we can determine only the intermittent track segments. Long-term continuous tracking is achieved when the ELM is combined with this method. In order to sufficiently reflect the track segment features, we associate the tracks on the basis of features extracted from track segments rather than comparing the distances of track segments point by point. Moreover, we present a new scheme to solve the problem of track segment association using an ELM network. The method decides whether the present track segment is associated with the former tracks through the ELM network, whose feature vectors in the training set are extracted from the track segments obtained by the IMMEKF. Hence, the long-term continuous tracking of multiple targets can be achieved via the cooperation between the IMMEKF and the ELM. Both the simulation and the field experiment results show that the proposed method has better tracking performance than conventional algorithms.

This paper is organized as follows. In Section 2, we briefly introduce the stereoscopic HFSWR station and the system model of the vessel target. In Section 3, we propose the long-term continuous tracking method with stereoscopic HFSWR based on IMMEKF combined with ELM. Furthermore, the simulation and the field data experiment results are presented to show the effectiveness of the proposed method in Section 4. In Section 5, the results are discussed, and areas for future potential research are considered. Finally, conclusions are drawn in Section 6.

2. System Model

In this paper, we examine the vessel target tracking for stereoscopic HFSWR [22], which is shown in Figure 1. We consider the T-R/R mode, with one transmitting station and two receiving radar stations working independently along the coast. For this mode, the target position is determined through geometric relation without measuring the target angle. The motion characteristics of the target, such as the radial range d and the radial velocity v, can be obtained. The distance between the two receiving radar stations is set as 2a.

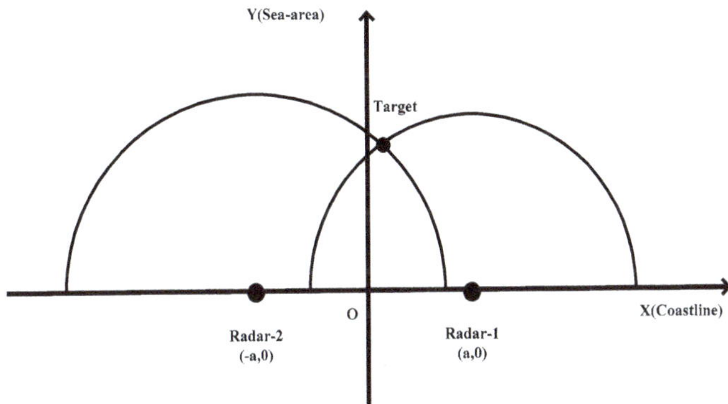

Figure 1. Schematic diagram of the stereoscopic High frequency surface wave radar (HFSWR) system.

At time point k, the state is described as

$$X(k) = [x(k), v_x(k), y(k), v_y(k)]^{\mathrm{T}} \tag{1}$$

where $x(k)$ and $y(k)$ describe the position of the vessel, and $v_x(k)$ and $v_y(k)$ represent the vessel velocity along the x and y direction, respectively. For the constant velocity (CV), constant acceleration (CA), and constant turn (CT) model [23], the state equation can easily be obtained:

$$X(k+1) = A(k)X(k) + B(k)W(k) \tag{2}$$

The observation vector is described as

$$Z(k) = [d_1(k), v_1(k), d_2(k), v_2(k)]^T \tag{3}$$

where $d_1(k)$ and $d_2(k)$ are the radial range, and $v_1(k)$ and $v_2(k)$ are the radial velocity. The measurement model is

$$Z(k) = H(X(k)) + V(k) \tag{4}$$

where the observation noise $V(k)$ is zero mean Gaussian white noise. Based on the geometrical relationship between the state and the measurement, the nonlinear measurement function $H(X(k))$ can be determined accordingly:

$$H(X(k)) = \begin{bmatrix} \dfrac{x-a}{\sqrt{(x-a)^2+y^2}} & 0 & \dfrac{y}{\sqrt{(x-a)^2+y^2}} & 0 \\[3mm] \dfrac{y^2 \cdot v_x - (x-a)\cdot y \cdot v_y}{\left[(x-a)^2+y^2\right]^{\frac{3}{2}}} & \dfrac{x-a}{\sqrt{(x-a)^2+y^2}} & \dfrac{x-a}{\sqrt{(x-a)^2+y^2}} & \dfrac{(x-a)^2 \cdot v_y - (x-a)\cdot y \cdot v_x}{\left[(x-a)^2+y^2\right]^{\frac{3}{2}}} \\[3mm] \dfrac{x+a}{\sqrt{(x+a)^2+y^2}} & 0 & \dfrac{y}{\sqrt{(x+a)^2+y^2}} & 0 \\[3mm] \dfrac{y^2 \cdot v_x - (x+a)\cdot y \cdot v_y}{\left[(x+a)^2+y^2\right]^{\frac{3}{2}}} & \dfrac{x+a}{\sqrt{(x-a)^2+y^2}} & \dfrac{y}{\sqrt{(x+a)^2+y^2}} & \dfrac{(x+a)^2 \cdot v_y - (x+a)\cdot y \cdot v_x}{\left[(x+a)^2+y^2\right]^{\frac{3}{2}}} \end{bmatrix} \tag{5}$$

The process of the target fusion and tracking algorithm based on IMM is shown in Figure 2, and the specific details of the algorithm are presented below.

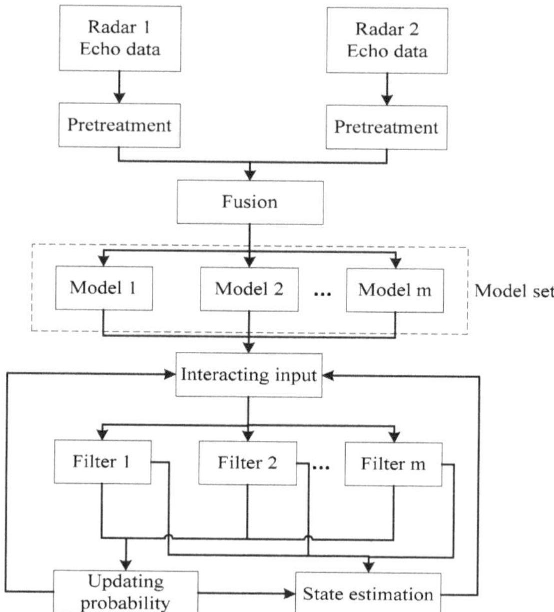

Figure 2. Flow graph of interacting multiple model extended Kalman filter (IMMEKF).

2.1. Interacting Input

$$\bar{c}_i = \sum_{j=1}^{m} u_j(k)p_{ji} \sum_{j=1}^{m} p_{ij} = 1, \ i,j = 1,\cdots,m \tag{6}$$

$$u_{ji}(k) = \frac{u_j(k)p_{ji}}{\bar{c}_i} \tag{7}$$

$$\hat{X}_{0i}(k) = \sum_{j=1}^{m} u_{ji}(k)\hat{X}_j(k) \tag{8}$$

$$P_{0i}(k) = \sum_{j=1}^{m} u_{ji}(k)\left\{P_j(k) + [\hat{X}_j(k) - \hat{X}_{0i}(k)][\hat{X}_j(k) - \hat{X}_{0i}(k)]^T\right\} \tag{9}$$

where $\bar{c}_i$ is the normalizing constant, u is the model probability, p is the model transition probability, m is the number of the motion model, $\hat{X}$ is the state estimation, and P is the residual covariance matrix variance.

2.2. Model Filtering

$$\hat{X}_i(k+1|k) = H_i(k+1|k)\hat{X}_{0i}(k) \tag{10}$$

$$P_i(k+1|k) = A_i(k+1|k)P_{0i}(k)A_i^T(k+1|k) + B_i(k)Q(k)B_i^T(k) \tag{11}$$

$$K_i(k+1) = P_i(k+1|k)H_i^T(k+1)\left[H_i(k+1)P_i(k+1|k)H_i^T(k+1) + R_i(k+1)\right]^{-1} \tag{12}$$

$$P_i(k+1) = [I - K_i(k+1)H_i(k+1)]P_i(k+1|k) \tag{13}$$

$$\hat{X}_i(k+1) = \hat{X}_i(k+1|k) + K_i(k+1)\left[Z_i(k+1) - H_i(k+1)\hat{X}_i(k+1|k)\right] \tag{14}$$

where $Q(k)$ is the covariance of the process noise and $R(k)$ is the covariance of the observation noise.

2.3. Updating Model Probability

$$u_i(k+1) = \frac{\Lambda_i(k+1)\bar{c}_i(k+1)}{\sum\limits_{i=1}^{m} \bar{c}_i\Lambda_i(k+1)} \tag{15}$$

$$\Lambda_i(k+1) = \frac{\exp\left\{-\frac{1}{2}(v_i(k+1))^T(S_i(k+1))^{-1}v_i(k+1)\right\}}{\sqrt{2\pi\det(S_i(k+1))}} \tag{16}$$

where v is measurement residuals and S is the residual covariance matrix.

2.4. Estimation Fusion

$$\hat{X}_i(k+1) = \sum_{i=1}^{m} \hat{X}_i(k+1)u_i(k+1) \tag{17}$$

$$P(k+1) = \sum_{i=1}^{m} u_i(k+1)\left\{P_j(k+1) + \left[\hat{X}_i(k+1) - \hat{X}(k+1)\right]\left[\hat{X}_i(k+1) - \hat{X}(k+1)\right]^T\right\} \tag{18}$$

3. Long-Term Continuous Tracking Method

For conventional tracking methods of HFSWR, the trajectory of vessel targets often breaks in segments due to interferences from other vessels, strong clutters, instant maneuvering, etc. In this section, we present a long-term continuous tracking method on the basis of the ELM network by extracting effective features to associate the track segments of the same vessel.

3.1. ELM Model

As is shown in Figure 3, the extreme learning machine (ELM) is a single hidden layer neural network with low computational complexity and good general performance [24,25]. Compared with the traditional neural network, ELM can randomly initialize the input weight and offset without adjusting during the training process [26], which make it simple and fast with the guarantee of accuracy [27].

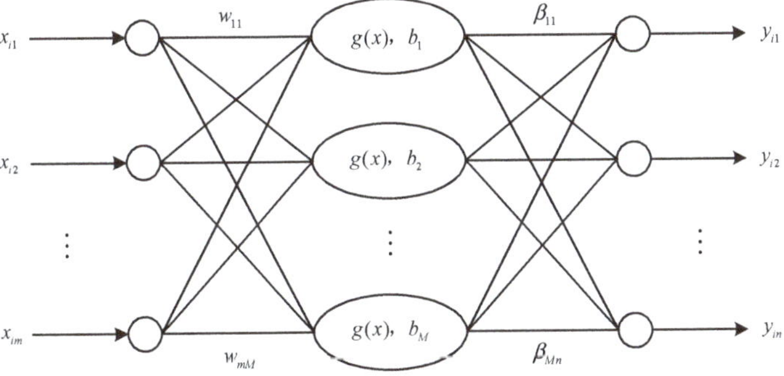

Figure 3. Extreme learning machine (ELM) network model.

Supposing there are N arbitrary data, the ELM network with M hidden layer nodes can be expressed as follows:

$$\sum_{j=1}^{M} \beta_j g(w_j \cdot x_i + b_j) = y_i (i = 1, 2, \cdots, N) \tag{19}$$

where $g(x)$ is the activation function, w_j is the input weight, β_j is the output weight, and b_j is the bias of the ith node.

The aim of neural network learning is minimizing the output error, which can be expressed as follows:

$$E(w, b, \beta) = \sum_{i=1}^{N} \|y_i - t_i\| \tag{20}$$

$$minE(w, b, \beta) = \min_{\omega, b, \beta} \|H(w_1, \cdots, w_M; b_1, \cdots, b_M; x_1, \cdots, x_N)\beta - T\| \tag{21}$$

$$H\beta = T \tag{22}$$

$$\hat{\beta} = H^+ T \tag{23}$$

where H is the output of the hidden layer, β is the output weight, T is the output of mathematical expectation, and H^+ is the Moore-Penrose generalized inverse of matrix H.

3.2. Feature Extraction

In order to sufficiently reflect the track segment features, we associate the tracks based on features extracted from track segments rather than comparing the distances of track segments point by point. To improve the tracking performance in a complex environment, more features should be extracted for track association. In this paper, we selected the average velocity, average curvature, ratio of the arc length to the chord length, and the wavelet coefficient as the feature vectors to train and test the ELM.

3.2.1. Average Velocity ($\bar{v}$)

Velocity as an inherent property varying with vessels can be reflected in the track segments, and thus we select the average velocity of the track segment as an important characteristic variable, which is defined as follows:

$$\bar{v} = \frac{\sum_{i=0}^{n} \sqrt{v_{xi}^2 + v_{yi}^2}}{n} \tag{24}$$

where v_{xi} is the velocity component of the vessel along the x axis, and v_{yi} is the velocity component of the vessel along the y axis.

3.2.2. Average Curvature ($\bar{k}$)

Curvature is the rate of change of the angle between the tangent of a point on the curve and the x axis relative to the arc length, which is defined by differentiation to indicate the degree of deviation of the curve from a straight line. The average curvature represents the deviation and regression degree of the course, which reflects the course correction ability varying with vessels under the interference of wind and waves. The average curvature ($\bar{k}$) is defined as follows:

$$\bar{k} = \frac{\sum_{i=0}^{n} \dfrac{|l_i''|}{[1+(l_i')^2]^{3/2}}}{n} \tag{25}$$

where l_i' is the first derivative of the arc length, and l_i'' is the second derivative of the arc length.

3.2.3. Ratio of the Arc Length to the Chord Length (R)

The arc length is the invariant of smooth curved motion, and the chord length has similar invariance. The ratio of the arc length to the chord length can roughly reflect the deviation and regression degree of course varying with vessels under the interference of wind and waves, which is similar to the average curvature. The ratio of the arc length to the chord length is defined as follows:

$$R = \frac{l}{d} \tag{26}$$

where d is the chord length, and l is the arc length.

3.2.4. Wavelet Coefficient L

Wavelet transform is a multi-resolution signal analysis method. It assumes that the measurement sequence is a non-stationary sequence $x(n)$ with polynomial trend. After a wavelet transformation, $x(n)$ can be reconstructed as

$$x(n) = \sum_k h(2n-k)x(k) + \sum_k g(2n-k)x(k) \tag{27}$$

where h is the high frequency coefficient reflecting the overall situation of the track and g is the low frequency coefficient reflecting the change of the target movement. In the frequency domain, the trend of the signal is represented by the low frequency part of the signal, which is represented by the low frequency coefficient in a wavelet analysis. The low frequency coefficient of a two-scale wavelet transformation based on the Haar function is taken as a track feature, which can represent the trend of the track segment.

3.3. Procedure of the Tracking Method Based on an IMMEKF Combined with an ELM

The procedure of the target tracking method based on an IMMEKF combined with an ELM is shown in Figure 4. Track segments are obtained by the IMMEKF when tracking the maneuvering target using the stereoscopic HFSWR data, and simultaneously, the ELM decides whether the newly obtained track segment associates with the former segment via feature classification. In this way, the method can achieve long-term continuous tracking of the vessel target.

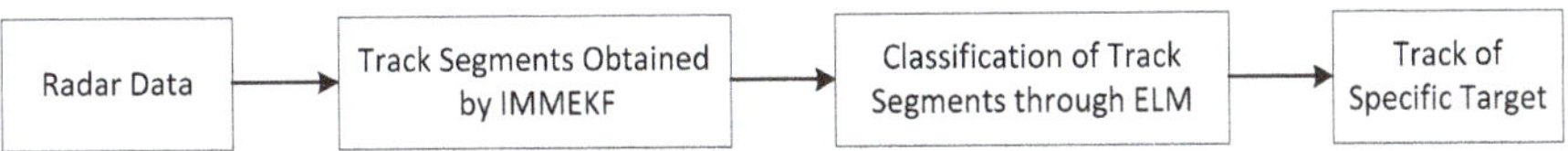

Figure 4. Flow graph of the continuous tracking method based on IMMEKF combined with ELM.

4. Experiment Results

4.1. Simulation Experiment

We assume that the radar monitoring area is approximately 20 km to 38 km in the x axis and approximately 20 km to 32 km in the y axis, where four vessels maneuvering with multiple models are simulated with MATLAB (MathWorks, Natick, MA, USA). For the conventional tracking method [13], the vessel track breaks into segments, as shown in Figure 5a. Combined with a trained ELM network, the proposed method realizes correct track segment association. As is shown in Figure 5b, different types of track segments are colored according to the classification results obtained by the ELM.

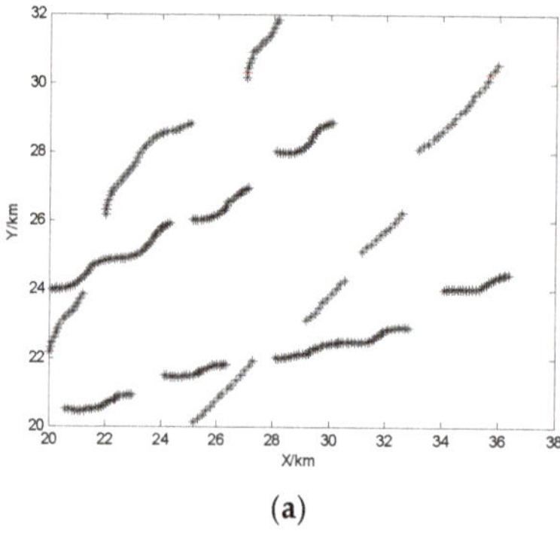 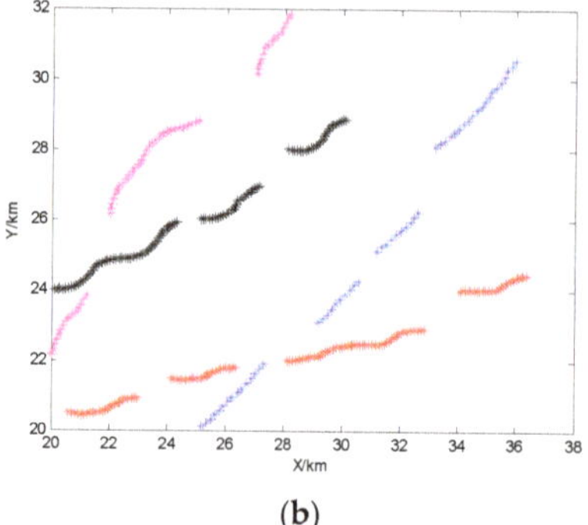

(a) **(b)**

Figure 5. Results of a simulation experiment: (**a**) track segments obtained by a conventional tracking method; (**b**) association results of the proposed method.

In order to verify the rationality and superiority of the proposed method, we carried out simulations on the basis of different combinations of features and different machine learning methods. To evaluate the association performance, the correct association probability (R_t), the error association probability (R_f) and the missing association probability (R_n) are defined as follows:

$$R_t = \frac{n_t}{n} \tag{28}$$

$$R_f = \frac{n_f}{n} \tag{29}$$

$$R_n = \frac{n_n}{n} \tag{30}$$

where n represents the total number of track segments in the experiment, n_t represents the number of correctly associated track segments, n_f represents the number of incorrectly associated track segments, and n_n represents the number of missed track segments.

4.1.1. Simulations Based on Different Combinations of Features

Among the four proposed features, $\bar{v}$ is an inherent property varying with vessels and L reflects the overall movement trend of the track segment, while both $\bar{k}$ and R reflect the degree of deviation and regression of vessels under the interference of wind and waves, respectively. We carried out simulations on the basis of different combinations of features. As is shown in Table 1, $\bar{k}$ is more effective than R in characterizing the features of track segments, and the correct association probability reaches the highest when they work together, which illustrates that $\bar{k}$ and R can reinforce one another in reflecting the degree of deviation and regression of vessels. Therefore, the proposed method in this paper selects all four of these features to work together for better association performance.

Table 1. Statistical results of track segment association based on different features (%).

	$\bar{v},L,\bar{k}$	$\bar{v},L,R$	$\bar{v},L,\bar{k},R$
R_t	92.5	89.3	94.7
R_f	6.7	9.8	4.6
R_n	0.8	0.9	0.7

4.1.2. Simulations Based on Different Machine Learning Methods

We carried out simulations based on different machine learning methods to show the performance of an ELM. As is shown in Table 2, the ELM has the fastest speed and the highest accuracy compared with the back propagation (BP) network and the SVM. Moreover, the ELM is more efficient than the BP and the SVM. Hence, we selected the ELM to combine with the IMMEKF in the proposed method in this paper to achieve long-term continuous tracking of multiple targets.

Table 2. Statistical results of track segment association based on different machine learning methods. BP: back propagation; SVM: support vector machine.

	ELM	BP	SVM
R_t (%)	94.7	82.3	89.7
R_f (%)	4.6	16.8	9.5
R_n (%)	0.7	0.9	0.8
Time (ms)	684	5103	1568

4.1.3. Comparison between the Conventional TSA Algorithm and the Proposed Method

The above simulations show that the extracted features of the track segment based on an IMMEKF can adequately reflect the vessel's attributes and motion characteristics; even better, the ELM can effectively learn the features of different track segments and accurately classify them. Furthermore, we carried out simulations comparing the conventional TSA algorithm and the proposed method. As is shown in Table 3, the proposed method has better performance than the conventional TSA algorithm.

Table 3. Statistical results of track segment association based on simulation (%). TSA: track segment association.

	Conventional TSA Algorithm	Proposed Method
R_t	76.5	94.7
R_f	12.1	4.6
R_n	11.4	0.7

4.2. Field Experiment

To compare the performance of the conventional TSA algorithm and the proposed method, we conducted extensive experiments basing the field data collected by a stereoscopic HFSWR system located on the northern shore of Weihai, China, on 21 July 2019. Data from an AIS (automatic identification system) is defined as the ground truth, which includes the number plates of the vessels of each track segment.

We selected the track as an example in Figure 6, where track segments x and y, marked with "*" and "+", respectively, are the tracks of vessel A, and segment z, marked with "○", is the track of vessel B. Although x is close to z, the ELM can determine that the track trend and characteristics between x and y are consistent, which are from the same target. The proposed method associated the track segment correctly in Figure 6c, yet the conventional TSA algorithm associated the track incorrectly in Figure 6b.

On the basis of 60 track segments obtained from field data, we calculated the average correct association probability, average error association probability, and average missing association probability of the two methods. The statistical results are shown in Table 4. Compared with the conventional TSA algorithm, the average correct association probability of the proposed method was increased by 21.7%, and the average error association probability and average missing association probability of the proposed method were reduced by 5.0% and 16.7%, respectively. Hence, the method proposed in this paper can effectively solve the problem of track segment association, which can effectively overcome the track breaking caused by strong sea clutter interference and interference from other vessels near dense channels, and realize the long-term continuous tracking of a specific vessel target with higher accuracy and stronger adaptability.

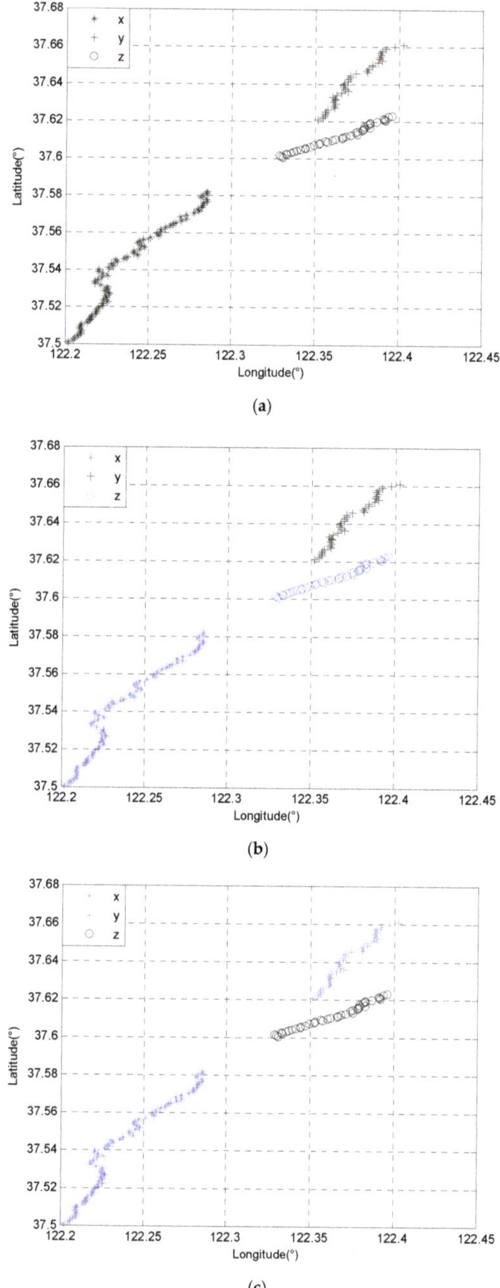

Figure 6. Comparison of track association results: (**a**) track segments from field data; (**b**) association results of the conventional TSA algorithm; (**c**) association results of the proposed method.

Table 4. Statistical results of track segment association based on radar data (%).

	Conventional TSA Algorithm	Proposed Method
R_t	69.5	91.2
R_f	13.1	8.1
R_n	17.4	0.7

5. Discussion

In Section 4.1.1, simulations were carried out on the basis of different combinations of features, which illustrates that the association performance is the best when the four extracted features of the tracks work together; in Section 4.1.2, we gave a comparison among different machine learning methods, which shows that the ELM had the fastest speed and the highest accuracy compared with the BP network and the SVM. Hence, we selected the ELM and all four of those features working together in the proposed method to realize track segment association. In Section 4.2, the proposed method was verified to have a better performance compared with the conventional TSA algorithm on the basis of radar data.

There is no doubt that the method proposed in this paper can effectively improve the track continuity of the target and realize the long-term continuous tracking of a specific vessel target. Moreover, the new method is easy for engineering implementation due to its generality, simple structure, reduced calculations, high learning speed, and high accuracy. In future research, we will consider further mining features of track segments and improving the network structure of the ELM to achieve better association accuracy.

6. Conclusions

We proposed a long-term continuous tracking method for vessel targets with stereoscopic HFSWR based on an IMMEKF combined with an ELM to solve the problem of trajectory breaking in large-scale marine surveillance. The IMMEKF is applied for the vessel target tracking, meanwhile the ELM network is combined to judge whether the present track is associated with the former track segments. For fully embodying the characteristics of track segments, we selected the average velocity, average curvature, ratio of the arc length to the chord length, and the wavelet coefficient as the feature vectors to train and test the ELM. The tracking algorithm IMMEKF and the track segments associating scheme of the ELM work simultaneously and iteratively. Both the simulation and the field experiment results showed that the proposed method has better tracking performance than the conventional algorithms, with an average correct track segment association rate of over 91.2%. Further, field experiment data from stereoscopic onshore HFSWR located in Weihai showed that the ELM-based track association method had higher accuracy, with lower error and missing association probability due to effective features extracted from the track segments. The method can effectively solve the problem of track fracture caused by rapid maneuvering, strong clutter, target occlusion, long sampling intervals, low detection probabilities, and interference from other vessels near channels, providing a new way for long-term continuous tracking of vessel targets in complex environments. Moreover, the new method is efficient and easy to implement due to its simple structure, which allows the real-time tracking of vessel targets for stereoscopic HFSWR.

Author Contributions: Conceptualization: L.Z.; Methodology: D.M. and L.Z.; Software: D.M. and J.N.; Validation: J.N. and D.M.; Investigation: D.M.; Resources: Y.J.; Data Curation: Y.J.; Writing—Original Draft Preparation: L.Z. and D.M.; Project Administration: Y.J.; Writing—Review and Editing: Q.M.J.W. All authors have read and agreed to the published version of the manuscript.

Funding: This project is sponsored by National Key R&D Program of China (No. 2017YFC1405202), and the National Natural Science Foundation of China (No. 51979256, No. 61671166, No. 41506114).

References

1. Grosdidier, S.; Baussard, A. Ship detection based on morphological component analysis of high-frequency surface wave radar images. *IET Radar Sonar Navig.* **2012**, *6*, 813–821. [CrossRef]
2. Zhang, L.; You, W.; Wu, Q.M.; Qi, S.B.; Ji, Y.G. Deep learning-based automatic clutter/interference detection for HFSWR. *Remote Sens.* **2018**, *10*, 1517. [CrossRef]
3. Pan, S.S.; Bao, Q.L.; Hou, W.B.; Che, Z.P. An improved track segment association algorithm using MM-GNN method. In Proceedings of the 2017 Progress in Electromagnetics Research Symposium Spring—(PIERS), St Petersburg, Russia, 22–25 May 2017; pp. 675–681.
4. Raghu, J.; Srihari, P.; Tharmarasa, R.; Kirubarajan, T. Comprehensive track segment association for improved track continuity. *IEEE Trans. Aerosp. Electron. Syst.* **2018**, *1*, 2463–2480. [CrossRef]
5. Bar-Shalom, Y. On the track-to-track correlation problem. *IEEE Trans. Autom. Control.* **1981**, *26*, 571–572. [CrossRef]
6. Yeom, S.W.; Kirubarajan, T.; Bar-Shalom, Y. Track segment association, fine-step IMM and initialization with Doppler for improved track performance. *IEEE Trans. Aerosp. Electron. Syst.* **2004**, *40*, 293–309. [CrossRef]
7. Zhang, S.; Bar-Shalom, Y. Track segment association for GMTI tracks of evasive move-stop-move maneuvering targets. *IEEE Trans. Aerosp. Electron. Syst.* **2011**, *47*, 1899–1914. [CrossRef]
8. Aybars, T.; Ali, K.H. A fast track to track association algorithm by sequence processing of target states. In Proceedings of the 2018 21st International Conference on Information Fusion (FUSION), Cambridge, UK, 10–13 July 2018; pp. 1280–1284.
9. Zhu, H.; Wang, C. Joint track-to-track association and sensor registration at the track level. *Digit. Signal. Process.* **2015**, *41*, 48–59. [CrossRef]
10. Ashraf, M.A.; Murali, T.; Roberto, C. Fuzzy logic data correlation approach in multisensor–multitarget tracking systems. *Signal. Process.* **1999**, *76*, 195–209.
11. Stubberud, S.C.; Kramer, K.A. Data association for multiple sensor types using fuzzy logic. In Proceedings of the IEEE Instrumentation & Measurement Technology Conference, Ottawa, ON, Canada, 17–19 May 2005; pp. 2154–2159.
12. Shao, H.; Japkowicz, N.; Abielmona, R.S. Vessel track correlation and association using fuzzy logic and Echo State Networks. In Proceedings of the 2014 IEEE Congress on Evolutionary Computation (CEC), Beijing, China, 6–11 July 2014; pp. 2322–2329.
13. Hong, S.X.; Peng, D.L.; Shi, Y.F. Track-to-track association using fuzzy membership function and clustering for distributed information fusion. In Proceedings of the 37th Chinese Control Conference, Wuhan, China, 25–27 July 2018; pp. 4028–4032.
14. Siwei, L.; Wujie, Y.; Aijun, Z. Performance of feedback BP networks. *J. Syst. Eng. Electron.* **1995**, *6*, 11–18.
15. Xie, J.; Zhou, J. Classification of urban building type from high spatial resolution remote sensing imagery using extended MRS and soft BP network. *IEEE J. Sel. Top. Appl. Earth Obs. Remote Sens.* **2017**, *1*, 3515–3528. [CrossRef]
16. Liu, X.; Tang, J. Mass classification in mammograms using selected geometry and texture features, and a new SVM-based feature selection method. *IEEE Syst. J.* **2014**, *8*, 910–920. [CrossRef]
17. Pal, M.; Mather, P.M. Support vector machines for classification in remote sensing. *Int. J. Remote Sens.* **2005**, *26*, 1007–1011. [CrossRef]
18. Lu, X.; Zou, H.; Zhou, H.; Xie, L.; Huang, G.B. Robust extreme learning machine with its application to indoor positioning. *IEEE Trans. Cybern.* **2016**, *46*, 194–205. [CrossRef] [PubMed]
19. Huang, G.B.; Wang, D.H.; Lan, Y. Extreme learning machines: A survey. *Int. J. Mach. Learn. Cybern.* **2011**, *2*, 107–122. [CrossRef]
20. Levine, S.; Pastor, P.; Krizhevsky, A.; Quillen, D. Learning hand-eye coordination for robotic grasping with deep learning and large-scale data collection. *Int. J. Robot. Res.* **2016**, *1*, 173–184. [CrossRef]
21. Shin, H.C.; Roth, H.R.; Gao, M.C.; Lu, L.; Xu, Z.; Nogues, I.; Yao, J.H.; Mollura, D.; Summers, R.M. Deep convolutional neural networks for computer-aided detection: CNN architectures, dataset characteristics and transfer learning. *IEEE Trans. Med. Imaging* **2016**, *35*, 1285–1298. [CrossRef] [PubMed]
22. Zhang, L.; Jiang, Y.; Li, Y.S.; Li, G.S.; Ji, Y.G. Adaptive maneuvering target tracking with 2-HFSWR multisensor surveillance system. *IEEE Aerosp. Electron. Syst. Mag.* **2017**, *32*, 70–76. [CrossRef]

23. Bahar, E.; Kurt, G.K. Location estimation in multi-carrier systems using extended Kalman based interacting multiple model and data fusion. In Proceedings of the 2011 IEEE Symposium on Computers and Communications (ISCC), Kerkyra, Greece, 28 June–1 July 2011; pp. 469–471.

24. Huang, G.B.; Zhu, Q.Y.; Siew, C.K. Extreme learning machine: A new learning scheme of feedforward neural networks. In Proceedings of the 2004 IEEE International Joint Conference on Neural Networks, Budapest, Hungary, 25–29 July 2004; pp. 489–501.

25. Huang, G.B.; Zhu, Q.Y.; Siew, C.K. Extreme learning machine: Theory and applications. *Neurocomputing* **2006**, *70*, 489–501. [CrossRef]

26. Zong, W.; Huang, G.B. Face recognition based on extreme learning machine. *Neurocomputing* **2011**, *74*, 2541–2551. [CrossRef]

27. Rong, H.J.; Huang, G.B.; Sundararajan, N. Online sequential fuzzy extreme learning machine for function approximation and classification problems. *IEEE Trans. Syst. Man Cybern. Part B* **2009**, *39*, 1067–1072. [CrossRef] [PubMed]

MDPI

St. Alban-Anlage 66

4052 Basel

Switzerland

Tel. +41 61 683 77 34

Fax +41 61 302 89 18

www.mdpi.com

Remote Sensing Editorial Office

E-mail: remotesensing@mdpi.com

www.mdpi.com/journal/remotesensing